CAMBRIDGE TRACTS IN MATHEMATICS

General Editors

163 Linear and Projective Representations of Symmetric Groups

T0291493

Cambridge Tracts in Mathematics
All the titles listed below can be obtained from good booksellers or from Cambridge University Press. For a complete series listing visit http://publishing.cambridge.org/stm/mathematics/ctm/

142. Harmonic Maps between Rienmannian Polyhedra. By J. Eells and B. Fuglede
143. Analysis on Fractals. By J. Kigami
144. Torsors and Rational Points. By A. Skorobogatov
145. Isoperimetric Inequalities. By I. Chavel
146. Restricted Orbit Equivalence for Actions of Discrete Amenable Groups. By J. W. Kammeyer and D. J. Rudolph
147. Floer Homology Groups in Yang–Mills Theory. By S. K. Donaldson
148. Graph Directed Markov Systems. By D. Mauldin and M. Urbanski
149. Cohomology of Vector Bundles and Syzygies. By J. Weyman
150. Harmonic Maps, Conservation Laws and Moving Frames. By F. Hélein
151. Frobenius Manifolds and Moduli Spaces for Singularities. By C. Hertling
152. Permutation Group Algorithms. By A. Seress
153. Abelian Varieties, Theta Functions and the Fourier Transform. By Alexander Polishchuk
156. Harmonic Mappings in the Plane. By Peter Duren
157. Affine Hecke Algebras and Orthogonal Polynomials. By I. G. MacDonald
158. Quasi-Frobenius Rings. By W. K. Nicholson and M. F. Yousif
159. The Geometry of Total Curvature on Complete Open Surfaces. By Katsuhiro Shiohama, Takashi Shioya and Minoru Tanaka
160. Approximation by Algebraic Numbers. By Yann Bugeaud
161. Equivalence and Duality for Module Categories with Tilting and Cotilting for Rings. By R. R. Colby and K. R. Fuller

Linear and Projective Representations of Symmetric Groups

ALEXANDER KLESHCHEV
University of Oregon

CAMBRIDGE
UNIVERSITY PRESS

CAMBRIDGE UNIVERSITY PRESS
Cambridge, New York, Melbourne, Madrid, Cape Town, Singapore, São Paulo, Delhi

Cambridge University Press
The Edinburgh Building, Cambridge CB2 8RU, UK

Published in the United States of America by Cambridge University Press, New York

www.cambridge.org
Information on this title: www.cambridge.org/9780521104180

First published 2005
This digitally printed version 2009

A catalogue record for this publication is available from the British Library

ISBN 978-0-521-83703-3 hardback
ISBN 978-0-521-10418-0 paperback

Contents

Preface

The subject of this book is representation theory of symmetric groups. We explain a new approach to this theory based on the recent work of Lascoux, Leclerc, Thibon, Ariki, Grojnowski, Brundan, Kleshchev, and others. We are mainly interested in *modular* representation theory, although everything works in arbitrary characteristic, and in case of characteristic 0 our approach is somewhat similar to the theory of Okounkov and Vershik [OV], described in Chapter 2 of this book. The methods developed here are quite general and they apply to a number of related objects: finite and affine Iwahori–Hecke algebras of type A, cyclotomic Hecke algebras, spin-symmetric groups, Sergeev algebras, Hecke–Clifford superalgebras, affine and cyclotomic Hecke–Clifford superalgebras, ... We concentrate on symmetric and spin-symmetric groups though.

We now outline some of the ideas which lead to the new approach. Let us concentrate on the modular case, as this is where things get really interesting. So let F be a field of characteristic $p > 0$, and S_n be the symmetric group. Irreducible FS_n-modules were classified by James. His approach is as follows (see [J] for details). Let S^λ be the Specht module corresponding to a partition λ of n (the Specht construction works over any field and even over $\mathbb{Z}$). This module has a canonical S_n-invariant bilinear form. The form is non-zero if and only if λ is *p-regular*, that is no non-zero part of λ is repeated p or more times. In this case the radical of the form, call it Q^λ, is a maximal proper submodule. Thus $D^\lambda := S^\lambda / Q^\lambda$ is an irreducible FS_n-module. Finally, $\{D^\lambda \mid \lambda \text{ is a } p\text{-regular partition of } n\}$ is a complete set of irreducible FS_n-modules.

The main unsolved problem in modular representation theory of symmetric groups is to find (Brauer) characters and in particular dimensions of irreducible modules. These will be known if any of the following two equivalent problems can be solved (see [K_6] for the proof of equivalence). Denote by $\mathcal{P}(n)$ (resp. $\mathcal{P}_p(n)$) the set of all partitions (resp. p-regular partitions) of n.

Decomposition numbers problem. Find the composition multiplicities

$$[S^\lambda : D^\mu] \tag{0.1}$$

for any $\lambda \in \mathcal{P}(n)$ and $\mu \in \mathcal{P}_p(n)$.

Branching problem. Determine the composition multiplicities of the restriction

$$[\mathrm{res}^{S_n}_{S_{n-1}} D^\lambda : D^\mu] \tag{0.2}$$

for any $\lambda \in \mathcal{P}_p(n)$ and $\mu \in \mathcal{P}_p(n-1)$.

Motivated by the second problem, the author [K_1] obtained certain partial *branching rules*, including an explicit description of the socle of the restriction $\mathrm{res}^{S_n}_{S_{n-1}} D^\lambda$, which is equivalent to the description of the spaces

$$\mathrm{Hom}_{FS_{n-1}}(D^\mu, \mathrm{res}^{S_n}_{S_{n-1}} D^\lambda) \qquad \big(\lambda \in \mathcal{P}_p(n),\ \mu \in \mathcal{P}_p(n-1)\big). \tag{0.3}$$

Another result from [K_1] describes the spaces

$$\mathrm{Hom}_{FS_{n-1}}(S^\mu, \mathrm{res}^{S_n}_{S_{n-1}} D^\lambda) \qquad \big(\lambda \in \mathcal{P}_p(n),\ \mu \in \mathcal{P}_p(n-1)\big). \tag{0.4}$$

It turns out that the spaces in (0.3) and (0.4) are at most 1-dimensional, which gives two different generalizations of the multiplicity freeness of the branching rule in characteristic 0. The solution is in terms of delicate combinatorial notions of a *normal node* and a *good node* of a Young diagram (see Chapter 11): the space (0.3) (resp. (0.4)) is non-trivial if and only if μ is obtained from λ by removing a good node (resp. normal node). As observed in [K_2], it follows from this description that all irreducible modules appearing in the socle of $\mathrm{res}^{S_n}_{S_{n-1}} D^\lambda$ belong to different blocks, the fact sometimes referred to as the *strong multiplicity freeness* of the branching rule. A number of further results on the modular branching problem was established in [K_5, K_4, BK_1]. For example, in [K_4] we have described the multiplicity (0.2) when μ is obtained from λ by removing a node. It turns out that this multiplicity is non-zero if and only if the node is normal, but it can be arbitrarily large.

Many applications of the branching rules were obtained, see for example [K_2, FK, BO_1, BK_4]. But the most important consequence was that they led to a discovery of deep connections between modular representation theory and the theory of crystal bases. The link was first made by Lascoux, Leclerc, and Thibon [LLT]. The nodes of Young diagrams come with residues, which are elements of $\mathbb{Z}/p\mathbb{Z}$, and we obtain a structure of a directed colored graph on the set of all p-regular partitions: by definition, there is an arrow from μ to λ of color $i \in \mathbb{Z}/p\mathbb{Z}$ if and only if μ is obtained from λ by removing a good node of residue i. Lascoux, Leclerc, and Thibon made the startling combinatorial observation that this "branching graph" coincides with the crystal graph of the basic representation of the affine Kac–Moody algebra $\mathfrak{g} = A^{(1)}_{p-1}$, determined explicitly by Misra and Miwa [MiMi]. It turned out later that the same observation applies to the branching graph for the associated complex Iwahori–Hecke algebras at a primitive $(\ell+1)$th root of unity and the crystal graph of the basic representation of the affine Kac–Moody algebra $\mathfrak{g} = A^{(1)}_{\ell}$, see [B]. In this latter case, Lascoux, Leclerc, and Thibon conjectured moreover that the coefficients of the canonical basis of the basic representation coincide with the decomposition numbers (0.1) for the Iwahori–Hecke algebras.

This conjecture was proved by Ariki [A_1] (see also Grojnowski [G_1]). More generally, Ariki established a similar result connecting the canonical basis of an arbitrary integrable highest weight module of $\mathfrak{g}$ to the representation theory of the corresponding cyclotomic Hecke algebra, as defined in [Ch, AK, BM]. Note that Ariki's work is concerned with the cyclotomic Hecke algebras over the ground field $\mathbb{C}$, but Ariki and Mathas [A_2, AM] were later able to extend the classification of the irreducible modules, but not the result on decomposition numbers, to arbitrary fields. For further developments related to the LLT conjecture, see [LT_1, VV, Sch, A_3].

Subsequently, Grojnowski and Vazirani [G_2, G_3, GV, V_2] have developed a powerful new approach to (among other things) the classification of the irreducible modules of the cyclotomic Hecke algebras. The approach is valid over an arbitrary ground field, and is entirely independent of the "Specht module theory" that plays an important role in Ariki's work. Branching rules are built in from the outset, resulting in an explanation and generalization of the link between modular branching rules and crystal graphs. The methods are purely algebraic, exploiting affine Hecke algebras in the spirit of [BZ, Z_1] and others. On the other hand, results on decomposition numbers do not follow, since that ultimately depends on the geometric work of Kazhdan, Lusztig, and Ginzburg.

In this book, we explain Grojnowski's approach to the theory in the case of degenerate affine Hecke algebra. In particular, we obtain an algebraic construction purely in terms of the representation theory of degenerate affine Hecke algebras of the positive part $U^+_{\mathbb{Z}}$ of the enveloping algebra of $\mathfrak{g} = A^{(1)}_{p-1}$, as well as of Kashiwara's highest weight crystals $B(\infty)$ and $B(\lambda)$ for each dominant weight λ. These emerge as the modular branching graphs of the appropriate algebras. As a consequence, a parametrization of irreducible FS_n-modules, classification of blocks ("Nakayama's conjecture"), and some of the modular branching rules mentioned above follow from the special case $\lambda = \Lambda_0$ of the main results.

Part II of the book deals with representation theory of Schur's double covers of symmetric groups. This is equivalent to studying projective (or spin) representations of symmetric groups. The spin theory in characteristic p was developed in [BK_3] (cf. [BMO, ABO_1, ABO_2, MoY_2])–it is parallel to the theory of linear representations described above, but with the role of the Kac–Moody algebra $\mathfrak{g} = A_{p-1}^{(1)}$ played by the twisted Kac–Moody algebra $\mathfrak{g} = A_{p-1}^{(2)}$. We note that the modular irreducible spin representations of S_n were first classified in [BK_2], following a conjecture in [LT_2], using a more classical approach via "Specht modules". However, that approach did not allow us to obtain any branching rules.

We hope that having both linear and spin theory under one cover will be useful for the reader. The two theories are actually very similar, if developed from the new point of view adopted in this book. In fact, a glance at the contents shows that many sections of Part II are exactly parallel to the corresponding sections of Part I.

Let us now describe the contents of the book in more detail. We note that each chapter has its own introduction where the results of the chapter are motivated and described, sometimes informally. Chapter 1 contains notation and some basic preliminary results. Chapter 2 is a presentation of the beautiful theory of Okounkov and Vershik in characteristic 0. It is not directly related to the rest of the book. However, it is a nice introduction to some ideas employed further on, and might be a good place to start for less-advanced readers. Also, it is perhaps the shortest way to some key results of the classical representation theory of symmetric groups in characteristic 0, such as classification of irreducible representations, Young's formulas, and the Murnaghan–Nakayama formula.

Degenerate affine Hecke algebras $\mathcal{H}_n$ are introduced in Chapter 3. Basis Theorem for $\mathcal{H}_n$ is proved and the center of $\mathcal{H}_n$ is described. We introduce parabolic subalgebras $\mathcal{H}_\mu \subset \mathcal{H}_n$ and the corresponding induction and restriction functors ind_μ^n and res_μ^n. "Mackey Theorem" is a result describing a filtration of $\mathrm{res}_\mu^n \mathrm{ind}_\nu^n$ by certain induced modules. An important result relating induction and duality is proved in Section 3.7.

In Chapter 4, we introduce the formal characters of finite dimensional $\mathcal{H}_n$-modules, discuss central characters and blocks, and then study in detail a remarkable irreducible $\mathcal{H}_n$-module, called Kato module, as well as its "covering modules".

The functors e_a and their versions Δ_a and Δ_{a^m}, which are affine analogues of Robinson's a-restriction functors for symmetric groups, are studied in

Chapter 5. The (affine version of the) strong multiplicity freeness of the branching rule is established in Corollary 5.1.7. This allows us to define the crystal operators $\tilde{e}_a$ and $\tilde{f}_a$ in Section 5.2 as the socle and the head of a certain direct summand of restriction and induction, respectively. We then prove that the formal characters of irreducible modules are linearly independent and use the crystal operators $\tilde{f}_a$ to label the irreducible $\mathcal{H}_n$-modules. Section 5.5 contains some further branching results.

In Chapter 6 we establish a sufficient condition for the irreducibility of the module, induced from an irreducible module over a parabolic subalgebra (this condition is far from necessary–see for example [LNT] for results on the subtle problem of finding necessary and sufficient condition). Then we calculate characters of some irreducible $\mathcal{H}_n$-modules for $n \leq 4$. These calculations provide the reader with concrete examples to play with, but also are important for the theory, as they will imply that the operators e_i on the Grothendieck group satisfy Serre relations. Finally, some higher crystal operators are introduced in Section 6.3–these play only a technical role.

Integral representations and (degenerate) cyclotomic Hecke algebras are treated in Chapter 7. We explain why it is enough to study integral representations and reveal their relations with cyclotomic Hecke algebras $\mathcal{H}_n^\lambda$. We introduce Lie theoretic notation related to the Kac–Moody algebra $\mathfrak{g}$ of type $A_{p-1}^{(1)}$, which will be used until the end of Part I. The main results of the chapter are Basis Theorem for cyclotomic Hecke algebras, Cyclotomic Mackey Theorem, and the fact that τ-duality commutes with induction for cyclotomic Hecke algebras.

In Chapter 8 we study the cyclotomic analogues e_i^λ of the functors e_i and introduce the "dual" functors f_i^λ and related notions. We also introduce the divided powers functors which generalize e_i^λ and f_i^λ. The main goal of the Chapter is to get the f_i^λ-analogues of the results on e_i obtained in Chapter 5, that is to get the results on induction similar to our previous results on restriction. This turns out to be quite a bit harder.

Key Chapter 9 begins by defining a Hopf algebra structure on the Grothendieck group $K(\infty)$ of integral $\mathcal{H}_n$-modules for all $n \geq 0$, operations coming from induction and restriction. We prove in Theorem 9.5.3 that the dual algebra $K(\infty)^*$ can be identified with the universal enveloping algebra of the positive part of the Kac–Moody algebra $\mathfrak{g}$. We also obtain a natural action of $K(\infty)^*$ on the Grothendieck group $K(\lambda)$ of finite dimensional $\mathcal{H}_n^\lambda$-modules for all $n \geq 0$–under this action the Chevalley generators e_i act as operators e_i^λ, which come from representation theory of $\mathcal{H}_n^\lambda$. This action is extended with the operators f_i^λ and h_i^λ to give the action of the full Kac–Moody algebra $\mathfrak{g}$, and the module $K(\lambda)$ is identified with the irreducible high weight module

$V(\lambda)$. Finally, the weight spaces of $V(\lambda)$ are interpreted in terms of blocks of $\mathcal{H}_n^\lambda$-modules, see Section 9.6.

In Chapter 10 we identify the crystal graph of $V(\lambda)$ as the socle branching graph of the cyclotomic Hecke algebras $\mathcal{H}_n^\lambda$. The consequences for symmetric groups are deduced in Chapter 11, where we specialize to the case $\lambda = \Lambda_0$ and use the explicit description of the corresponding crystal graph to label the irreducible modules by p-regular partitions, to describe the blocks, and to deduce some of the branching rules. We also get useful results on formal characters.

Part II starts with Chapter 12, which is a review on superalgebras and their representations. This will be needed for the spin theory, as it turns out extremely convenient to consider the twisted group algebra $\mathcal{T}_n$ of S_n as a superalgebra and then work in the category of $\mathcal{T}_n$-supermodules. In fact, it is even more convenient to work with the Sergeev superalgebra $\mathcal{Y}_n$, which is defined in Chapter 13. We also prove in this Chapter that $\mathcal{Y}_n$ is "almost Morita equivalent" to $\mathcal{T}_n$. Chapters 14–22 are all parallel to the corresponding chapters of Part I, as indicated in the beginning of each of them, so we do not review them here in detail.

I have benefited greatly from collaboration with Jon Brundan. I was greatly influenced by the work of Jantzen and Seitz; Lascoux, Leclerc, and Thibon; Ariki; and, especially, Grojnowski. The idea to write this book appeared when I lectured on the topic at the University of Wisconsin-Madison during my sabbatical leave from Oregon. I am very grateful to Georgia Benkart, Ken Ono, and especially Arun Ram for their hospitality. I should also thank the editors–without their repeated (patient) emails I would never have finished. Most of all, I am indebted to my family (without whom this book might have been written faster, but that's not the point...).

PART I

Linear representations

1
Notation and generalities

Throughout the book: $\mathbb{Z}_+$ is the set of non-negative integers and F is an algebraically closed field of characteristic $p \geq 0$. Throughout Part I

$$I := \mathbb{Z} \cdot 1 \subset F. \tag{1.1}$$

If $p = \operatorname{char} F > 0$ then I is identified with $\{0, 1, \dots, p-1\}$, and if $p = 0$ then $I = \mathbb{Z}$.

If $\mathcal{A}$ is an associative F-algebra we denote by $\mathcal{A}$-mod the category of all finite dimensional left $\mathcal{A}$-modules and by $\mathcal{A}$-proj $\subset \mathcal{A}$-mod the full subcategory of all projective $\mathcal{A}$-modules. We write $K(\mathcal{A}\text{-mod})$, $K(\mathcal{A}\text{-proj})$ for the corresponding Grothendieck groups. The embedding $\mathcal{A}$-proj $\subset \mathcal{A}$-mod induces the natural *Cartan map*

$$\omega : K(\mathcal{A}\text{-proj}) \to K(\mathcal{A}\text{-mod}).$$

Note that in general ω does not have to be injective.

Let $M \in \mathcal{A}$-mod. The *socle* of M, written $\operatorname{soc} M$, is the largest completely reducible submodule of M, and the *head* of M, written $\operatorname{hd} M$, is the largest completely reducible quotient module of M. If V is an irreducible $\mathcal{A}$-module, we write $[M : V]$ for the multiplicity of V as a composition factor of M.

For algebras $\mathcal{A}$, $\mathcal{B}$, an $\mathcal{A}$-module M, and a $\mathcal{B}$-module N, we write $M \boxtimes N$ for the outer tensor product, that is the tensor product of vector spaces $M \otimes N$ considered as an $\mathcal{A} \otimes \mathcal{B}$-module in the usual way.

If $\mathcal{B}$ is a subalgebra of $\mathcal{A}$, and M is a $\mathcal{B}$-module we write $\operatorname{ind}_{\mathcal{B}}^{\mathcal{A}} M$ or $\operatorname{ind}^{\mathcal{A}}$ for the induced module $\mathcal{A} \otimes_{\mathcal{B}} M$. We may consider $\operatorname{ind}_{\mathcal{B}}^{\mathcal{A}}$ as a functor from the category of $\mathcal{B}$-modules to the category of $\mathcal{A}$-modules. This functor is left adjoint to the restriction functor $\operatorname{res}_{\mathcal{B}}^{\mathcal{A}}$ (or $\operatorname{res}_{\mathcal{B}}$) going in the other direction. If $\mathcal{A}$ is free as a right $\mathcal{B}$-module the induction functor is exact.

We denote by $\mathcal{Z}(\mathcal{A})$ the center of $\mathcal{A}$. By a *central character* of A we mean a (unital) algebra homomorphism $\chi : \mathcal{Z}(\mathcal{A}) \to F$. For a central character χ the corresponding *block* is the full subcategory $A\text{-mod}[\chi]$ of $\mathcal{A}$-mod consisting of all modules $M \in \mathcal{A}$-mod such that $(z-\chi(z))^k M = 0$ for $k \gg 0$. We have a decomposition

$$\mathcal{A}\text{-mod} = \bigoplus_{\chi} \mathcal{A}\text{-mod}[\chi]$$

as χ runs over all central characters. If $\mathcal{A}$ is finite dimensional, then two non-isomorphic irreducible $\mathcal{A}$-modules L and M belong to the same block if and only if there exists a chain $L \cong L_0, L_1, \dots, L_m \cong M$ of irreducible $\mathcal{A}$-modules with either $\operatorname{Ext}^1_{\mathcal{A}}(L_i, L_{i+1}) \neq 0$ or $\operatorname{Ext}^1_{\mathcal{A}}(L_{i+1}, L_i) \neq 0$ for each i.

Let $\mathcal{B}$ be a subalgebra of an F-algebra $\mathcal{A}$ and $\mathcal{C}$ be the centralizer of $\mathcal{B}$ in $\mathcal{A}$. If V is an $\mathcal{A}$-module and W is a $\mathcal{B}$-module then $\operatorname{Hom}_{\mathcal{B}}(W, \operatorname{res}_{\mathcal{B}} V)$ is naturally a $\mathcal{C}$-module with respect to the action $(cf)(w) = cf(w)$ for $w \in W, f \in \operatorname{Hom}_{\mathcal{B}}(W, \operatorname{res}_{\mathcal{B}} V), c \in \mathcal{C}$.

Lemma 1.0.1 *Let $\mathcal{B} \subseteq \mathcal{A}$ be semisimple finite dimensional F-algebras. If V is irreducible over $\mathcal{A}$ and W is irreducible over $\mathcal{B}$ then*

$$\operatorname{Hom}_{\mathcal{B}}(W, \operatorname{res}_{\mathcal{B}} V)$$

is irreducible over $\mathcal{C}$.

Proof By Wedderburn–Artin, we may assume that $\mathcal{A} = \operatorname{End}(V)$. Decompose $\operatorname{res}_{\mathcal{B}} V = W^{\oplus k} \oplus X$, where W is not a composition factor of X. Then the algebra $\operatorname{End}_{\mathcal{B}}(W^{\oplus k})$, naturally contained in $\mathcal{C}$, acts on the space $\operatorname{Hom}_{\mathcal{B}}(W, \operatorname{res}_{\mathcal{B}} V)$ as the full endomorphism algebra. □

For any $n \geq 0$, let $\alpha = (\alpha_1, \alpha_2, \dots)$ be a *partition of* n, that is a non-increasing sequence of non-negative integers summing to n. If $p > 0$, the partition α is called *p-regular* if for any $k > 0$ we have

$$\sharp\{j \mid \alpha_j = k\} < p.$$

By definition, every partition is 0-regular. Let $\mathcal{P}(n)$ (resp. $\mathcal{P}_p(n)$) denote the set of all (resp. all p-regular) partitions of n. Thus $\mathcal{P}(n) = \mathcal{P}_0(n)$. Set

$$\mathcal{P} := \bigcup_{n \geq 0} \mathcal{P}(n) \quad \text{and} \quad \mathcal{P}_p := \bigcup_{n \geq 0} \mathcal{P}_p(n).$$

We identify a partition α with its *Young diagram*

$$\alpha = \{(r, s) \in \mathbb{Z}_{>0} \times \mathbb{Z}_{>0} \mid s \leq \alpha_r\}.$$

Elements $(r, s) \in \mathbb{Z}_{>0} \times \mathbb{Z}_{>0}$ are called *nodes*. We label the nodes of α with *residues*, which are elements of I. By definition, the residue of the node (r, s)

is $s-r \pmod p$ if p is positive, and simply $s-r$ if $p=0$. The residue of the node A is denoted $\operatorname{res} A$. Define the *residue content* of α to be the tuple

$$\operatorname{cont}(\alpha) = (\gamma_i)_{i\in I}, \tag{1.2}$$

where for each $i \in I$, γ_i is the number of nodes of residue i contained in the diagram α.

Let $i \in I$ be some fixed residue. A node $A = (r, s) \in \alpha$ is called *i-removable* (resp. *i-addable*) for α if $\operatorname{res} A = i$ and $\alpha_A := \alpha \setminus \{A\}$ (resp. $\alpha^A := \alpha \cup \{A\}$) is a Young diagram of a partition. A node is called *removable* (resp. *addable*) if it is i-removable (resp. i-addable) for some i. Thus, for example, a removable node is always of the form (m, α_m) with $\alpha_m > \alpha_{m+1}$.

Throughout the book, S_n denotes the symmetric group on n letters. The permutations act on numbers $1, \dots, n$ on the *left* so that for the product we have, for example, $(1, 2)(2, 3) = (1, 2, 3)$. S_n also acts on n-tuples of objects by place permutations on the right:

$$(a_1, a_2, \dots, a_n) \cdot w = (a_{w1}, \dots, a_{wn})$$

or on the left:

$$w \cdot (a_1, a_2, \dots, a_n) = (a_{w^{-1}1}, \dots, a_{w^{-1}n}).$$

The length function on S_n in the sense of Coxeter groups is denoted by ℓ. The number $\ell(w)$ can be characterized as the number of inversions in the permutation w.

Finally we recall one classical result. Let

$$\mathcal{P}_n = F[x_1, \dots, x_n]$$

be the polynomial algebra in n indeterminates,

$$\mathcal{Z}_n := F[x_1, \dots, x_n]^{S_n}$$

be the ring of symmetric polynomials, and $\mathcal{Z}_n^+ \subset \mathcal{Z}_n$ be the symmetric polynomials without free term. The following fact is well known over $\mathbb{C}$. That it holds over $\mathbb{Z}$, and hence over F, is proved in [St].

Theorem 1.0.2 *$\mathcal{P}_n$ is a free module of rank $n!$ over $\mathcal{Z}_n$. Moreover we can take the set*

$$B := \{x_1^{a_1} \dots x_n^{a_n} \mid 0 \le a_i < i \text{ for all } 1 \le i \le n\}$$

as a basis. In particular, the cosets of elements of B form a basis of the algebra $\mathcal{P}_n/\mathcal{P}_n\mathcal{Z}_n^+$.

A slightly more general result easily follows:

Corollary 1.0.3 *Let* $r \leq n$. *Then* $\mathcal{P}_n^{S_r}$ *is a free module of rank* $n!/r!$ *over* $\mathcal{Z}_n$. *Moreover we can take the set*

$$B := \{x_{r+1}^{a_{r+1}} \dots x_n^{a_n} \mid 0 \leq a_i < i \textit{ for all } r+1 \leq i \leq n\}$$

as a basis.

Proof It suffices to prove that elements of B generate $\mathcal{P}_n^{S_r}$ as a module over $\mathcal{Z}_n$. Let $f \in \mathcal{P}_n^{S_r}$. In view of Theorem 1.0.2, we can write

$$f = \sum f_{\underline{a}} x_1^{a_1} \dots x_n^{a_n}, \tag{1.3}$$

where the summation is over all n-tuples $\underline{a} = (a_1, \dots, a_n)$ with $0 \leq a_i < i$, and $f_{\underline{a}} \in \mathcal{Z}_n$. Using Theorem 1.0.2 with $n = r$ we can also see that $\mathcal{P}_n$ is a free $\mathcal{P}_n^{S_r}$-module on basis $\{x_1^{a_1} \dots x_r^{a_r} \mid 0 \leq a_i < i \text{ for all } 1 \leq i \leq r\}$. Now note that the polynomials $f_{\underline{a}} x_{r+1}^{a_{r+1}} \dots x_n^{a_n}$ are in $\mathcal{P}_n^{S_r}$, so, since f is also in $\mathcal{P}_n^{S_r}$, it follows that $f_{\underline{a}} = 0$ in (1.3) unless $a_1 = \dots = a_r = 0$. This completes the proof. □

2
Symmetric groups I

In order to illustrate the theory we are trying to develop let us start from an "easy" special case, namely the case of complex representations of the symmetric group S_n. We explain the beautiful elementary approach of Okounkov and Vershik [OV] (see also [DG]). The idea of this approach is not new: to study all symmetric groups at once. However, it is rather amazing that in this way the whole theory can be built quickly from scratch using only the classical Maschke and Wedderburn–Artin Theorems.

We will obtain the following well-known results: labeling the irreducible $\mathbb{C}S_n$-modules by partitions of n, construction of Young's orthogonal bases in irreducible modules, explicit description of matrices of simple transpositions with respect to these bases, and the Murnaghan–Nakayama formula for irreducible characters.

2.1 Gelfand–Zetlin bases

Define the kth *Jucys–Murphy elements* (JM-element for short) $L_k \in FS_n$ as follows:

$$L_k := \sum_{1 \leq m < k} (m, k). \tag{2.1}$$

These elements were introduced in [Ju], [Mu$_1$]. Note that $L_1 = 0$ and L_k commutes with S_{k-1}. As $L_k \in FS_k$, it follows that the JM-elements commute. Here and below, if $m < n$, the default embedding of S_m into S_n is with respect to the *first m* letters. A copy of S_m embedded with respect to the *last m* letters is denoted by S'_m.

Denote by $\mathcal{Z}_n$ the center of the group algebra FS_n. Also let

$$\mathcal{Z}_{n,m} := (FS_{n+m})^{S_n}$$

be the centralizer of FS_n in FS_{n+m}. It is clear that $\mathcal{Z}_{n,m}$ is spanned by the class sums corresponding to the S_n-conjugacy classes in S_{n+m}. These conjugacy classes can be thought of as cycle shapes with "fixed positions" for $n+1, n+2, \dots, n+m$ – we call them *marked cycle shapes*. For example, the symbol

$$(*, *, *, *, *)(*, *)(*)(*)(12, *, 13, 14, *)(15) \tag{2.2}$$

corresponds to the S_{11}-conjugacy class in S_{15}, which consists of all permutations whose cycle presentation is obtained by inserting the numbers 1 through 11 instead of asterisks.

Proposition 2.1.1 [O_1] *The algebra $\mathcal{Z}_{n,m}$ is generated by S'_m, $\mathcal{Z}_n$, and $L_{n+1}, \dots, L_{n+m}$.*

Proof It is clear that S'_m, $\mathcal{Z}_n$, and $L_{n+1}, \dots, L_{n+m}$ are contained in $\mathcal{Z}_{n,m}$, so they generate a subalgebra $\mathcal{A} \subseteq \mathcal{Z}_{n,m}$. Conversely, let us filter $\mathcal{Z}_{n,m}$ so that the ith filtered component $\mathcal{Z}^i_{n,m}$ is the span of the class sums which correspond to the marked cycle shapes moving at most i elements. For example the class sum corresponding to (2.2) belongs to $\mathcal{Z}^{12}_{11,4}$, but not $\mathcal{Z}^{11}_{11,4}$. We prove by induction on $i = 0, 1, \dots$ that $\mathcal{Z}^i_{n,m} \subseteq \mathcal{A}$. For $i = 0$ and 1, we have $\mathcal{Z}^i_{n,m} = F \cdot 1 \subseteq \mathcal{A}$. We explain the inductive step on example. Let $z \in \mathcal{Z}^{12}_{11,4}$ be the class sum corresponding to the marked cycle shape from (2.2). Let $c \in \mathcal{Z}_{11}$ denote the sum of all elements of S_{11} whose cycle shape is

$$(*, *, *, *, *)(*, *)(*)(*).$$

Also, let

$$x = (12, 13)L_{12}(13, 14)(L_{14} - (12, 14) - (13, 14)) \in \mathcal{A}.$$

(Note that L_{12} is the class sum corresponding to $(*, 12)$, and $(L_{14} - (12, 14) - (13, 14))$ is the class sum corresponding to $(*, 14)$.) Then xc is equal to z modulo lower layers of our filtration. □

From now on until the end of Chapter 2 we assume that $F = \mathbb{C}$.

The following key multiplicity-freeness result is well known – it is a special case of the branching rule, which describes the restriction of an irreducible $\mathbb{C}S_n$-module to S_{n-1}. However, usually the branching rule is proved after some theory has been developed and irreducible modules have been studied. In the approach explained here the multiplicity-freeness result is proved from scratch and then used to develop a theory.

Theorem 2.1.2 *Let V be an irreducible $\mathbb{C}S_n$-module. Then the restriction* $\mathrm{res}_{S_{n-1}} V$ *is multiplicity free.*

Proof It follows from Proposition 2.1.1 that the centralizer of $\mathbb{C}S_{n-1}$ in $\mathbb{C}S_n$ is commutative. So the theorem comes from Lemmas 1.0.1. □

We now define the *branching graph* $\mathbb{B}$ whose vertices are isomorphism classes of irreducible $\mathbb{C}S_n$-modules for all $n \geq 0$ (by agreement $\mathbb{C}S_0 = \mathbb{C}$); we have a directed edge $W \to V$ from (an isoclass of) an irreducible $\mathbb{C}S_n$-module W to (an isoclass of) an irreducible $\mathbb{C}S_{n+1}$-module V if and only if W appears as a composition factor of $\mathrm{res}_{S_n} V$; there are no other edges. Our main goal is to find an explicit combinatorial description of the branching graph. This will give us a labeling of the irreducible $\mathbb{C}S_n$-modules *for all n*. This will also yield the branching rule. To achieve this goal we will actually do more.

Let V be an irreducible $\mathbb{C}S_n$-module. Pick an S_n-invariant inner product $(\cdot,\cdot)$ on V (it is unique up to a scalar). Theorem 2.1.2 implies that the decomposition

$$\mathrm{res}_{S_{n-1}} V = \bigoplus_{W \to V} W$$

is canonical. Decomposing each W on restriction to S_{n-2}, and continuing inductively all the way to S_0, we get a canonical decomposition

$$\mathrm{res}_{S_0} V = \bigoplus_T V_T$$

into irreducible $\mathbb{C}S_0$-modules, that is 1-dimensional subspaces V_T, where T runs over all paths $W_0 \to W_1 \to \cdots \to W_n = V$ in $\mathbb{B}$. Note that

$$\mathbb{C}S_k \cdot V_T = W_k \qquad (0 \leq k \leq n). \tag{2.3}$$

Choosing a vector $v_T \in V_T$, we get a basis $\{v_T\}$ of V called *Gelfand–Zetlin basis* (or GZ-basis). Vectors of GZ-basis are defined uniquely up to scalars. Moreover, if $\varphi : V \to V'$ is an isomorphism of irreducible modules then φ moves a GZ-basis of V to a GZ-basis of V'. Note also, for example using (2.3), that a GZ-basis is orthogonal with respect to $(\cdot,\cdot)$.

Now decompose the algebra $\mathbb{C}S_n$ according to the Wedderburn–Artin Theorem

$$\mathbb{C}S_n = \bigoplus_V \mathrm{End}_{\mathbb{C}}(V), \tag{2.4}$$

where the sum is over the representatives of the isoclasses of irreducible $\mathbb{C}S_n$-modules. This decomposition is canonical. Let us pick a GZ-basis in each V. Then we also identify

$$\mathbb{C}S_n = \bigoplus_V M_{\dim V}(\mathbb{C}). \tag{2.5}$$

Define the *GZ-subalgebra* $\mathcal{A}_n \subseteq \mathbb{C}S_n$ as the subalgebra which consists of all elements of $\mathbb{C}S_n$, which are diagonal with respect to a GZ-basis in every irreducible $\mathbb{C}S_n$-module. In terms of the decomposition (2.5), $\mathcal{A}_n$ consists of all diagonal matrices. In particular:

Lemma 2.1.3 *$\mathcal{A}_n$ is a maximal commutative subalgebra of $\mathbb{C}S_n$. Also, $\mathcal{A}_n$ is a semisimple algebra.*

We give two more descriptions of the GZ-subalgebra.

Lemma 2.1.4

(i) *$\mathcal{A}_n$ is generated by the subalgebras $\mathcal{Z}_0, \mathcal{Z}_1, \dots, \mathcal{Z}_n \subseteq \mathbb{C}S_n$.*
(ii) *$\mathcal{A}_n$ is generated by the JM-elements $L_1, L_2, \dots, L_n$.*

Proof (i) Let $e_V \in \mathcal{Z}_n$ be the central idempotent of $\mathbb{C}S_n$, which acts as identity on V and as zero on any irreducible $\mathbb{C}S_n$-module $V' \not\cong V$ (in terms of (2.4); e_V is the identity endomorphism in the V-component and zero endomorphism in other components). If $T = W_0 \to W_1 \to \dots \to W_n = V$ is a path in $\mathbb{B}$ then

$$e_{W_0} e_{W_1} \dots e_{W_n} \in \mathcal{Z}_0 \mathcal{Z}_1 \dots \mathcal{Z}_n$$

acts as the projection to V_T along $\oplus_{S \neq T} V_S$ and as zero on any irreducible $\mathbb{C}S_n$-module $V' \not\cong V$. So the subalgebra generated by $\mathcal{Z}_1, \mathcal{Z}_2, \dots, \mathcal{Z}_n$ contains $\mathcal{A}_n$. As this subalgebra is commutative and $\mathcal{A}_n$ is a maximal commutative subalgebra of $\mathbb{C}S_n$, the two must coincide.

(ii) Note that L_k is the sum of all transpositions in S_k minus the sum of all transpositions in S_{k-1}, that is L_k is a difference of a central element in S_k and a central element in S_{k-1}. So by (i), the JM-elements do belong to $\mathcal{A}_n$. To prove that they generate $\mathcal{A}_n$, proceed by induction on n, the inductive base being trivial. By (i), $\mathcal{A}_n$ is generated by $\mathcal{A}_{n-1}$ and $\mathcal{Z}_n$. In view of the inductive assumption, it suffices to prove that $\mathcal{A}_{n-1}$ and L_n generate $\mathcal{Z}_n$. But this follows from the obvious embedding $\mathcal{Z}_n \subseteq \mathcal{Z}_{n-1,1}$ and Proposition 2.1.1, as $\mathcal{Z}_{n-1} \subseteq \mathcal{A}_{n-1}$. □

Now, we will try to have the GZ-subalgebra play a role of a Cartan subalgebra in Lie Theory. As $\mathcal{A}_n$ is semisimple we can decompose every irreducible $\mathbb{C}S_n$-module V as a direct sum of simultaneous eigenspaces for the elements $L_1, \dots, L_n$. If $\lambda = (\lambda_1, \dots, \lambda_n) \in \mathbb{C}^n$ and V_λ is the simultaneous eigenspace for the $L_1, \dots, L_n$ corresponding to the eigenvalues $\lambda_1, \dots, \lambda_n$, respectively, then we say that λ is a *weight* of V and V_λ is the λ-*weight space* of V.

It follows from definitions that vectors of a GZ-basis are weight vectors. Also, since in terms of (2.5), $\mathcal{A}_n$ consists of *all* diagonal matrices, each weight space is 1-dimensional. Thus the weight spaces are precisely the spans of the elements of a GZ-basis. It also follows that if λ is a weight of an irreducible module V, then it is *not* a weight of irreducible $V' \not\cong V$. Thus, via GZ-bases, we get a one-to-one correspondence between all possible weights (for all symmetric groups) and all paths in $\mathbb{B}$. The weight corresponding to a path T will be denoted λ_T and a path corresponding to a weight λ will be denoted T_λ. We will also write v_λ for v_{T_λ}.

A path T ends at a vertex V if and only if the corresponding weight λ_T is a weight of V. It is clear now that in order to understand $\mathbb{B}$ it suffices to describe the sets

$$W(n) = \{\lambda \in \mathbb{C}^n \mid \lambda \text{ is a weight of a } \mathbb{C}S_n\text{-module}\} \quad (n \geq 0) \tag{2.6}$$

and the equivalence relation

$$\lambda \approx \mu \;\Leftrightarrow\; \lambda \text{ and } \mu \text{ are weights of the same irreducible } \mathbb{C}S_n\text{-module} \tag{2.7}$$

on $W(n)$. Indeed, note that

$$\mathbb{B} = \bigsqcup_{n \geq 0} (W(n)/\approx)$$

and for equivalence classes $[\lambda]$ and $[\mu]$ of $\lambda \in W(n-1)$ and $\mu \in W(n)$ we have $[\lambda] \to [\mu]$ if and only if $\lambda = (\nu_1, \dots, \nu_{n-1})$ for some $\nu \approx \mu$.

Remark 2.1.5 For the reader who is spoiled by knowing what the final answer should be: yes, the elements of the set $W(n)/\approx$ will be labeled by the partitions α of n, and the elements of the set $W(n)$ will be labeled by the standard α-tableaux for all such α, with two tableaux being equivalent if and only if they have the same shape α. To be more precise, if t is a standard α-tableaux, then the corresponding weight λ is obtained as follows: λ_i is the residue of the box in α which is occupied by i in the α-tableaux t $(1 \leq i \leq n)$.

The following notation will be convenient: if $\lambda = (\lambda_1, \dots, \lambda_n) \in W(n)$, we write $V(\lambda)$ for an irreducible $\mathbb{C}S_n$-module, which has λ as its weight.

The weight λ determines $V(\lambda)$ uniquely up to isomorphism, but $V(\lambda) \cong V(\mu)$ if and only if $\lambda \approx \mu$ Now, (2.3) can now be restated as follows:

$$\mathbb{C}S_k \cdot v_\lambda = V(\lambda_1, \ldots, \lambda_k) \qquad (0 \le k \le n). \tag{2.8}$$

2.2 Description of weights

We write $s_i := (i, i+1) \in S_n$, $1 \le i < n$, for basic transpositions. Note important relations

$$s_i L_i = L_{i+1} s_i - 1, \qquad s_i L_j = L_j s_i \quad (j \ne i, i+1). \tag{2.9}$$

The second relation immediately implies:

Lemma 2.2.1 *Let $\lambda = (\lambda_1, \ldots, \lambda_n) \in W(n)$, and $1 \le k < n$. Then $s_k v_\lambda$ is a linear combination of vectors v_μ such that $\mu_i = \lambda_i$ for $i \ne k, k+1$.*

If the role of Cartan subalgebra is played by $\mathcal{A}_n$, then the role of sl_2-subalgebras will be played by $\mathcal{B}_i := \langle L_i, L_{i+1}, s_i \rangle$. In view of (2.9), every $\mathcal{B}_i$ is a quotient of the *rank one degenerate affine Hecke algebra* (of which we will see much more later in this book):

$$\mathcal{H}_2 = \langle s, x, y \mid xy = yx, s^2 = 1, sx = ys - 1 \rangle.$$

We now construct some $\mathcal{H}_2$-modules. Fix a pair of numbers $a, b \in \mathbb{C}$. If $b = a + 1$, let $L(a, b) = \mathbb{C}v$ with the action of the generators

$$xv = av, \;\; yv = bv, \; sv = v.$$

Clearly, the relations are satisfied, so we have a well-defined action of $\mathcal{H}_2$. Similarly, if $b = a - 1$, we have $L(a, b) = \mathbb{C}v$ with

$$xv = av, \;\; yv = bv, \; sv = -v.$$

Finally, assume that $a \ne b \pm 1$. Let $L(a, b)$ be 2-dimensional with the action of the generators x, y, s, given, respectively, by the matrices

$$\begin{pmatrix} a & -1 \\ 0 & b \end{pmatrix}, \quad \begin{pmatrix} b & 1 \\ 0 & a \end{pmatrix}, \quad \begin{pmatrix} 0 & 1 \\ 1 & 0 \end{pmatrix}. \tag{2.10}$$

Note that, if $a = b$, then x and y do not act on $L(a, b)$ semisimply, while, if $a \ne b, b \pm 1$, then we can simultaneously diagonalize x and y, so that the matrices of x, y, s are

$$\begin{pmatrix} a & 0 \\ 0 & b \end{pmatrix}, \quad \begin{pmatrix} b & 0 \\ 0 & a \end{pmatrix}, \quad \begin{pmatrix} (b-a)^{-1} & 1 - (b-a)^{-2} \\ 1 & (a-b)^{-1} \end{pmatrix}. \tag{2.11}$$

To achieve this, change basis from $\{v_1, v_2\}$ to $\{v_1, v_2 - (b-a)^{-1}v_1\}$. If instead we change to

$$\{v_1, (1-(b-a)^{-2})^{-1/2}(v_2 - (b-a)^{-1}v_1)\}, \tag{2.12}$$

the matrix of s becomes orthogonal:

$$\begin{pmatrix} (b-a)^{-1} & \sqrt{1-(b-a)^{-2}} \\ \sqrt{1-(b-a)^{-2}} & (a-b)^{-1} \end{pmatrix}. \tag{2.13}$$

Proposition 2.2.2

(i) *Every irreducible $\mathcal{H}_2$-module is isomorphic to some $L(a, b)$.*
(ii) *If $a \neq b \pm 1$, then $L(a, b) \cong L(b, a)$, and there are no other isomorphic pairs among $\{L(a, b) \mid a, b \in \mathbb{C}\}$.*

Proof (i) Let V be an irreducible $\mathcal{H}_2$-module. There exists $v \in V$ which is a simultaneous eigenvector for x and y. So $xv = av$, $yv = bv$ for some $a, b \in \mathbb{C}$. If sv is proportional to v, then $V = \mathbb{C}v$, and we must have $sv = \pm v$, as $s^2 = 1$. This immediately leads to $b = a \pm 1$ and $V = L(a, b)$. If sv is not proportional to v, then $\{v, sv\}$ must be a basis of V, which leads to the formulas (2.10), but these formulas determine an irreducible module only if $a \neq b \pm 1$.

(ii) That no other pairs are isomorphic is clear, because if $L(a, b)$ and $L(c, d)$ are isomorphic, then their restrictions to $\langle x, y\rangle$ are isomorphic. Finally, if $a \neq b, b \pm 1$, it is easy to write down an explicit isomorphism between $L(a, b)$ and $L(b, a)$ using the formulas (2.11). □

Corollary 2.2.3 *Let $\lambda \in W(n)$, $V = V(\lambda)$, $1 \leq i < n$, and*

$$\mu = s_i\lambda = (\lambda_1, \dots, \lambda_{i-1}, \lambda_{i+1}, \lambda_i, \lambda_{i+2}, \dots, \lambda_n).$$

Then:

(i) $\lambda_i \neq \lambda_{i+1}$.
(ii) *If $\lambda_{i+1} = \lambda_i \pm 1$ then $s_i v_\lambda = \pm v_\lambda$ and μ is not a weight of V.*
(iii) *Let $\lambda_{i+1} \neq \lambda_i \pm 1$. Then μ is a weight of V. Moreover, the vector $w := (s_i - (\lambda_{i+1} - \lambda_i)^{-1})v_\lambda$ is a non-zero vector of weight μ, the elements s_i, L_i, L_{i+1} leave $X := \operatorname{span}(v_\lambda, w)$ invariant, and act in the basis $\{v_\lambda, w\}$ of X with matrices (2.11), respectively.*

Proof By (2.8), $\mathbb{C}S_{i+1} \cdot v_\lambda \cong V(\lambda_1, \dots, \lambda_{i+1})$. Consider

$$M := \operatorname{Hom}_{S_{i-1}}(V(\lambda_1, \dots, \lambda_{i-1}), V(\lambda_1, \dots, \lambda_{i+1}))$$

as a module over $\mathcal{Z}_{i-1,2} = \langle \mathcal{B}_i, \mathcal{Z}_{i-1} \rangle$, see Proposition 2.1.1. This module is irreducible by Lemma 1.0.1. By Schur's Lemma, $\mathcal{Z}_{i-1}$ acts on M with scalars, so M is irreducible even as a $\mathcal{B}_i$-module. Note that the $\mathcal{B}_i$-module M is isomorphic to the B_i-submodule $N := \mathcal{B}_i \cdot v_\lambda \subseteq V$. Inflating along the surjection $\mathcal{H}_2 \to \mathcal{B}_i$, N becomes an irreducible $\mathcal{H}_2$-module, with λ_i and λ_{i+1} appearing as eigenvalues of x and y, respectively. Hence $N \cong L(\lambda_i, \lambda_{i+1})$, see Proposition 2.2.2. Now the result follows from the classification of irreducible $\mathcal{H}_2$-modules obtained above, noting for (i) that x and y do not act semisimply of $L(a, a)$, so this case is impossible. □

Corollary 2.2.4 *Let $\lambda = (\lambda_1, \dots, \lambda_n) \in \mathbb{C}^n$. If $\lambda_i = \lambda_{i+2} = \lambda_{i+1} \pm 1$ for some i, then $\lambda \notin W(n)$.*

Proof Otherwise, Corollary 2.2.3(ii) gives $s_i v_\lambda = \pm v_\lambda$ and $s_{i+1} v_\lambda = \mp v_\lambda$, which contradicts the braid relation $s_i s_{i+1} s_i = s_{i+1} s_i s_{i+1}$. □

Lemma 2.2.5 *Let $\lambda \in W(n)$. Then:*

(i) $\lambda_1 = 0$.
(ii) $\{\lambda_i - 1, \lambda_i + 1\} \cap \{\lambda_1, \dots, \lambda_{i-1}\} \neq \varnothing$ *for all* $1 < i \leq n$.
(iii) *If* $\lambda_i = \lambda_j = a$ *for some* $i < j$ *then*

$$\{a-1, a+1\} \subseteq \{\lambda_{i+1}, \dots, \lambda_{j-1}\}.$$

Proof (i) is clear as $L_1 = 0$.

If (ii) fails, apply Corollary 2.2.3(iii) repeatedly to swap λ_i with λ_{i-1}, then with λ_{i-2}, etc., all the way to the second position. Now, if $\lambda_i = 0$, we get a weight which starts with two 0s, which contradicts Corollary 2.2.3(i). Otherwise, again by Corollary 2.2.3(iii), we can move λ_i to the first position, which contradicts (i).

If (iii) fails, let us pick i, j with the minimal $j - i$ for which this happens. By Corollaries 2.2.3(i), (iii) and 2.2.4, we have

$$\lambda = (\dots, a, a \pm 1, \dots, a \pm 1, a, \dots).$$

By the choice of $j - i$, there must be both $a \pm 1 + 1$ and $a \pm 1 - 1$ between the two entries $a \pm 1$ in λ. So there is an entry equal to a in between, which again contradicts the minimality of $j - i$. □

Recall partition notation introduced in Chapter 1. Define the *Young graph* $\mathbb{Y}$ as a directed graph with the set $\mathcal{P}$ of all partitions as its set of vertices; moreover, for $\alpha, \beta \in \mathcal{P}$ we have $\beta \to \alpha$ if and only if $\beta = \alpha_A$ for some removable node A for α. A path in $\mathbb{Y}$ ending in α will be referred to as an *α-path*. Thus an α-path T can be thought of as a sequence of nodes $T_1, \dots, T_n$ of α such that T_n is removable for α, T_{n-1} is removable for α_{T_n}, etc. If, for all $1 \le i \le n$, we substitute the number i for the box T_i, we get an *α-tableau*, that is an array of integers $1, \dots, n$ of shape α and such that the numbers increase from top to bottom along the columns and from left to right along the rows. In this way we get a one-to-one correspondence between the α-paths in $\mathbb{Y}$ and the α-tableaux. We will not distinguish between the two. Thus, if T is an α-tableaux, T_i is the box occupied by i in T. To each α-tableau T, we associate the n-tuple

$$\lambda_T = (\operatorname{res} T_1, \dots, \operatorname{res} T_n) \in \mathbb{Z}^n.$$

Example 2.2.6 If $\alpha = (4, 2, 1)$, an example of an α-tableau is given by

$$T = \begin{array}{|c|c|c|c|}\hline 1 & 2 & 4 & 5 \\ \hline 3 & 7 \\ \cline{1-2} 6 \\ \cline{1-1}\end{array}$$

In this case $\lambda_T = (0, 1, -1, 2, 3, -2, 0)$.

Set

$$W'(n) := \{\lambda_T \mid T \text{ is an } \alpha\text{-tableau for some } \alpha \in \mathcal{P}(n)\}. \tag{2.14}$$

Note that the shape α of T can be recovered from the weight λ_T: the amount of as among the λ_is is the amount of nodes on the ath diagonal of the Young diagram α. So the n-tuples $\lambda, \mu \in W'(n)$ come from tableaux of the same shape if and only if λ can be obtained from μ by a place permutation, in which case we write $\lambda \sim \mu$.

Lemma 2.2.7 *The set $W'(n)$ is precisely the set of all n-tuples $\lambda \in \mathbb{C}^n$ which satisfy the properties (i)–(iii) of Lemma 2.2.5. In particular, $W(n) \subseteq W'(n)$.*

Proof Easy combinatorial exercise. □

If $\nu = (\nu_1, \dots, \nu_n) \in \mathbb{C}^n$, and $\nu_i \neq \nu_{i+1} \pm 1$, then a place permutation which swaps ν_i and ν_{i+1} will be called an *admissible transposition*. If $\nu = \lambda_T$ for a tableau T, then an admissible transposition amounts to swapping i and $i+1$ that do not lie on adjacent diagonals in T. It is clear that such a swap always transforms an α-tableaux to an α-tableaux.

Let $\alpha = (\alpha_1 \ge \alpha_2 \ge \dots \ge \alpha_k) \in \mathcal{P}(n)$. We define the corresponding *canonical α-tableau* $T(\alpha)$ to be the α-tableau obtained by filling in the numbers

$1, 2, \ldots, n$ from left to right along the rows, starting from the first row and going down.

Lemma 2.2.8 *Let $\alpha \in \mathcal{P}(n)$. If T is an α-tableau, then there is a series of admissible transpositions which moves T to $T(\alpha)$. Moreover, these transpositions $s_{i_1}, s_{i_2}, \ldots, s_{i_\ell}$ can be chosen in such a way that $\ell = \ell(s_{i_1} s_{i_2} \ldots s_{i_\ell})$.*

Proof Let A be the last node of the last row of α. In $T(\alpha)$, A is occupied by n. In T, A is occupied by some number i. Note also that in T, $i+1$ and i do not lie on adjacent diagonals. So we can apply an admissible transposition to swap i and $i+1$, then to swap $i+1$ and $i+2$, etc. As a result, we get a new α-tableau in which A is occupied by n. Next, forget about A, and take care of $(n-1)$. Continuing this way we will get a chain of admissible transpositions which transform T to $T(\alpha)$. Finally, note that this chain yields a reduced word. □

Lemma 2.2.9 *If $\lambda \in W'(n)$ and $\lambda \sim \mu$ for some $\mu \in W(n)$, then $\lambda \in W(n)$ and $\lambda \approx \mu$.*

Proof Let $\lambda = \lambda_T$. By Lemma 2.2.7, $\mu = \lambda_S$. As $\lambda \sim \mu$, the tableaux S and T have the same shape. In view of Corollary 2.2.3(iii), it suffices to show that we can go from λ_S to λ_T by a chain of admissible transpositions. But this follows from Lemma 2.2.8. □

The following is the main result of this chapter.

Theorem 2.2.10 *We have $W(n) = W'(n)$. Moreover, $\lambda_T \approx \lambda_S \Leftrightarrow$ the tableaux T and S have the same shape $\Leftrightarrow \lambda_T \sim \lambda_S$. In particular, the branching graph $\mathbb{B}$ is isomorphic to the Young graph $\mathbb{Y}$.*

Proof By Lemma 2.2.7, $W(n) \subseteq W'(n)$. As the number of (isoclasses of) irreducible $\mathbb{C}S_n$-modules equals the number of conjugacy classes of S_n, which are labeled by partitions of n, we have

$$|W(n)/\approx| = |\mathcal{P}(n)| = |W'(n)/\sim|. \tag{2.15}$$

Now, let $\lambda \in W'(n)$. In view of Lemma 2.2.9, the $\sim$-equivalence class of λ either contains no elements of $W(n)$ or is a subset of $\approx$-equivalence class of $W(n)$. In view of (2.15), this now implies $W(n) = W'(n)$ and $\sim$ is equivalent to $\approx$. □

Now to every irreducible $\mathbb{C}S_n$-module V we can associate a partition $\alpha \in \mathcal{P}(n)$. Indeed, if $\lambda \in W(n)$ is a weight of V then $\lambda = \lambda_T$ for some tableaux T, and we associate to V the shape α of T. We will write $V = V^\alpha$. This notation is better than $V(\lambda)$ for $\lambda \in W(n)$, because we have a one-to-one correspondence between the isoclasses of irreducible $\mathbb{C}S_n$-module and partitions of n. The weights of V^α are precisely $\{(\text{res}\, T_1, \dots, \text{res}\, T_n) \mid T \text{ is an } \alpha\text{-tableau}\}$.

Example 2.2.11 (i) If $\alpha = (n)$, the only α-tableau is $\boxed{1}\boxed{2}\boxed{3}\cdots\boxed{n}$. So $V^{(n)}$ is 1-dimensional, and its only weight is $(0, 1, \dots, n-1)$. Similarly, $V^{(1^n)}$ is 1-dimensional with the only weight $(0, -1, \dots, -n)$. It is clear from this information that $V^{(n)}$ is the trivial and $V^{(1^n)}$ is the sign modules over S_n.

(ii) Let $\alpha = (n-1, 1)$. Then the α-tableaux are $T(i) := \begin{array}{|c|c|c|c|}\hline 1 & 2 & \cdots & n \\ \hline i \\ \cline{1-1}\end{array}$ for $2 \le i \le n$, and the corresponding weights are

$$\lambda(i) := (0, 1, \dots, i-2, -1, i-1, \dots, n-2) \qquad (2 \le i \le n).$$

Note what we have done so far. We have started from a nested family of algebras $\mathbb{C}S_0 \subset \mathbb{C}S_1 \subset \dots$, proved the multiplicity-freeness of the branching rule from scratch, defined the branching graph $\mathbb{B}$, and tried to learn enough facts about $\mathbb{B}$, so that we could identify it with some known graph. This have lead to a classification of irreducible $\mathbb{C}S_n$-modules for all n and a description of the branching rule at the same time. On the way we have obtained other useful results about irreducible modules. We are going to follow this scheme again and again in this book for various families of algebras, although to realize it we will need more sophisticated tools.

2.3 Formulas of Young and Murnaghan–Nakayama

Formulas of Young describe explicitly the matrices of simple transpositions s_i with respect to a nice choice of a GZ-basis. The formulas come more or less from (2.11) and (2.13). But we need to scale elements of a GZ-basis in a consistent way.

In order to do this, fix $\alpha \in \mathcal{P}(n)$. Pick a basis vector $v_{T(\alpha)} \in V^\alpha_{T(\alpha)}$ corresponding to the canonical α-tableau. Let T be an arbitrary α-tableau. Write $T = w \cdot T(\alpha)$ for $w \in S_n$. Define $\ell(T)$ to be $\ell(w)$. Denote by π_T the projection to V^α_T along $\oplus_{S \neq T} V^\alpha_S$, and set

$$v_T = \pi_T(w v_{T(\alpha)}). \tag{2.16}$$

By Lemma 2.2.8, there is a reduced decomposition $w = s_{i_1} \dots s_{i_\ell}$ with all simple transpositions being admissible. So Corollary 2.2.3(iii) implies

$$wv_{T(\alpha)} = v_T + \sum_{S:\,\ell(S)<\ell(T)} c_S v_S, \tag{2.17}$$

and $v_T \neq 0$.

Theorem 2.3.1 (Young's seminormal form) *Let $\alpha \in \mathcal{P}(n)$, $\{v_T\}$ be the GZ-basis of V^α defined in (2.16), and $1 \leq i < n$. Then the action of the simple transposition $s_i \in S_n$ is given as follows:*

(i) *If* $\operatorname{res} T_{i+1} = \operatorname{res} T_i \pm 1$, *then* $s_i v_T = \pm v_T$.
(ii) *Let* $\rho := (\operatorname{res} T_{i+1} - \operatorname{res} T_i)^{-1} \neq \pm 1$ *and set* $S = s_i T$. *Then*

$$s_i v_T = \begin{cases} \rho v_T + v_S, & \text{if } \ell(S) > \ell(T), \\ -\rho v_T + (1-\rho^2) v_S, & \text{if } \ell(S) < \ell(T). \end{cases}$$

Proof If $\operatorname{res} T_{i+1} = \operatorname{res} T_i \pm 1$, the result follows from Corollary 2.2.3(ii). Otherwise s_i is an admissible transposition for T. We may assume that $\ell(S) > \ell(T)$. As weight spaces of V^α are 1-dimensional, Corollary 2.2.3(iii) implies that v_S equals $s_i v_T - \rho v_T$ up to a scalar multiple, and, using (2.17), we see that the scalar is 1. □

Corollary 2.3.2 *Irreducible representations of S_n are defined over $\mathbb{Q}$ and are self-dual.*

Theorem 2.3.3 (Young's orthogonal form) *Let $\alpha \in \mathcal{P}(n)$. There exists a GZ-basis $\{w_T\}$ of V^α such that the action of an arbitrary simple transposition $s_i \in S_n$ is given by*

$$s_i w_T = \rho w_T + \sqrt{1-\rho^2}\, w_{s_i T}$$

where $\rho := (\operatorname{res} T_{i+1} - \operatorname{res} T_i)^{-1}$ *(note that when $\rho = \pm 1$, the coefficient of $w_{s_i T}$ is zero, so this term should be omitted).*

Proof Let $\{v_T\}$ be the basis of Theorem 2.3.1, and set

$$w_T = v_T / \sqrt{(v_T, v_T)}.$$

Let $S = s_i T$. We may assume that s_i is an admissible transposition and $\ell(S) > \ell(T)$. As s_i preserves $(\cdot,\cdot)$, the formulas of Theorem 2.3.1(ii) imply $(v_S, v_S) = (1-\rho^2)(v_T, v_T)$, whence $w_S = v_S / (\sqrt{(v_T, v_T)}\sqrt{1-\rho^2})$. Now, the result follows from (2.12) and (2.13). □

Example 2.3.4 Let $\alpha = (n-1, n)$. Using the notation of Example 2.2.11(ii) and writing v_j for $v_{T(j)}$, $2 \le j \le n$, the formulas of Young's orthogonal form become

$$s_i v_j = \begin{cases} v_j, & \text{if } j \neq i, i+1, \\ \frac{1}{i} v_i + \sqrt{1 - \frac{1}{i^2}} v_{i+1}, & \text{if } j = i, \\ \sqrt{1 - \frac{1}{i^2}} v_i - \frac{1}{i} v_{i+1}, & \text{if } j = i+1. \end{cases} \tag{2.18}$$

Let M be the natural permutation $\mathbb{C}S_n$-module with basis $e_1, \dots, e_n$. It has the irreducible submodule $N = \{\sum_i a_i e_i \in M \mid \sum_i a_i = 0\}$. Set

$$v_j := \frac{1}{\sqrt{j(j-1)}}(e_1 + \dots + e_{j-1} - (j-1)e_j) \qquad (2 \le j \le n).$$

Then $\{v_2, v_3, \dots, v_n\}$ is a basis of N with respect to which the simple permutations act by formulas (2.18).

Let $\alpha \in \mathcal{P}(n)$ and $\beta \in \mathcal{P}(n-k)$. Set

$$V^{\alpha/\beta} := \operatorname{Hom}_{S_{n-k}}(V^\beta, \operatorname{res}_{S_{n-k}} V^\alpha).$$

It is clear from the branching rule that $V^{\alpha/\beta} \neq 0$ if and only if the Young diagram β is contained in the Young diagram α, in which case we denote the complement by α/β. A set of nodes of this form will be called a *skew shape*. The number of nodes in α/β will be denoted $|\alpha/\beta|$. The number of rows occupied by $\lambda/\mu - 1$ will be denoted by $L(\alpha/\beta)$. A skew shape is called a *skew hook* if it is connected and does not have two boxes on the same diagonal (equivalently, if the residues of the nodes of the shape form a segment of integers).

We know that $V^{\alpha/\beta}$ is an irreducible $\mathcal{Z}_{n-k,k}$-module. On restriction to $S'_k \subset \mathcal{Z}_{n-k,k}$ it becomes a (not necessarily irreducible) $\mathbb{C}S_k$-module. Let $\chi^{\alpha/\beta}$ be the character of this $\mathbb{C}S_k$-module. If $\beta = \varnothing$ we get the character χ^α of V^α. The results on GZ-bases and Young's canonical forms can be easily generalized to skew shapes. For example, define an *α/β-path* to be any path which connects β with α. We will not distinguish between α/β-paths and α/β-tableaux (defined in the obvious way). Then Theorem 2.3.3 implies:

Proposition 2.3.5 (Young's orthogonal form for skew shapes) *Let α/β be a skew shape with $|\alpha/\beta| = k$, $|\beta| = n-k$. There exists a basis $\{w_T \mid T$ is an α/β-tableau$\}$ of $V^{\alpha/\beta}$ such that the action of an arbitrary simple transposition $s_i \in S_k$ is given by*

$$s_i w_T = \rho w_T + \sqrt{1-\rho^2}\, w_{s_i T}$$

where $\rho := (\text{res}\, T_{i+1} - \text{res}\, T_i)^{-1}$ *(note that when* $\rho = \pm 1$*, the coefficient of* $w_{s_i T}$ *is zero, so this term should be omitted). Moreover, each vector* w_T *is a simultaneous eigenvector for* $L_{n-k+1}, \dots, L_n \in \mathcal{Z}_{n-k,k}$ *with eigenvalues* $\text{res}\, T_1, \dots, \text{res}\, T_k$*, respectively.*

The final main result of this section is:

Theorem 2.3.6 *Let* α/β *be a skew shape with* $|\alpha/\beta| = k$*. Then*

$$\chi^{\alpha/\beta}\big((1, 2, \dots, k)\big) = \begin{cases} (-1)^{L(\alpha/\beta)}, & \text{if } \alpha/\beta \text{ is a skew hook,} \\ 0, & \text{otherwise.} \end{cases}$$

The following is a very effective way to evaluate an irreducible character on a given element.

Corollary 2.3.7 (Murnaghan–Nakayama rule) *Let* α/β *be a skew shape with* $|\alpha/\beta| = k$*, and* c *be an element of* S_k *whose cycle shape corresponds to a partition* $\rho = (\rho_1 \geq \cdots \geq \rho_l > 0) \in \mathcal{P}(k)$*. Then*

$$\chi^{\alpha/\beta}(c) = \sum_H (-1)^{L(H)},$$

where the sum is over all sequences H *of partitions*

$$\beta = \alpha(0) \subset \alpha(1) \subset \cdots \subset \alpha(l) = \alpha$$

such that $\alpha(i)/\alpha(i-1)$ *is a skew hook with* $|\alpha(i)/\alpha(i-1)| = \rho_i$ *for all* $1 \leq i \leq l$*, and* $L(H) = \sum_{i=1}^{l} L(\alpha(i)/\alpha(i-1))$*.*

Proof By the branching rule for $m < k$ we have

$$\text{res}_{S_m \times S_{k-m}} V^{\alpha/\beta} = \oplus_\gamma V^{\gamma/\beta} \boxtimes V^{\alpha/\gamma},$$

where the sum is over all partitions γ with $\beta \subset \gamma \subset \alpha$, such that $|\gamma/\beta| = m$. More generally,

$$\text{res}_{S_{\rho_1} \times \cdots \times S_{\rho_l}} V^{\alpha/\beta} = \oplus_H V^{\alpha(1)/\alpha(0)} \boxtimes \cdots \boxtimes V^{\alpha(l)/\alpha(l-1)},$$

where the summation is over all H as in the statement of the corollary. Now the result follows from Theorem 2.3.6. □

We proceed to prove Theorem 2.3.6. Fix a skew shape α/β with $|\alpha/\beta| = k$.

Lemma 2.3.8 *Theorem 2.3.6 is true for $\beta = \varnothing$.*

Proof It is easy to see that $L_2L_3 \dots L_k$ is the sum of all k-cycles in S_k. If $v \in V^\alpha$ is a weight vector of weight λ, then $L_2L_3 \dots L_k v = \lambda_2\lambda_3 \dots \lambda_n v$, which is zero unless α is a hook, see Theorem 2.2.10. However, if $\alpha = (k-b, 1^b)$ is a hook with $L(\alpha) = b$, then, again by Theorem 2.2.10, we have $\lambda_2 \dots \lambda_n = (-1)^b b!(k-b-1)!$ and $\dim V^\alpha = \binom{k-1}{b}$. Now the result follows from the fact that there are $(k-1)!$ k-cycles in S_k. □

Lemma 2.3.9 *In the notation of Proposition 2.3.5, $\mathbb{C}S_k \cdot w_T = V^{\alpha/\beta}$ for any T.*

Proof As $V^{\alpha/\beta}$ is irreducible over $\mathcal{Z}_{n-k,k}$, we have $\mathcal{Z}_{n-k,k} \cdot w_T = V^{\alpha/\beta}$. However, in view of (2.9), every element of $\mathcal{Z}_{n-k,k}$ can be written as gxz, where $g \in \mathbb{C}S_k$, $x \in \langle L_{n-k+1}, \dots, L_n \rangle$, $z \in \mathcal{Z}_{n-k}$. As x and z act on w_T by multiplication with scalars, the result follows. □

Lemma 2.3.10 *If α/β is not connected, then $\chi^{\alpha/\beta}((1, 2, \dots, k)) = 0$.*

Proof Let $\alpha/\beta = \gamma \cup \delta$, where γ and δ are skew shapes disconnected from each other, that is $|\mathrm{res}\, C - \mathrm{res}\, D| > 1$ for any $C \in \gamma$ and $D \in \delta$. Let $c := |\gamma|$ and $d := |\delta|$. There exists an α/β-tableau T such that $T_1, \dots, T_c \in \gamma$ and $T_{c+1}, \dots, T_k \in \delta$. By Proposition 2.3.5, the subspace of $V^{\alpha/\beta}$, spanned by vectors w_T for all such tableaux T, is invariant with respect to $S_c \times S_d < S_k$, and, as a $\mathbb{C}[S_c \times S_d]$-module, it is isomorphic to $V^\gamma \boxtimes V^\delta$. By Lemma 2.3.9 and Frobenius reciprocity, we get a surjective homomorphism $\mathrm{ind}^{S_k}(V^\gamma \boxtimes V^\delta) \to V^{\alpha/\beta}$. But, using Proposition 2.3.5, we see that the dimensions of both modules are equal to $\binom{k}{c} \dim V^\gamma \dim V^\delta$. So $V^{\alpha/\beta} \cong \mathrm{ind}^{S_k}(V^\gamma \boxtimes V^\delta)$. Now the lemma follows from the following standard general fact: if H is a subgroup of a finite group G, $g \in G$ is not conjugate to an element of H, and V is a $\mathbb{C}G$-module induced from H, then the character of V on g is zero. □

Lemma 2.3.11 *If α/β has two nodes on the same diagonal, and $\gamma = (a, 1^{k-a})$ be an arbitrary hook with k-boxes, then V^γ is not a composition factor of $V^{\alpha/\beta}$. In particular, $\chi^{\alpha/\beta}((1, 2, \dots, k)) = 0$.*

Proof The second statement follows from the first by Lemma 2.3.8. By assumption a 2×2 square ⊞ is contained in α/β. It follows from Proposition 2.3.5 that $V^{(2,2)}$ is an S_4-submodule of $V^{\alpha/\beta}$ (for S_4 embedded not necessarily with respect to the first four letters, but such S_4 is conjugate to the canonical one anyway). By Frobenius reciprocity and Lemma 2.3.9, there is a surjection $\text{ind}^{S_k} V^{(2,2)} \to V^{\alpha/\beta}$, and the result now follows from the branching rule. □

Lemma 2.3.12 *Let α/β be a skew hook, and $\gamma = (k-b, 1^b)$. Then V^γ appears as a composition factor of $V^{\alpha/\beta}$ if and only if $b = L(\alpha/\beta)$, in which case its multiplicity is one.*

Proof It follows from Proposition 2.3.5 that translation of α/β does not change the corresponding S_k-module. So we may assume that α and β are minimal possible, as in the picture

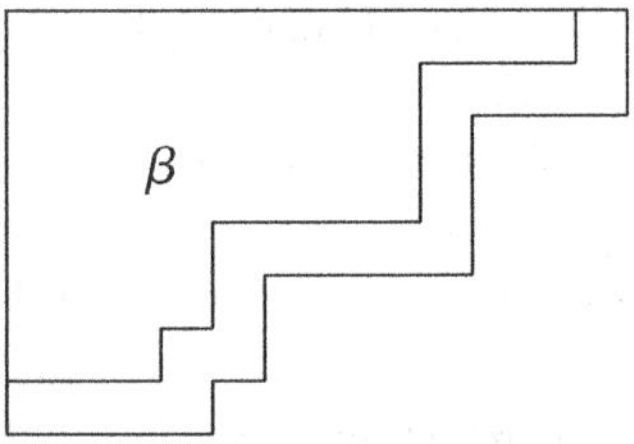

Now, if $b \neq L(\alpha/\beta)$, then $\gamma \not\subseteq \alpha$, so, by the branching rule, V^γ does not appear as a composition factor of $\text{res}_{S_k} V^\alpha$, hence it does not appear in $V^{\alpha/\beta}$ either.

Let $b = L(\alpha/\beta)$. Note that α/γ has shape β. So it follows from Proposition 2.3.5 that $V^{\alpha/\gamma}$ and V^β are isomorphic as $\mathbb{C}S_{n-k}$-modules. So $[\text{res}_{S_k \times S_{n-k}} V^\alpha : V^\gamma \boxtimes V^\beta] = 1$. Then $[\text{res}_{S_{n-k} \times S_k} V^\alpha : V^\beta \boxtimes V^\gamma] = 1$. It remains to note that $[V^{\alpha/\beta} : V^\gamma] = [\text{res}_{S_{n-k} \times S_k} V^\alpha : V^\beta \boxtimes V^\gamma]$. □

Theorem 2.3.6 follows from Lemmas 2.3.8 and 2.3.10–2.3.12.

Remark 2.3.13 We sketch another interpretation of the graph $\mathbb{Y}$ referring the reader to [KR, Lecture 4] for details. In fact, this interpretation is closer to what we are going to do later in the book. Let $\mathfrak{g} = \mathfrak{gl}_\infty(\mathbb{C})$ be the Lie algebra of all $\mathbb{Z} \times \mathbb{Z}$-matrices over $\mathbb{C}$ with only finitely many non-zero entries. Thus, the matrix units $\{E_{ij} \mid i, j \in \mathbb{Z}\}$ form a basis of $\mathfrak{g}$. The Lie algebra $\mathfrak{g}$ acts on the *Fock space* $\mathcal{F}$, which is the complex vector space, whose basis consists of the formal semi-infinite wedges $v_{i_0} \wedge v_{i_1} \wedge v_{i_2} \wedge \cdots$ such that $i_0 > i_1 > \dots$ and $i_k = -k$ for $k \gg 0$. To write down the action we follow the usual rules for the action of Lie algebra on a wedge power of a module. For example,

$$E_{2,-1} \cdot v_0 \wedge v_{-1} \wedge v_{-2} \wedge \cdots = v_0 \wedge v_2 \wedge v_{-2} \wedge \cdots = -v_2 \wedge v_0 \wedge v_{-2} \wedge \cdots .$$

In fact, more than just $\mathfrak{g}$ acts on $\mathcal{F}$. Let $a_k = \sum_{j-i=k} E_{i,j}$ be the kth diagonal. Even though a_k is not an element of $\mathfrak{g}$, we can still extend the action of $\mathfrak{g}$ to it, at least if $k \neq 0$. For example,

$$a_{-2} \cdot v_0 \wedge v_{-1} \wedge v_{-2} \wedge \cdots = v_2 \wedge v_{-1} \wedge v_{-2} \wedge \cdots - v_1 \wedge v_0 \wedge v_{-2} \wedge \cdots .$$

It is convenient to label semi-infinite wedges by partitions: to a partition $\alpha = (\alpha_1 \geq \alpha_2 \geq \ldots)$ we associate the vector $v_\alpha := v_{i_0} \wedge v_{i_1} \wedge v_{i_2} \wedge \cdots$ with $i_j = \alpha_j - j$. For example, $v_\varnothing = v_0 \wedge v_{-1} \wedge v_{-2} \wedge \cdots$. Then $\{v_\alpha \mid \alpha \in \mathcal{P}\}$ is a basis of $\mathcal{F}$, and we have in some sense recovered the vertices of $\mathbb{Y}$. For the edges, note that $E_{i,i+1} v_\alpha = v_\beta$ where β is obtained from α by removing a removable node of residue i, if it exists, and otherwise v_β is interpreted as 0. Similarly, $E_{i+1,i} v_\alpha = v_{\geq}$, where $\geq$ is obtained from α by adding an addable node of residue i, if it exists, and otherwise $v_{\geq}$ is interpreted as 0. Thus the action of the Chevalley generators of $\mathfrak{g}$ on the basis vectors $\{v_\alpha\}$ recovers the edges of $\mathbb{Y}$. Is it possible to explain this remarkable coincidence of two graphs, one coming from representation theory of S_n and the other from (completely different) representation theory of $\mathfrak{gl}_\infty(\mathbb{C})$? If you want to know the answer, keep reading this book...

We make one more observation along these lines. It is easy to see that for $i > j$ we have $E_{i,j} v_\alpha = \varepsilon v_\beta$, where β is obtained from α by removing a skew hook of length $j - i$, starting at the node of residue i and ending at the node of residue $j - 1$; if no such hook exists, interpret v_β as 0. Moreover, $\varepsilon = (-1)^{L(\alpha/\beta)}$. It follows that for $k > 0$ we have

$$a_k v_\alpha = \sum (-1)^{L(\alpha/\beta)} v_\beta$$

where the sum is over all β such that α/β is a skew hook with $|\alpha/\beta| = k$. So the Murnaghan–Nakayama rule can be interpreted as follows: the value $\chi^\alpha(c_\rho)$ of the irreducible character χ^α on an element c_ρ with cycle-shape $(\rho_1, \rho_2, \ldots, \rho_l)$ is equal to the coefficient of $v_\varnothing$ in $a_{\rho_1} a_{\rho_2} \ldots a_{\rho_l} v_\alpha$. Or better yet:

$$\chi^\alpha(c_\rho) = (v_\alpha\,,\, a_{-\rho_1} a_{-\rho_2} \cdots a_{-\rho_l} v_\varnothing), \tag{2.19}$$

where $(\cdot, \cdot)$ is the contravariant form on $\mathcal{F}$ normalized so that $(v_\varnothing, v_\varnothing) = 1$.

3
Degenerate affine Hecke algebra

In this chapter we define the degenerate affine Hecke algebra $\mathcal{H}_n$. As a vector space, $\mathcal{H}_n$ is the tensor product $FS_n \otimes F[x_1, \ldots, x_n]$ of the group algebra FS_n and the free commutative polynomial algebra $F[x_1, \ldots, x_n]$. Moreover, $FS_n \otimes 1$ and $1 \otimes F[x_1, \ldots, x_n]$ are subalgebras of $\mathcal{H}_n$ isomorphic to FS_n and $F[x_1, \ldots, x_n]$, respectively. Furthermore, there exists an algebra homomorphism $\mathcal{H}_n \to FS_n$, which is the "identity" on the subalgebra FS_n, that is sends $w \otimes 1$ to w, see Chapter 7. However, the relations between the elements of FS_n and $F[x_1, \ldots, x_n]$ are *not quite* the ones coming from the natural action of S_n on $F[x_1, \ldots, x_n]$ – they are those modulo some "garbage", which often can be kept under control, see for example Lemma 3.2.1. In particular, the center of $\mathcal{H}_n$ is what we would like it to be: the ring of symmetric polynomials $F[x_1, \ldots, x_n]^{S_n}$.

We introduce parabolic subalgebras of $\mathcal{H}_n$, the corresponding induction and restriction functors, and prove a Mackey-type theorem, which as usual is a result on induction followed by restriction. The only difference with Mackey Theorem for S_n is that here we have a filtration instead of a direct sum. Some important antiautomorphisms of $\mathcal{H}_n$ are defined next. Informally, the automorphism σ swaps the roles of the left- and the right-hand sides. For example, if we have proved something about the action of x_n on $\mathcal{H}_n$-modules, we may then apply σ and get a similar result about x_1. The antiautomorphism τ, however, will be used to define a dual of an $\mathcal{H}_n$-module. It will turn out later (see Chapter 5) that the irreducible $\mathcal{H}_n$-modules are self-dual. A non-trivial Theorem 3.7.5 establishes a nice relation between duality and induction. Finally, we introduce certain intertwining elements Φ_w for $w \in S_n$, whose main property is that they commute with the variables x_i according to the natural action of the symmetric group element w (no "garbage" this time!)

3.1 The algebras

Let

$$\mathcal{P}_n = F[x_1, \dots, x_n]$$

be the algebra of polynomials in $x_1, \dots, x_n$. If $\alpha = (\alpha_1, \dots, \alpha_n) \in \mathbb{Z}_+^n$, set

$$x^\alpha := x_1^{\alpha_1} \dots x_n^{\alpha_n}.$$

We record the relations

$$x_i x_j = x_j x_i, \tag{3.1}$$

for all $1 \le j < i \le n$ (the precise form of the relations is used in the proof of Theorem 3.2.2).

The group algebra FS_n of S_n over F will be denoted by $\mathcal{G}_n$. Thus, $\mathcal{G}_n$ is generated by the basic transpositions $s_1, \dots, s_{n-1}$ subject to relations

$$s_i^2 = 1, \tag{3.2}$$

$$s_i s_j = s_j s_i, \quad s_i s_{i+1} s_i = s_{i+1} s_i s_{i+1}, \tag{3.3}$$

for all admissible i, j with $|i - j| > 2$. The corresponding Bruhat ordering on S_n is denoted by $\le$. We define a left action of S_n on $\mathcal{P}_n$ by algebra automorphisms so that

$$w \cdot x_i = x_{wi}, \tag{3.4}$$

for each $w \in S_n$, $i = 1, \dots, n$.

Now we define the main object of study: the *(degenerate) affine Hecke algebra* $\mathcal{H}_n$. This was introduced by Drinfeld [D] and Lusztig [L]. The associative algebra $\mathcal{H}_n$ is given by generators $x_1, \dots, x_n$ and $s_1, \dots, s_{n-1}$, subject to the same relations as $\mathcal{P}_n$ (3.1), and as $\mathcal{G}_n$ (3.2), (3.3), together with

$$s_i x_j = x_j s_i, \tag{3.5}$$

$$s_i x_i = x_{i+1} s_i - 1, \tag{3.6}$$

for all admissible i, j with $j \ne i, i+1$. We will call $x_1, \dots, x_n$ *polynomial generators* and $s_1, \dots, s_{n-1}$ *Coxeter generators*. Note that $\mathcal{H}_1 \cong \mathcal{P}_1 \cong F[x]$. Also, *by agreement*, $\mathcal{H}_0 \cong F$. The relation (3.6) implies

$$s_i x_{i+1} = x_i s_i + 1. \tag{3.7}$$

By induction, we deduce a convenient general formula

$$s_i f = (s_i \cdot f) s_i + \frac{f - s_i \cdot f}{x_{i+1} - x_i} \tag{3.8}$$

for $f \in F[x_1, \dots, x_n]$. In particular, for $j \geq 1$ we have

$$s_i x_i^j = x_{i+1}^j s_i - \sum_{k=0}^{j-1} x_i^k x_{i+1}^{j-1-k}, \tag{3.9}$$

$$s_i x_{i+1}^j = x_i^j s_i + \sum_{k=0}^{j-1} x_i^k x_{i+1}^{j-1-k}. \tag{3.10}$$

3.2 Basis Theorem

Our first goal is to construct a "PBW-type" basis of $\mathcal{H}_n$. There are obvious homomorphisms $\varphi : \mathcal{P}_n \to \mathcal{H}_n$ and $\psi : \mathcal{G}_n \to \mathcal{H}_n$ under which the x_i or s_j map to the corresponding elements of $\mathcal{H}_n$. We also write x^α for the image under φ of the basis element $x^\alpha \in \mathcal{P}_n$, and w for the image under ψ of the basis element $w \in S_n \subset \mathcal{G}_n$. This notation will be justified shortly, when we show that φ and ψ are both algebra monomorphisms. The following lemma is obvious from the defining relations:

Lemma 3.2.1 *Let $f \in \mathcal{P}_n$, $w \in S_n$. Then in $\mathcal{H}_n$ we have*

$$wf = (w \cdot f)w + \sum_{u<w} f_u u, \qquad fw = w(w^{-1} \cdot f) + \sum_{u<w} u f'_u$$

for some $f_y, f'_y \in \mathcal{P}_n$ of degrees less than the degree of f.

It follows easily from this lemma that that $\mathcal{H}_n$ is at least *spanned* by all $x^\alpha w$, $\alpha \in \mathbb{Z}_+^n$, $w \in S_n$. We wish to prove that these elements are linearly independent too:

Theorem 3.2.2 *The $\{x^\alpha w \mid \alpha \in \mathbb{Z}_+^n,\ w \in S_n\}$ form a basis for $\mathcal{H}_n$.*

Proof We give two proofs of this important result.
First proof. Consider instead the algebra $\tilde{\mathcal{H}}_n$ given by generators $\tilde{x}_i, \tilde{s}_j$, $1 \leq i \leq n$, $1 \leq j < n$, subject to the relations (3.1), (3.2), (3.5), (3.6), and (3.7). Thus we have all the relations of $\mathcal{H}_n$, *except for* the braid relations (3.3). Using these relations as the reduction system in Bergman's diamond lemma [Be, 1.2] we see that $\tilde{\mathcal{H}}_n$ has a basis given by all $\tilde{x}^\alpha \tilde{w}$ for all $\alpha \in \mathbb{Z}_+^n$ and all words $\tilde{w}$ in the $\tilde{s}_j$ which do not involve subword of the form $\tilde{s}_j^2$. Hence, the subalgebra $\tilde{\mathcal{P}}_n$ of $\tilde{\mathcal{H}}_n$ generated by the $\tilde{x}_i$ is isomorphic to $\mathcal{P}_n$. Also let $\tilde{\mathcal{G}}_n$ denote the subalgebra of $\tilde{\mathcal{H}}_n$ generated by the $\tilde{s}_j$, so that $\tilde{\mathcal{G}}_n$ is isomorphic to the algebra on generators $\tilde{s}_1, \dots, \tilde{s}_{n-1}$ subject to relations $\tilde{s}_j^2 = 1$ for each j.

Now, $\mathcal{H}_n$ is the quotient of $\tilde{\mathcal{H}}_n$ by the two-sided ideal $\mathcal{I}$ generated by the elements

$$a_{i,j} = \tilde{s}_i\tilde{s}_j - \tilde{s}_j\tilde{s}_i, \quad b_i = \tilde{s}_i\tilde{s}_{i+1}\tilde{s}_i - \tilde{s}_{i+1}\tilde{s}_i\tilde{s}_{i+1}$$

for i, j as in (3.3). Let $\mathcal{J}$ be the two-sided ideal of $\tilde{\mathcal{G}}_n$ generated by the same elements $a_{i,j}, b_i$ for all i, j. Then $\tilde{\mathcal{G}}_n/\mathcal{J} \cong \mathcal{G}_n$, and to prove the theorem it suffices to show that $\mathcal{I} = \widetilde{P}_n\mathcal{J}$ in $\widetilde{\mathcal{H}}_n$. In turn, this follows if we can show that $r\tilde{x}_k \in \tilde{\mathcal{P}}_n r$ for each $k = 1, \dots, n$, and $r \in \{a_{i,j}, b_i\}$. This is an easy check. For example,

$$(\tilde{s}_{i+1}\tilde{s}_i\tilde{s}_{i+1} - \tilde{s}_i\tilde{s}_{i+1}\tilde{s}_i)\tilde{x}_{i+2} = \tilde{x}_i(\tilde{s}_{i+1}\tilde{s}_i\tilde{s}_{i+1} - \tilde{s}_i\tilde{s}_{i+1}\tilde{s}_i),$$

using only the relations in $\tilde{\mathcal{H}}_n$.

Second proof. By verifying the defining relations of $\mathcal{H}_n$, we can see that $\mathcal{H}_n$ acts on the polynomials $F[y_1, \dots, y_n]$ in n variables by the following formulas:

$$x_i \circ f = y_i f, \quad s_j \circ f = (s_j \cdot f) + \frac{f - s_j \cdot f}{y_{j+1} - y_j}$$

for all admissible i, j and all $f \in F[y_1, \dots, y_n]$ (secretly, we consider the action of $\mathcal{H}_n$ on the module $\mathrm{ind}_{FS_n}^{\mathcal{H}_n} \mathbf{1}_{S_n}$ induced from the trivial FS_n-module). Now, to see that the elements $x^\alpha w$ are linearly independent, it suffices to show that they act by linearly independent linear transformations on $F[y_1, \dots, y_n]$. But this is clear if we consider the action on an element of the form $y_1^N y_2^{2N} \dots y_n^{nN}$ for $N \gg 0$. □

By Theorem 3.2.2, we have a right from now on to *identify* $\mathcal{P}_n$ and $\mathcal{G}_n$ with the corresponding subalgebras of $\mathcal{H}_n$. Then $\mathcal{H}_n$ is a free right $\mathcal{G}_n$-module on a basis $\{x^\alpha \mid \alpha \in \mathbb{Z}^n\}$. As another consequence, if $m \le n$, we can consider $\mathcal{H}_m$ as the subalgebra of $\mathcal{H}_n$ generated by $x_1, \dots, x_m, s_1, \dots, s_{m-1}$.

Finally, let us point out that there is the obvious variant of Theorem 3.2.2: $\mathcal{H}_n$ also has $\{wx^\alpha \mid w \in S_n, \ \alpha \in \mathbb{Z}_+^n\}$ as a basis. This follows using Lemma 3.2.1.

3.3 The center of $\mathcal{H}_n$

The following simple description of the center is very important.

Theorem 3.3.1 *The center of $\mathcal{H}_n$ consists of all symmetric polynomials in $x_1, \dots, x_n$.*

Proof That the symmetric polynomials are indeed central is easily verified using (3.9). Conversely, take a central element $z = \sum_{w \in S_n} f_w w \in \mathcal{H}_n$ where each $f_w \in \mathcal{P}_n$. Let w be maximal with respect to the Bruhat order such that $f_w \neq 0$. Assume $w \neq 1$. Then there exists $i \in \{1, \ldots, n\}$ with $wi \neq i$. By Lemma 3.2.1, $x_i z - z x_i$ looks like $f_w(x_i - x_{wi})w$ plus a linear combination of terms of the form $f'_u u$ for $f'_u \in \mathcal{P}_n$ and $u \in S_n$ with $u \not\geq w$ in the Bruhat order. So, in view of Theorem 3.2.2, z is not central, giving a contradiction.

Hence, we must have that $z \in \mathcal{P}_n$. To see that z is actually a symmetric polynomial, write $z = \sum_{i,j \geq 0} a_{i,j} x_1^i x_2^j$, where the coefficients $a_{i,j}$ lie in $F[x_3, \ldots, x_n]$. Applying Lemma 3.2.1 to $s_1 z = z s_1$ now gives that $a_{i,j} = a_{j,i}$ for each i, j, hence z is symmetric in x_1 and x_2. Similar argument shows that z is symmetric in x_i and x_{i+1} for all $i = 1, \ldots, n-1$. $\square$

3.4 Parabolic subalgebras

Suppose that $\mu = (\mu_1, \ldots, \mu_r)$ is a composition of n. Let

$$S_\mu \cong S_{\mu_1} \times \cdots \times S_{\mu_r}$$

denote the corresponding Young subgroup of S_n. Then the subalgebra $\mathcal{G}_\mu$ of $\mathcal{G}_n$ generated by the s_j for which $s_j \in S_\mu$ is isomorphic to

$$FS_\mu \cong \mathcal{G}_{\mu_1} \otimes \cdots \otimes \mathcal{G}_{\mu_r}.$$

We define the *parabolic subalgebra* $\mathcal{H}_\mu$ of the degenerate affine Hecke algebra $\mathcal{H}_n$ in a similar way: it is the subalgebra generated by $\mathcal{P}_n$ and all s_j for which $s_j \in S_\mu$. It follows easily from Theorem 3.2.2 that the elements

$$\{x^\alpha w \mid \alpha \in \mathbb{Z}_+^n,\ w \in S_\mu\}$$

form a basis for $\mathcal{H}_\mu$. In particular,

$$\mathcal{H}_\mu \cong \mathcal{H}_{\mu_1} \otimes \cdots \otimes \mathcal{H}_{\mu_r}.$$

Note that the parabolic subalgebra $\mathcal{H}_{(1,1,\ldots,1)}$ is precisely the subalgebra $\mathcal{P}_n$.

We will use the induction and restriction functors between $\mathcal{H}_n$ and $\mathcal{H}_\mu$. These will be denoted simply

$$\mathrm{ind}_\mu^n : \mathcal{H}_\mu\text{-mod} \to \mathcal{H}_n\text{-mod}, \qquad \mathrm{res}_\mu^n : \mathcal{H}_n\text{-mod} \to \mathcal{H}_\mu\text{-mod}, \tag{3.11}$$

the former being the tensor functor $\mathcal{H}_n \otimes_{\mathcal{H}_\mu} ?$, which is left adjoint to res_μ^n. More generally, we will consider induction and restriction between nested parabolic subalgebras, with obvious notation. We will also occasionally consider the restriction functor

$$\mathrm{res}^n_{n-1} : \mathcal{H}_n\text{-mod} \to \mathcal{H}_{n-1}\text{-mod}. \tag{3.12}$$

3.5 Mackey Theorem

Let μ, ν be compositions of n. Denote by D_ν the set of minimal length left S_ν-coset representatives in S_n, and by D_μ^{-1} the set of minimal length right S_μ-coset representatives. Then $D_{\mu,\nu} := D_\mu^{-1} \cap D_\nu$ is the set of minimal length (S_μ, S_ν)-double coset representatives in S_n. We recall three known properties, see for example [DJ, Section 1],

(1) For $x \in D_{\mu,\nu}$, $S_\mu \cap xS_\nu x^{-1}$ and $x^{-1}S_\mu x \cap S_\nu$ are Young subgroups of S_n. So we can define compositions $\mu \cap x\nu$ and $x^{-1}\mu \cap \nu$ of n from

$$S_\mu \cap xS_\nu x^{-1} = S_{\mu\cap x\nu} \quad \text{and} \quad x^{-1}S_\mu x \cap S_\nu = S_{x^{-1}\mu\cap\nu}.$$

(2) For $x \in D_{\mu,\nu}$, the map $w \mapsto x^{-1}wx$ restricts to a length preserving isomorphism

$$S_{\mu\cap x\nu} \to S_{x^{-1}\mu\cap\nu}.$$

(3) For $x \in D_{\mu,\nu}$, every $w \in S_\mu x S_\nu$ can be written as $w = uxv$ for unique elements $u \in S_\mu$ and $v \in S_\nu \cap D^{-1}_{x^{-1}\mu\cap\nu}$. Moreover, $S_\nu \cap D^{-1}_{x^{-1}\mu\cap\nu}$ is the set of minimal length right coset representatives of $S_{x^{-1}\mu\cap\nu}$ in S_ν.

Now fix some total order $\prec$ refining the Bruhat order $<$ on $D_{\mu,\nu}$. For $x \in D_{\mu,\nu}$, set

$$\mathcal{B}_{\preceq x} = \bigoplus_{y\in D_{\mu,\nu},\ y\preceq x} \mathcal{H}_\mu y \mathcal{G}_\nu, \tag{3.13}$$

$$\mathcal{B}_{\prec x} = \bigoplus_{y\in D_{\mu,\nu},\ y\prec x} \mathcal{H}_\mu y \mathcal{G}_\nu, \tag{3.14}$$

$$\mathcal{B}_x = \mathcal{B}_{\preceq x}/\mathcal{B}_{\prec x}. \tag{3.15}$$

It follows from Lemma 3.2.1 that $\mathcal{B}_{\preceq x}$ and $\mathcal{B}_{\prec x}$ are invariant under right multiplication by $\mathcal{P}_n$. Hence, since $\mathcal{X}_\nu = \mathcal{G}_\nu \mathcal{P}_n$, we have defined a filtration of $\mathcal{X}_n$ as an $(\mathcal{X}_\mu, \mathcal{X}_\nu)$-bimodule. We want to describe the quotients $\mathcal{B}_x$ explicitly. To this end, note using the property (2) above that for each $y \in D_{\mu,\nu}$, there exists an algebra isomorphism

$$\varphi_{y^{-1}} : \mathcal{H}_{\mu\cap y\nu} \to \mathcal{H}_{y^{-1}\mu\cap\nu}$$

with $\varphi_{y^{-1}}(w) = y^{-1}wy$ and $\varphi_{y^{-1}}(x_i) = x_{y^{-1}i}$ for $w \in S_{\mu\cap y\nu}$, $1 \le i \le n$. If N is a left $\mathcal{H}_{y^{-1}\mu\cap\nu}$-module, then by twisting the action with the isomorphism $\varphi_{y^{-1}}$ we get a left $\mathcal{H}_{\mu\cap y\nu}$-module, which will be denoted ${}^yN$.

Lemma 3.5.1 *Let us view $\mathcal{H}_\mu$ as an $(\mathcal{H}_\mu, \mathcal{H}_{\mu\cap x\nu})$-bimodule and $\mathcal{H}_\nu$ as an $(\mathcal{H}_{x^{-1}\mu\cap\nu}, \mathcal{H}_\nu)$-bimodule in the natural ways. Then ${}^x\mathcal{H}_\nu$ is an $(\mathcal{H}_{\mu\cap x\nu}, \mathcal{H}_\nu)$-bimodule, and*

$$\mathcal{B}_x \cong \mathcal{H}_\mu \otimes_{\mathcal{H}_{\mu\cap x\nu}} {}^x\mathcal{H}_\nu$$

is an $(\mathcal{H}_\mu, \mathcal{H}_\nu)$-bimodule.

Proof We define a bilinear map $\mathcal{H}_\mu \times {}^x\mathcal{H}_\nu \to \mathcal{B}_x = \mathcal{B}_{\preceq x}/\mathcal{B}_{\prec x}$ by

$$(h, h') \mapsto hxh' + \mathcal{B}_{\prec x}.$$

The map is checked to be $\mathcal{H}_{\mu\cap x\nu}$-balanced, so it yields an $(\mathcal{H}_\mu, \mathcal{H}_\nu)$-bimodule map

$$\Phi : \mathcal{H}_\mu \otimes_{\mathcal{H}_{\mu\cap x\nu}} {}^x\mathcal{H}_\nu \to \mathcal{B}_x.$$

To prove that Φ is bijective, note by the Property (3) above that

$$\{x^\alpha u \otimes v \mid \alpha \in \mathbb{Z}_+^n,\ u \in S_\mu,\ v \in S_\nu \cap D_{x^{-1}\mu\cap\nu}^{-1}\}$$

is a basis of the induced module $\mathcal{H}_\mu \otimes_{\mathcal{H}_{\mu\cap x\nu}} {}^x\mathcal{H}_\mu$ as a vector space, and the image of these elements under Φ is a basis of $\mathcal{B}_x$. □

Now we can prove the Mackey Theorem.

Theorem 3.5.2 ("Mackey Theorem") *Let M be an $\mathcal{H}_\nu$-module. Then* $\mathrm{res}^n_\mu \mathrm{ind}^n_\nu M$ *admits a filtration with subquotients isomorphic to*

$$\mathrm{ind}^\mu_{\mu\cap x\nu} {}^x(\mathrm{res}^\nu_{x^{-1}\mu\cap\nu} M),$$

one for each $x \in D_{\mu,\nu}$. Moreover, the subquotients can be taken in any order refining the Bruhat order on $D_{\mu,\nu}$, in particular $\mathrm{ind}^\mu_{\mu\cap\nu}\mathrm{res}^\nu_{\mu\cap\nu} M$ *appears as a submodule.*

Proof This follows from Lemma 3.5.1 and the isomorphism

$$(\mathcal{H}_\mu \otimes_{\mathcal{H}_{\mu\cap x\nu}} {}^x\mathcal{H}_\nu) \otimes_{\mathcal{H}_\nu} M \cong \mathrm{ind}^{\mathcal{H}_\mu}_{\mathcal{H}_{\mu\cap x\nu}} {}^x(\mathrm{res}^{\mathcal{H}_\nu}_{\mathcal{H}_{x^{-1}\mu\cap\nu}} M),$$

which is easy to check. □

3.6 Some (anti) automorphisms

A check of relations shows that $\mathcal{H}_n$ possesses an automorphism σ and an antiautomorphism τ defined on the generators as follows:

$$\sigma : s_i \mapsto -s_{n-i}, \quad x_j \mapsto x_{n+1-j}; \tag{3.16}$$

$$\tau : s_i \mapsto s_i, \quad x_j \mapsto x_j, \tag{3.17}$$

for all $i = 1, \dots, n-1, j = 1, \dots, n$.

If M is a finite dimensional $\mathcal{H}_n$-module, we can use τ to make the dual space M^* into an $\mathcal{H}_n$-module denoted M^τ. Note τ leaves invariant every parabolic subalgebra of $\mathcal{H}_n$, so also induces a duality on finite dimensional $\mathcal{H}_\mu$-modules for each composition μ of n.

Instead, given any $\mathcal{H}_n$-module M, we can twist the action with σ to get a new module denoted M^σ. Moreover, for any composition $\nu = (\nu_1, \dots, \nu_r)$ of n we denote by ν^* the composition with the same non-zero parts but taken in the opposite order. For example $(3, 2, 1)^* = (1, 2, 3)$. Then σ induces an isomorphism of parabolic subalgebras $\mathcal{H}_{\nu^*} \to \mathcal{H}_\nu$. So if M is an $\mathcal{H}_\nu$-module, we can inflate through σ to get an $\mathcal{H}_{\nu^*}$-module denoted M^σ. If $M = M_1 \boxtimes \cdots \boxtimes M_r$ is an outer tensor product module over $\mathcal{H}_\nu = \mathcal{H}_{\nu_1} \otimes \cdots \otimes \mathcal{H}_{\nu_r}$ then $M^\sigma \cong M_r^\sigma \boxtimes \cdots \boxtimes M_1^\sigma$. These observations imply:

Lemma 3.6.1 *Let $M \in \mathcal{H}_m$-mod and $N \in \mathcal{H}_n$-mod. Then*

$$(\mathrm{ind}_{m,n}^{m+n} M \boxtimes N)^\sigma \cong \mathrm{ind}_{n,m}^{m+n} N^\sigma \boxtimes M^\sigma.$$

3.7 Duality

Throughout this section, μ is a composition of n and $\nu = \mu^*$. Let d be the longest element of $D_{\mu,\nu}$. Note that $\mu \cap d\nu = \mu$ and $d^{-1}\mu \cap \nu = \nu$, so $S_\mu d S_\nu = S_\mu d = d S_\nu$. There is an isomorphism

$$\varphi = \varphi_{d^{-1}} : \mathcal{H}_\mu \to \mathcal{H}_\nu, \tag{3.18}$$

see Section 3.5, and for an $\mathcal{H}_\nu$-module M, ${}^dM$ denotes the $\mathcal{H}_\mu$-module obtained by pulling back the action through φ. We begin by considering the situation for $\mathcal{G}_n$.

Lemma 3.7.1 *Define a linear map $\theta' : \mathcal{G}_n \to {}^d\mathcal{G}_\nu$ by*

$$\theta'(w) = \begin{cases} d^{-1}w & \text{if } w \in dS_\nu, \\ 0 & \text{otherwise,} \end{cases}$$

for each $w \in S_n$. *Then:*

(i) θ' *is a homomorphism of* $(\mathcal{G}_\mu, \mathcal{G}_\nu)$*-bimodules;*
(ii) $\ker \theta'$ *contains no non-zero left ideals of* $\mathcal{G}_n$*;*
(iii) *the map*

$$f' : \mathcal{G}_n \to \mathrm{Hom}_{\mathcal{G}_\mu}(\mathcal{G}_n, {}^d\mathcal{G}_\nu), \qquad h \mapsto h\theta'$$

is an isomorphism of $(\mathcal{G}_n, \mathcal{G}_\nu)$*-bimodules.*

Proof (i) Follows easily using $S_\mu d = dS_\nu$.

(ii) It is enough to show that $\theta'(\mathcal{G}_n t) \neq 0$ for any $t \in \mathcal{G}_n$. By multiplying t with an appropriate group element on the left, we may assume that $t = \sum_{y \in D_\nu} y h_y$ with each $h_y \in \mathcal{G}_\nu$ and $h_d \neq 0$. Now $\theta'(t) = h_d \neq 0$, as required.

(iii) We remind that $h\theta' : \mathcal{G}_n \to {}^d\mathcal{G}_\mu$ denotes the map with $(h\theta')(t) = \theta'(th)$. Given (i), it is straightforward to check that f' is a homomorphism of $(\mathcal{G}_n, \mathcal{G}_\nu)$-bimodules. To see that it is an isomorphism, it suffices by dimension to show that it is injective. Suppose t lies in the kernel, then $(f'(t))(h) = \theta'(ht) = 0$ for all $h \in \mathcal{G}_n$. Hence $t = 0$ by (ii). □

Now we extend this result to $\mathcal{H}_n$.

Lemma 3.7.2 *Define a linear map* $\theta : \mathcal{H}_n \to {}^d\mathcal{H}_\nu$ *by*

$$\theta(fw) = \begin{cases} \varphi(f)d^{-1}w & \text{if } w \in dS_\nu, \\ 0 & \text{otherwise,} \end{cases}$$

for each $f \in \mathcal{P}_n$, $w \in S_n$. *Then:*

(i) θ *is a homomorphism of* $(\mathcal{H}_\mu, \mathcal{H}_\nu)$*-bimodules;*
(ii) *the map*

$$f : \mathcal{H}_n \to \mathrm{Hom}_{\mathcal{H}_\mu}(\mathcal{H}_n, {}^d\mathcal{H}_\nu), \qquad h \mapsto h\theta$$

is an isomorphism of $(\mathcal{H}_n, \mathcal{H}_\nu)$*-bimodules.*

Proof (i) According to a special case of Lemma 3.5.1, the top factor $\mathcal{B}_d$ in the bimodule filtration of $\mathcal{H}_n$ defined in (3.15) is isomorphic to ${}^d\mathcal{H}_\nu$ as an $(\mathcal{H}_\mu, \mathcal{H}_\nu)$-bimodule. The map θ is simply the composite of this isomorphism with the quotient map $\mathcal{H}_n \to \mathcal{B}_d$.

(ii) Recall that $\{w \mid w \in D_\mu^{-1}\}$ forms a basis for $\mathcal{H}_n$ as a free left $\mathcal{H}_\mu$-module, and ${}^d\mathcal{H}_\nu$ is isomorphic to $\mathcal{H}_\mu$ as a left $\mathcal{H}_\mu$-module. It follows that the maps $\{\psi_w \mid w \in D_\mu^{-1}\}$ form a basis for $\mathrm{Hom}_{\mathcal{H}_\mu}(\mathcal{H}_n, {}^d\mathcal{H}_\nu)$ as a free right $\mathcal{H}_\nu$-module, where $\psi_w : \mathcal{H}_n \to {}^d\mathcal{H}_\nu$ is the unique left $\mathcal{H}_\mu$-module homomorphism with $\psi_w(u) = \delta_{w,u}.1$ for all $u \in D_\mu^{-1}$.

Given (i), it is straightforward to check that f is a homomorphism of $(\mathcal{H}_n, \mathcal{H}_\nu)$-bimodules. So, to prove that f is an isomorphism, it suffices to find a basis of $\mathcal{H}_n$ as a free right $\mathcal{H}_\nu$-module, which is mapped by f to the basis $\{\psi_w \mid w \in D_\mu^{-1}\}$ of $\mathrm{Hom}_{\mathcal{H}_\mu}(\mathcal{H}_n, {}^d\mathcal{H}_\nu)$.

The analogous maps $\psi'_w \in \mathrm{Hom}_{\mathcal{G}_\mu}(\mathcal{G}_n, {}^d\mathcal{G}_\nu)$ defined by $\psi'_w(u) = \delta_{w,u}.1$ for $u \in D_\mu^{-1}$ form a basis for $\mathrm{Hom}_{\mathcal{G}_\mu}(\mathcal{G}_n, {}^d\mathcal{G}_\nu)$ as a free right $\mathcal{G}_\nu$-module. So, in view of Lemma 3.7.1(iii), we can find a basis $\{a_w \mid w \in D_\mu^{-1}\}$ for $\mathcal{G}_n$ viewed as a right $\mathcal{G}_\nu$-module such that $f'(a_w) = \psi'_w$ for each $w \in D_\mu^{-1}$, that is

$$\theta'(ua_w) = \begin{cases} 1 & \text{if } u = w, \\ 0 & \text{otherwise} \end{cases}$$

for every $u \in D_\mu^{-1}$. But $\mathcal{H}_\nu = \mathcal{G}_\nu \mathcal{P}_n$, so the elements $\{a_w \mid w \in D_\mu^{-1}\}$ also form a basis for $\mathcal{H}_n$ as a right $\mathcal{H}_\nu$-module, and $f(a_w) = \psi_w$ since $\theta = \theta'$ on $\mathcal{G}_n$. □

Corollary 3.7.3 *There is a natural isomorphism*

$$\mathrm{Hom}_{\mathcal{H}_\mu}(\mathcal{H}_n, {}^d M) \cong \mathcal{H}_n \otimes_{\mathcal{H}_\nu} M$$

of $\mathcal{H}_n$-modules, for every left $\mathcal{H}_\nu$-module M.

Proof Let $f : \mathcal{H}_n \to \mathrm{Hom}_{\mathcal{H}_\mu}(\mathcal{H}_n, {}^d\mathcal{H}_\nu)$ be the bimodule isomorphism constructed in Lemma 3.7.2. Then there are natural isomorphisms

$$\begin{aligned} \mathcal{H}_n \otimes_{\mathcal{H}_\nu} M &\xrightarrow{f\otimes \mathrm{id}} \mathrm{Hom}_{\mathcal{H}_\mu}(\mathcal{H}_n, {}^d\mathcal{H}_\nu) \otimes_{\mathcal{H}_\nu} M \\ &\cong \mathrm{Hom}_{\mathcal{H}_\mu}(\mathcal{H}_n, {}^d\mathcal{H}_\nu \otimes_{\mathcal{H}_\nu} M) \cong \mathrm{Hom}_{\mathcal{H}_\mu}(\mathcal{H}_n, {}^d M), \end{aligned}$$

the second isomorphism depending on the fact that $\mathcal{H}_n$ is a free left $\mathcal{H}_\mu$-module, see for example [AF, 20.10]. □

Corollary 3.7.4 *There is a natural isomorphism*

$$\mathrm{ind}_\nu^n(M^\tau) \cong (\mathrm{ind}_\mu^n({}^d M))^\tau$$

for every finite dimensional $\mathcal{H}_\nu$-module M.

Proof The functor $\tau \circ \mathrm{ind}_\mu^n \circ \tau : \mathcal{H}_\mu\text{-mod} \to \mathcal{H}_n\text{-mod}$ is right adjoint to res_μ^n. Hence it is isomorphic to $\mathrm{Hom}_{\mathcal{H}_\mu}(\mathcal{H}_n, ?)$ by uniqueness of adjoint functors.

Now combine this natural isomorphism with Corollary 3.7.3 (with μ and ν swapped and d replaced by d^{-1}). □

We finally record an important special case:

Theorem 3.7.5 *For $M \in \mathcal{H}_m$-mod and $N \in \mathcal{H}_n$-mod, we have*

$$(\mathrm{ind}_{m,n}^{m+n} M \boxtimes N)^\tau \cong \mathrm{ind}_{n,m}^{m+n}(N^\tau \boxtimes M^\tau).$$

3.8 Intertwining elements

We will need certain elements of $\mathcal{H}_n$ which go back Cherednik. Given $1 \leq i < n$, define

$$\Phi_i := s_i(x_i - x_{i+1}) + 1. \tag{3.19}$$

A straightforward calculation gives:

$$\Phi_i^2 = (x_i - x_{i+1} - 1)(x_{i+1} - x_i - 1), \qquad 1 \leq i < n, \tag{3.20}$$

$$\Phi_i x_i = x_{i+1}\Phi_i, \; \Phi_i x_{i+1} = x_i \Phi_i, \; \Phi_i x_j = x_j \Phi_i, \qquad j \neq i, i+1, \tag{3.21}$$

$$\Phi_i \Phi_j = \Phi_j \Phi_i, \; \Phi_i \Phi_{i+1} \Phi_i = \Phi_{i+1} \Phi_i \Phi_{i+1}, \qquad |i-j| > 1. \tag{3.22}$$

Property (3.22) means that for every $w \in S_n$ we obtain a well-defined element $\Phi_w \in \mathcal{H}_n$, namely,

$$\Phi_w := \Phi_{i_1} \dots \Phi_{i_m}$$

where $w = s_{i_1} \dots s_{i_m}$ is any reduced expression for w. According to (3.21), these elements have the property that

$$\Phi_w x_i = x_{wi} \Phi_w, \tag{3.23}$$

for all $w \in S_n$ and $1 \leq i \leq n$. We note that only the properties (3.20) and (3.21) will be essential in what follows.

4

First results on $\mathcal{H}_n$-modules

The polynomial subalgebra $\mathcal{P}_n$ of $\mathcal{H}_n$ is a maximal commutative subalgebra, and so we may try to "do Lie Theory" using $\mathcal{P}_n$ as an analogue of Cartan subalgebra. The main difference however is that $\mathcal{P}_n$ is not semisimple, so we have to consider generalized eigenspaces, rather than usual eigenspaces. We define the formal character of an $\mathcal{H}_n$-module M as the generating function for the dimensions of simultaneous generalized eigenspaces of the elements $x_1, \dots, x_n$ on M. In Chapter 5 we will prove that the formal characters of irreducible $\mathcal{H}_n$-modules are linearly independent (as any reasonable formal characters should be). The "Shuffle Lemma", which is a special case of the Mackey Theorem, gives a transparent description of what induction "does" to the formal characters.

Our knowledge of the center of $\mathcal{H}_n$ allows us to develop an easy theory of blocks. The central characters of $\mathcal{H}_n$ (and so the blocks too) are labeled by the S_n-orbits on the n-tuples of scalars.

Next we study the properties of what can be considered as one of the main technical tools of the theory: the so-called *Kato module* (cf. [Kt]). Miraculously, if we take the 1-dimensional module over the polynomial algebra $\mathcal{P}_n$ on which every x_i acts with the *same* scalar, and then induce it to $\mathcal{H}_n$, we get an irreducible module. This is not hard to prove once you believe it is true. As a module over symmetric group, any Kato module is just the regular module, and thus we deal with a remarkable extension of the regular FS_n-module to $\mathcal{H}_n$, which is irreducible. The simple definition of Kato modules as induced modules allows us to investigate them in great detail. We conclude the chapter with a study of some "finite dimensional approximations" of projective covers of Kato modules.

4.1 Formal characters

Let $a \in F$. Denote by $L(a)$ the 1-dimensional $\mathcal{P}_1$-module with $x_1 v = av$ for $v \in L(a)$. As $\mathcal{P}_n \cong \mathcal{P}_1 \otimes \cdots \otimes \mathcal{P}_1$, we obtain the irreducible $\mathcal{P}_n$-modules by taking outer tensor products $L(a_1) \boxtimes \cdots \boxtimes L(a_n)$, for $a_1, \ldots, a_n \in F$.

Now take any $M \in \mathcal{P}_n$-mod. For any $\underline{a} = (a_1, \ldots, a_n) \in F^n$, let $M_{\underline{a}}$ be the largest submodule of M, all of whose composition factors are isomorphic to $L(a_1) \boxtimes \cdots \boxtimes L(a_n)$. Alternatively, $M_{\underline{a}}$ is the simultaneous generalized eigenspace for the commuting operators $x_1, \ldots, x_n$ corresponding to the eigenvalues $a_1, \ldots, a_n$, respectively. Hence:

Lemma 4.1.1 *For any $M \in \mathcal{P}_n$-mod, we have $M = \bigoplus_{\underline{a} \in F^n} M_{\underline{a}}$ as a $\mathcal{P}_n$-module.*

Note, for $M \in \mathcal{P}_n$-mod, knowledge of the dimensions of the spaces $M_{\underline{a}}$ for all $\underline{a}$ is equivalent to knowing the coefficients $r_{\underline{a}}$ when the class $[M]$ of M in the Grothendieck group $K(\mathcal{P}_n\text{-mod})$ is expanded as

$$[M] = \sum_{\underline{a} \in F^n} r_{\underline{a}} [L(a_1) \boxtimes \cdots \boxtimes L(a_n)]$$

in terms of the basis $\{[L(a_1) \boxtimes \cdots \boxtimes L(a_n)] \mid \underline{a} \in F^n\}$.

Now let $M \in \mathcal{H}_n$-mod. Recall that $\mathcal{H}_{1,\ldots,1} = \mathcal{P}_n$. We define the *formal character* of M by:

$$\operatorname{ch} M := [\operatorname{res}^n_{1,\ldots,1} M] \in K(\mathcal{P}_n\text{-mod}). \tag{4.1}$$

Since the functor $\operatorname{res}^n_{1,\ldots,1}$ is exact, ch induces a homomorphism

$$\operatorname{ch} : K(\mathcal{H}_n\text{-mod}) \to K(\mathcal{P}_n\text{-mod})$$

at the level of Grothendieck groups. We will later see that this map is actually injective (Theorem 5.3.1). We will also occasionally consider characters of modules over parabolic subalgebras $\mathcal{H}_\mu$. The definitions are modified in this case in obvious ways.

Lemma 4.1.2 *Let $\underline{a} = (a_1, \ldots, a_n) \in F^n$. Then*

$$\operatorname{ch} \operatorname{ind}^n_{1,\ldots,1} L(a_1) \boxtimes \cdots \boxtimes L(a_n) = \sum_{w \in S_n} [L(a_{w^{-1}1}) \boxtimes \cdots \boxtimes L(a_{w^{-1}n})]$$

Proof This follows from the Mackey Theorem with $\mu = \nu = (1^n)$. □

Lemma 4.1.3 ("Shuffle Lemma") *Let $n = m + k$, and let $M \in \mathcal{H}_m$-mod, $K \in \mathcal{H}_k$-mod. Assume*

$$\operatorname{ch} M = \sum_{\underline{a} \in F^m} r_{\underline{a}} [L(a_1) \boxtimes \cdots \boxtimes L(a_m)],$$

$$\operatorname{ch} K = \sum_{\underline{b} \in F^k} s_{\underline{b}} [L(b_1) \boxtimes \cdots \boxtimes L(b_k))].$$

Then

$$\operatorname{ch} \operatorname{ind}_{m,k}^n M \boxtimes K = \sum_{\underline{a} \in F^m} \sum_{\underline{b} \in F^k} r_{\underline{a}} s_{\underline{b}} (\sum_{\underline{c}} L(c_1) \boxtimes \cdots \boxtimes L(c_n)),$$

where the last sum is over all $\underline{c} = (c_1, \dots, c_n) \in F^n$ which are obtained by shuffling $\underline{a}$ and $\underline{b}$, that is there exist $1 \le u_1 < \cdots < u_m \le n$ such that $(c_{u_1}, \dots, c_{u_m}) = (a_1, \dots, a_m)$, and $(c_1, \dots, \widehat{c}_{u_1}, \dots, \widehat{c}_{u_m}, \dots, c_n) = (b_1, \dots, b_k)$.

Proof Apply the Mackey Theorem with $\mu = (1^n)$ and $\nu = (m, k)$. □

4.2 Central characters

Recall by Theorem 3.3.1 that every element z of the center $Z(\mathcal{H}_n)$ of $\mathcal{H}_n$ can be written as a symmetric polynomial $f(x_1, \dots, x_n)$. Given $\underline{a} \in F^n$, we associate the *central character*

$$\chi_{\underline{a}} : Z(\mathcal{H}_n) \to F, \quad f(x_1, \dots, x_n) \mapsto f(a_1, \dots, a_n). \tag{4.2}$$

Consider the action of S_n on F^n by place permutation. We write $\underline{a} \sim \underline{b}$ if $\underline{a}$ and $\underline{b}$ lie in the same orbit with respect to this action. The following lemma is immediate.

Lemma 4.2.1 *For $\underline{a}, \underline{b} \in F^n$, $\chi_{\underline{a}} = \chi_{\underline{b}}$ if and only if $\underline{a} \sim \underline{b}$.*

Thus the central characters of $\mathcal{H}_n$ are actually labeled by the set $F^n/\sim$ of S_n-orbits on F^n. If γ is such an orbit we set

$$\chi_\gamma := \chi_{\underline{a}}$$

for any $\underline{a} \in \gamma$.

Now let M be a finite dimensional $\mathcal{H}_n$-module and $\gamma \in F^n/\sim$. We let $M[\gamma]$ denote the generalized eigenspace of M over $Z(\mathcal{H}_n)$ that corresponds to the central character χ_γ, that is

$$M[\gamma] = \{v \in M \mid (z - \chi_\gamma(z))^k v = 0 \text{ for all } z \in Z(\mathcal{H}_n) \text{ and } k \gg 0\}.$$

Observe this is an $\mathcal{H}_n$-submodule of M. Now, for any $\underline{a} \in F^n$ with $\underline{a} \in \gamma$, $Z(\mathcal{H}_n)$ acts on $L(a_1) \boxtimes \cdots \boxtimes L(a_n)$ via the central character χ_γ. So applying Lemma 4.2.1, we see that

$$M[\gamma] = \bigoplus_{\underline{a} \in \gamma} M_{\underline{a}},$$

recalling the decomposition of M as a $\mathcal{P}_n$-module from Lemma 4.1.1. Therefore:

Lemma 4.2.2 *Any* $M \in \mathcal{H}_n$-mod *decomposes as*

$$M = \bigoplus_{\gamma \in F^n/\sim} M[\gamma]$$

as an $\mathcal{H}_n$*-module.*

Thus the $\{\chi_\gamma \mid \gamma \in F^n/\sim\}$ exhaust the possible central characters that can arise in a finite dimensional $\mathcal{H}_n$-module, while Lemma 4.1.2 shows that every such central character does arise in some finite dimensional $\mathcal{H}_n$-module.

If $\gamma \in F^n/\sim$, let us denote by $\mathcal{H}_n$-mod$[\gamma]$ for the full subcategory of $\mathcal{H}_n$-mod, consisting of all modules M with $M[\gamma] = M$. Then Lemma 4.2.2 implies that there is an equivalence of categories

$$\mathcal{H}_n\text{-mod} \cong \bigoplus_{\gamma \in F^n/\sim} \mathcal{H}_n\text{-mod}[\gamma]. \tag{4.3}$$

We say that $\mathcal{H}_n$-mod$[\gamma]$ is the *block* of $\mathcal{H}_n$-mod corresponding to γ (or to the central character χ_γ). If $M \in \mathcal{H}_n$-mod$[\gamma]$, we say that M *belongs* to the block corresponding to γ. If $M \neq 0$ is indecomposable then $M \in \mathcal{H}_n$-mod$[\gamma]$ for a unique $\gamma \in F^n/\sim$.

4.3 Kato's Theorem

Let $a \in F$. Introduce the *Kato module*

$$L(a^n) := \mathrm{ind}^n_{1,\dots,1} L(a) \boxtimes \cdots \boxtimes L(a). \tag{4.4}$$

By Lemma 4.1.2, we know immediately that

$$\mathrm{ch}\, L(a^n) = n![L(a) \boxtimes \cdots \boxtimes L(a)].$$

In particular, for each $k = 1, \dots, n$, the only eigenvalue of the element x_k on $L(a^n)$ is a.

Lemma 4.3.1 *Let $a \in F$. Set $L = L(a) \boxtimes \cdots \boxtimes L(a)$, so $L(a^n) = \mathcal{H}_n \otimes_{\mathcal{P}_n} L$. The common a-eigenspace of the operators $x_1, \dots, x_{n-1}$ on $L(a^n)$ is precisely $1 \otimes L$. Moreover, all Jordan blocks of x_1 on $L(a^n)$ are of size n.*

Proof Note $L(a^n) = \bigoplus_{w \in S_n} w \otimes L$, since by Theorem 3.2.2 we know that $\mathcal{H}_n$ is a free right $\mathcal{P}_n$-module on basis $\{w \mid w \in S_n\}$.

We first claim that the eigenspace of x_1 is a sum of the subspaces of the form $y \otimes L$, where $y \in S'_{n-1} = \langle s_2, \dots, s_{n-1} \rangle$. Well, any w can be written as $y s_1 s_2 \dots s_j$ for some $y \in S'_{n-1}$ and $0 \le j < n$. Note that $(x_{j+1} - a)v = 0$ for any $v \in L$. Now the defining relations of $\mathcal{H}_n$ imply

$$(x_1 - a) y s_1 s_2 \dots s_j \otimes v = -y s_1 \dots s_{j-1} \otimes v + (*),$$

where $(*)$ stands for terms which belong to subspaces of the form

$$y' s_1 \dots s_k \otimes L$$

for $y' \in S'_{n-1}$ and $0 \le k < j - 1$. Now assume that a linear combination

$$z := \sum_{y \in S'_{n-1}} \sum_{0 \le j < n} c_{y,j} y s_1 s_2 \dots s_j \otimes v$$

is an eigenvector for x_1. Choose the maximal j for which the coefficient $c_{y,j}$ is non-zero. Then the calculation above shows that $(x_1 - a)z \neq 0$ unless $j = 0$. This proves our claim.

Now apply the same argument to see that the common eigenspace of x_1 and x_2 is spanned by $y \otimes L$ for $y \in \langle s_3, \dots, s_{n-1} \rangle$, and so on, yielding the first claim of the lemma. Finally, define

$$V(m) := \{z \in L(a^n) \mid (x_1 - a)^m z = 0\}.$$

It follows by induction from the calculation above that

$$V(m) = \operatorname{span}\{y s_1 s_2 \dots s_j \otimes L \mid y \in S'_{n-1},\ j < m\},$$

giving the second claim. □

Now we prove the main theorem on the structure of the Kato module $L(a^n)$, compare [Kt].

Theorem 4.3.2 *Let $a \in F$ and $\mu = (\mu_1, \dots, \mu_r)$ be a composition of n:*

(i) *$L(a^n)$ is irreducible, and it is the only irreducible module in its block.*
(ii) *All composition factors of $\operatorname{res}^n_\mu L(a^n)$ are isomorphic to*

$$L(a^{\mu_1}) \boxtimes \cdots \boxtimes L(a^{\mu_r}),$$

and $\operatorname{soc} \operatorname{res}^n_\mu L(a^n)$ is irreducible.

(iii) $\operatorname{soc} \operatorname{res}^n_{n-1} L(a^n) \cong L(a^{n-1})$.

Proof Denote $L(a) \boxtimes \cdots \boxtimes L(a)$ by L.

(i) Let M be a non-zero $\mathcal{H}_n$-submodule of $L(a^n)$. Then $\mathrm{res}^n_{1,\dots,1} M$ must contain a $\mathcal{P}_n$-submodule N isomorphic to L. But the commuting operators $x_1, \dots, x_n$ act on L as scalars, giving that N is contained in their common eigenspace on $L(a^n)$. But by Lemma 4.3.1, this implies that $N = 1 \otimes L$. This shows that M contains $1 \otimes L$, but $1 \otimes L$ generates the whole of $L(a^n)$ over $\mathcal{H}_n$. So $M = L(a^n)$. To see that $L(a^n)$ is the only irreducible in its block use Frobenius reciprocity and the fact just proved that $L(a^n)$ is irreducible.

(ii) The fact that all composition factors of $\mathrm{res}^n_\mu L(a^n)$ are isomorphic to $L(a^{\mu_1}) \boxtimes \cdots \boxtimes L(a^{\mu_r})$ follows by formal characters and (i). To see that $\mathrm{soc}\,\mathrm{res}^n_\mu L(a^n)$ is irreducible, note that the submodule $\mathcal{H}_\mu \otimes L$ of $\mathrm{res}_{\mathcal{H}_\mu} L(a^n)$ is isomorphic to $L(a^{\mu_1}) \boxtimes \cdots \boxtimes L(a^{\mu_l})$. This module is irreducible, and so it is contained in the socle. Conversely, let M be an irreducible $\mathcal{H}_\mu$-submodule of $L(a^n)$. Then using Lemma 4.3.1 as in the proof of (i), we see that M must contain $1 \otimes L$, hence $\mathcal{H}_\mu \otimes L$.

(iii) By part (ii), $L(a^n)$ has a unique $\mathcal{H}_{n-1,1}$-submodule isomorphic to $L(a^{n-1}) \boxtimes L(a)$, namely $\mathcal{H}_{n-1,1} \otimes L$, which contributes a copy of $L(a^{n-1})$ to $\mathrm{soc}\,\mathrm{res}^n_{n-1} L(a^n)$. Conversely, take any irreducible $\mathcal{H}_{n-1}$-submodule M of $L(a^n)$. The common a-eigenspace of $x_1, \dots, x_{n-1}$ on M must lie in $1 \otimes L$ by Lemma 4.3.1. Hence, $M \subseteq \mathcal{H}_{n-1,1} \otimes L$, which completes the proof. □

4.4 Covering modules

Fix $a \in F$ and $n \geq 1$ throughout the section. We will construct for each $m \geq 1$ an $\mathcal{H}_n$-module $L_m(a^n)$ with irreducible head isomorphic to $L(a^n)$. Let $\mathcal{J}(a^n)$ denote the annihilator in $\mathcal{H}_n$ of $L(a^n)$. Introduce the quotient algebra

$$\mathcal{R}_m(a^n) := \mathcal{H}_n / \mathcal{J}(a^n)^m \tag{4.5}$$

for each $m \geq 1$. Obviously $\mathcal{J}(a^n)$ contains $(x_k - a)^{n!}$ for each $k = 1, \dots, n$, whence each algebra $\mathcal{R}_m(a^n)$ is finite dimensional. Moreover, by Theorem 4.3.2, $L(a^n)$ is the unique irreducible $\mathcal{R}_m(a^n)$-module up to isomorphism. Let $L_m(a^n)$ denote a projective cover of $L(a^n)$ in the category $\mathcal{R}_m(a^n)$-mod (for convenience, we also define $L_0(a^n) = \mathcal{R}_0(a^n) = 0$). Then:

Lemma 4.4.1 *For each $m \geq 1, \mathcal{R}_m(a^n) \cong L_m(a^n)^{\oplus n!}$*

There are obvious surjections

$$\mathcal{R}_1(a^n) \twoheadleftarrow \mathcal{R}_2(a^n) \twoheadleftarrow \dots, \tag{4.6}$$

$$L(a^n) = L_1(a^n) \twoheadleftarrow L_2(a^n) \twoheadleftarrow \dots, \tag{4.7}$$

where $R_m(a^n)$ and $L_m(a^n)$ are considered as $\mathcal{H}_n$-modules by inflation.

By an argument involving lifting idempotents (see for example [La]) we may assume that these surjections agree with the decompositions $\mathcal{R}_m(a^n) \cong L_m(a^n)^{\oplus n!}$ for different ms.

Lemma 4.4.2 *Let M be an $\mathcal{H}_n$-module annihilated by $\mathcal{J}(a^n)^k$ for some k. Then for all $m \geq k$ there is a natural isomorphism of $\mathcal{H}_n$-modules*

$$\operatorname{Hom}_{\mathcal{H}_n}(\mathcal{R}_m(a^n), M) \cong M.$$

Moreover, there is an isomorphism of functors

$$\varinjlim_m \operatorname{Hom}_{\mathcal{H}_n}(\mathcal{R}_m(a^n),\ ?) \cong \varinjlim_m \operatorname{Hom}_{\mathcal{H}_n}(L_m(a^n),\ ?)^{\oplus n!}$$

from the category of $\mathcal{H}_n$-module annihilated by some power of $\mathcal{J}(a^n)$ to the category of vector spaces.

Proof The assumption implies that M is the inflation of an $\mathcal{R}_m(a^n)$-module. So

$$\operatorname{Hom}_{\mathcal{H}_n}(\mathcal{R}_m(a^n), M) \cong \operatorname{Hom}_{\mathcal{R}_m(a^n)}(\mathcal{R}_m(a^n), M) \cong M,$$

all isomorphisms being the natural ones. The second statement follows from the remark preceding this lemma. □

Now let **1** and **sgn** be the trivial and the sign $\mathcal{G}_n$-modules, respectively, and set

$$P := \operatorname{ind}_{\mathcal{G}_n}^{\mathcal{H}_n} \mathbf{1}, \quad Q := \operatorname{ind}_{\mathcal{G}_n}^{\mathcal{H}_n} \mathbf{sgn}.$$

The following lemma gives another description of the modules $L_m(a^n)$.

Lemma 4.4.3 *We have $L_m(a^n) \cong P/\mathcal{J}(a^n)^m P \cong Q/\mathcal{J}(a^n)^m Q$.*

Proof We prove the result for P, the argument for Q being similar. By definition of $L(a^n)$, its restriction from $\mathcal{H}_n$ to $\mathcal{G}_n$ is the regular module. So, using Frobenius reciprocity, we get

$$\begin{aligned}\operatorname{Hom}_{\mathcal{H}_n}(P/\mathcal{J}(a^n)^m P, L(a^n)) &\cong \operatorname{Hom}_{\mathcal{H}_n}(P, L(a^n))\\ &\cong \operatorname{Hom}_{\mathcal{G}_n}(\mathbf{1}, \operatorname{res}_{\mathcal{G}_n}^{\mathcal{H}_n} L(a^n)) \cong F,\end{aligned}$$

whence $P/\mathcal{J}(a^n)^m P$ is an $\mathcal{R}_m(a^n)$-module with irreducible head $L(a^n)$. Therefore $P/\mathcal{J}(a^n)^m P$ is a quotient of the principal indecomposable module $L_m(a^n)$.

However, let $JL_m(a^n)$ be the radical of $L_m(a^n)$, and let

$$\pi : L_m(a^n) \to L_m(a^n)/JL_m(a^n) \cong L(a^n)$$

be the natural projection. Recall again that on restriction to $\mathcal{G}_n$, $L(a^n)$ is the regular module, so it has a non-zero S_n-invariant vector v. Moreover, as the regular module splits out, there exists an S_n-invariant vector $w \in L_m(a^n)$ such that $\pi(w) = v$. This vector w gives rise to the $\mathcal{G}_n$-homomorphism $\mathbf{1} \to L_m(a^n)$, whose image is Fw. By Frobenius reciprocity, this homomorphism yields an $\mathcal{H}_n$-homomorphism $P \to L_m(a^n)$ and hence an $\mathcal{R}_m(a^n)$-homomorphism

$$P/\mathcal{J}(a^n)^m P \to L_m(a^n),$$

which must be surjective by the choice of w. □

Let us finally consider two important special cases: $n = 1$ and $m = 1$. If $n = 1$, we easily check that $R_m(a) = L_m(a)$ is just the Jordan block of size m with the eigenvalue a, that is the vector space on basis $w_1, \dots, w_m$ with $x_1 w_k = a w_k + w_{k+1}$, interpreting w_{m+1} as 0. We can also describe the map $L_m(a) \twoheadleftarrow L_{m+1}(a)$ from (4.7) explicitly: it is the identity on $w_1, \dots, w_m$ but maps w_{m+1} to zero.

If $m = 1$, we know by Lemma 4.4.1 that $R_1(a^n)$ has dimension $(n!)^2$. Let $\chi_{(a^n)}$ be the central character of $L(a^n)$, cf. (4.2), and $\mathcal{Z}_{(a^n)}$ be the kernel of $\chi_{(a^n)}$. By Theorem 3.3.1, $\mathcal{Z}_{(a^n)}$ consists of all symmetric polynomials in the $x_i - a$ without free term. By Theorem 1.0.2, $\mathcal{P}_m/\mathcal{P}_m\mathcal{Z}_{(a^n)}$ has dimension $n!$, and a basis which consists of the cosets of

$$\{(x_1 - a)^{a_1} \dots (x_n - a)^{a_n} \mid 0 \leq a_i < i \text{ for all } 1 \leq i \leq n\}.$$

Now, in view of Theorem 3.2.2, $\mathcal{H}_n/\mathcal{H}_n\mathcal{Z}_{(a^n)}$ has dimension $(n!)^2$ and a basis which consists of the cosets of

$$\{w(x_1 - a)^{a_1} \dots (x_n - a)^{a_n} \mid 0 \leq a_i < i \text{ for all } 1 \leq i \leq n,\ w \in S_n\}.$$

As $\mathcal{H}_n\mathcal{Z}_{(a^n)}$ annihilates $L(a^n)$, we must have

$$J(a^n) = \mathcal{H}_n\mathcal{Z}_{(a^n)} \tag{4.8}$$

by dimensions.

5
Crystal operators

In this key chapter we begin to study branching rules for affine Hecke algebras. These are results about restrictions of irreducible $\mathcal{H}_n$-modules to some large natural subalgebras, such as $\mathcal{H}_{n-1}$ or $\mathcal{H}_{n-1} \otimes \mathcal{H}_1$. The idea of "Robinson's a-restriction", which comes from symmetric groups, is that when you restrict an irreducible module from $\mathcal{H}_n$ to $\mathcal{H}_{n-1}$ the blocks which can occur are of very limited form, and all possible blocks can be labeled by just one scalar parameter $a \in F$. So we can denote the corresponding block component of the restriction $\mathrm{res}^{\mathcal{H}_n}_{\mathcal{H}_{n-1}} M$ by $e_a M$ (of course $e_a M$ could be 0). There is a better definition though which works for any module and shows that e_a is actually a functor from $\mathcal{H}_n$-modules to $\mathcal{H}_{n-1}$-modules: $e_a M$ is the generalized a-eigenspace of x_n on M (considered as an $\mathcal{H}_{n-1}$-module on restriction).

An important result, which is a subtle generalization of the multiplicity freeness of the branching rule for symmetric groups in characteristic 0, states that the socle of the $\mathcal{H}_{n-1}$-module $e_a M$ is irreducible when M is an irreducible $\mathcal{H}_n$-module. For affine Hecke algebras this was first proved by Grojnowski and Vazirani [GV], following earlier result [K_2] for symmetric groups. Let us write $\tilde{e}_a$ for the socle of $e_a M$. Then $\tilde{e}_a$ is a map from {iso-classes of irreducible $\mathcal{H}_n$-modules} to {iso-classes of irreducible $\mathcal{H}_{n-1}$-modules}$\cup\{0\}$. We also define an important function ε_a on iso-classes of irreducible $\mathcal{H}_n$-modules. One of the equivalent definitions is: $\varepsilon_a(M) = \max\{m \geq 0 \mid \tilde{e}_a^m M \neq 0\}$. We prove that this function has many important representation theoretic interpretations: $\varepsilon_a(M)$ is the multiplicity of $\tilde{e}_a M$ as a composition factor of $e_a M$, it also is the maximal size of a Jordan block of x_n on M with eigenvalue a, and, finally, it is the dimension of the endomorphism algebra $\mathrm{End}_{\mathcal{H}_{n-1}}(e_a M)$.

If we try to use induction instead of restriction to define the analogues $f_a, \tilde{f}_a, \varphi_a$ of $e_a, \tilde{e}_a, \varepsilon_a$, we run into a problem. The trouble is that induction from $\mathcal{H}_n$ to $\mathcal{H}_{n+1}$ does not preserve finite dimensionality, while inducing

$M \otimes L(a)$ from $\mathcal{H}_n \otimes \mathcal{H}_1$ to $\mathcal{H}_{n+1}$, called ∇_a below, is not "large enough", and is quite "defective" in many respects. For example, it is not self-dual, while $e_a M$ is (for irreducible M). We will give "correct" definitions of f_a and φ_a in Chapter 8. Still, ∇_a applied to an irreducible module does produce a module with irreducible head (always non-zero, which is another difference with e_a), and so we use it to define the map $\tilde{f}_a$ from {iso-classes of irreducible $\mathcal{H}_n$-modules} to {iso-classes of irreducible $\mathcal{H}_{n+1}$-modules}. It has the following nice property: $\tilde{f}_a M = N$ if and only if $\tilde{e}_a N = M$. We use these operations $\tilde{f}_a$ to get a notation for irreducible $\mathcal{H}_n$-modules. This cannot be called labeling yet, because one irreducible can have several different notations.

5.1 Multiplicity-free socles

Let $M \in \mathcal{H}_n$-mod and $a \in F$. Define $\Delta_a M$ to be the generalized a-eigenspace of x_n on M. Equivalently

$$\Delta_a M = \bigoplus_{\underline{a} \in F^n,\ a_n = a} M_{\underline{a}}, \tag{5.1}$$

recalling the decomposition from Lemma 4.1.1. Note since x_n is central in the parabolic subalgebra $\mathcal{H}_{n-1,1}$ of $\mathcal{H}_n$, $\Delta_a M$ is invariant under this subalgebra. So, in fact, Δ_a can be viewed as an exact functor

$$\Delta_a : \mathcal{H}_n\text{-mod} \to \mathcal{H}_{n-1,1}\text{-mod}, \tag{5.2}$$

being defined on morphisms simply as restriction. Slightly more generally, given $m \geq 0$, define

$$\Delta_{a^m} : \mathcal{H}_n\text{-mod} \to \mathcal{H}_{n-m,m}\text{-mod} \tag{5.3}$$

so that $\Delta_{a^m} M$ is the simultaneous generalized a-eigenspace of the commuting operators x_k for $k = n-m+1, \dots, n$. In view of Theorem 4.3.2(i), $\Delta_{a^m} M$ can also be characterized as the largest submodule of $\mathrm{res}^n_{n-m,m} M$ all of whose composition factors are of the form $N \boxtimes L(a^m)$ for irreducible $N \in \mathcal{H}_{n-m}$-mod. The definition of Δ_{a^m} implies functorial isomorphisms

$$\begin{aligned} &\mathrm{Hom}_{\mathcal{H}_{n-m,m}}(N \boxtimes L(a^m), \Delta_{a^m} M) \\ &\cong \mathrm{Hom}_{\mathcal{H}_n}(\mathrm{ind}^n_{n-m,m} N \boxtimes L(a^m), M) \end{aligned} \tag{5.4}$$

for $N \in \mathcal{H}_{n-m}$-mod, $M \in \mathcal{H}_n$-mod. Also from definitions we get:

Lemma 5.1.1 *Let $M \in \mathcal{H}_n$-mod with*

$$\mathrm{ch}\, M = \sum_{\underline{a} \in F^n} r_{\underline{a}} [L(a_1) \boxtimes \cdots \boxtimes L(a_n)].$$

Then we have

$$\operatorname{ch}\Delta_{a^m} M = \sum_{\underline{b}} r_{\underline{b}}[L(b_1) \boxtimes \cdots \boxtimes L(b_n)],$$

summing over all $\underline{b} \in F^n$ *with* $b_{n-m+1} = \cdots = b_n = a$.

Now for $a \in F$ and $M \in \mathcal{H}_n$-mod, define

$$\varepsilon_a(M) = \max\{m \geq 0 \mid \Delta_{a^m} M \neq 0\}. \tag{5.5}$$

Lemma 5.1.1 shows that $\varepsilon_a(M)$ can be worked out just from knowledge of the character $\operatorname{ch} M$: it is the length of the "longest a-tail".

Lemma 5.1.2 *Let* $M \in \mathcal{H}_n$-mod *be irreducible,* $a \in F$, $\varepsilon = \varepsilon_a(M)$. *If* $N \boxtimes L(a^m)$ *is an irreducible submodule of* $\Delta_{a^m} M$ *for some* $0 \leq m \leq \varepsilon$, *then* $\varepsilon_a(N) = \varepsilon - m$.

Proof The definitions imply immediately that $\varepsilon_a(N) \leq \varepsilon - m$. For the reverse inequality, the property (5.4) and the irreducibility of M imply that M is a quotient of $\operatorname{ind}^n_{n-m,m} N \boxtimes L(a^m)$. So applying the exact functor Δ_{a^ε}, we see that $\Delta_{a^\varepsilon} M \neq 0$ is a quotient of

$$\Delta_{a^\varepsilon}(\operatorname{ind}^n_{n-m,m} N \boxtimes L(a^m)).$$

In particular, $\Delta_{a^\varepsilon}(\operatorname{ind}^n_{n-m,m} N \boxtimes L(a^m)) \neq 0$. Now we get that $\varepsilon_a(N) \geq \varepsilon - m$ applying the Shuffle Lemma and Lemma 5.1.1. □

Lemma 5.1.3 *Let* $m \geq 0$, $a \in F$ *and* $N \in \mathcal{H}_n$-mod *be irreducible with* $\varepsilon_a(N) = 0$. *Set* $M = \operatorname{ind}^{n+m}_{n,m} N \boxtimes L(a^m)$. *Then:*

(i) $\Delta_{a^m} M \cong N \boxtimes L(a^m)$;
(ii) $\operatorname{hd} M$ *is irreducible with* $\varepsilon_a(\operatorname{hd} M) = m$;
(iii) *all other composition factors* L *of* M *have* $\varepsilon_a(L) < m$.

Proof (i) Clearly a copy of $N \boxtimes L(a^m)$ appears in $\Delta_{a^m} M$. But by the Shuffle Lemma and Lemma 5.1.1, $\dim(\Delta_{a^m} M) = \dim(N \boxtimes L(a^m))$, hence $\Delta_{a^m} M \cong N \boxtimes L(a^m)$.

(ii) By (5.4), a copy of $N \boxtimes L(a^m)$ appears in $\Delta_{a^m} Q$ for any non-zero quotient Q of M, in particular for any constituent of $\operatorname{hd} M$. But by (i), $N \boxtimes L(a^m)$ only appears once in $\Delta_{a^m} M$, hence $\operatorname{hd} M$ must be irreducible.

(iii) We have shown that $\Delta_{a^m} M = \Delta_{a^m}(\operatorname{hd} M)$. Hence, $\Delta_{a^m} L = 0$ for any other composition factor of M by exactness of Δ_{a^m}. □

Lemma 5.1.4 *Let* $M \in \mathcal{H}_n$-mod *be irreducible,* $a \in F$*, and* $\varepsilon = \varepsilon_a(M)$*. Then* $\Delta_{a^\varepsilon} M$ *is isomorphic to* $N \boxtimes L(a^\varepsilon)$ *for some irreducible* $\mathcal{H}_{n-\varepsilon}$*-module* N *with* $\varepsilon_a(N) = 0$.

Proof Pick an irreducible submodule $N \boxtimes L(a^\varepsilon)$ of $\Delta_{a^\varepsilon} M$. Then $\varepsilon_a(N) = 0$ by Lemma 5.1.2. By (5.4) and the irreducibility of M, M is a quotient of $\mathrm{ind}^n_{n-\varepsilon,\varepsilon} N \boxtimes L(a^\varepsilon)$. Hence, $\Delta_{a^\varepsilon} M$ is a quotient of

$$\Delta_{a^\varepsilon} \mathrm{ind}^n_{n-\varepsilon,\varepsilon} N \boxtimes L(a^\varepsilon).$$

But this is isomorphic to $N \boxtimes L(a^\varepsilon)$ by Lemma 5.1.3(i). This shows that $\Delta_{a^\varepsilon} M \cong N \boxtimes L(a^\varepsilon)$. □

Lemma 5.1.5 *Let* $m \geq 0$*,* $a \in F$ *and* $N \in \mathcal{H}_n$-mod *be irreducible. Set*

$$M = \mathrm{ind}^{n+m}_{n,m} (N \boxtimes L(a^m)).$$

Then $\mathrm{hd}\, M$ *is irreducible with* $\varepsilon_a(\mathrm{hd}\, M) = \varepsilon_a(N) + m$*, and all other composition factors* L *of* M *have* $\varepsilon_a(L) < \varepsilon_a(N) + m$.

Proof Let $\varepsilon = \varepsilon_a(N)$. By Lemma 5.1.4, we have that

$$\Delta_{a^\varepsilon} N = K \boxtimes L(a^\varepsilon)$$

for an irreducible $K \in \mathcal{H}_{n-\varepsilon}$-mod with $\varepsilon_a(K) = 0$. By (5.4) and the irreducibility of N, N is a quotient of $\mathrm{ind}^n_{n-\varepsilon,\varepsilon} K \boxtimes L(a^\varepsilon)$. So the transitivity of induction implies that $\mathrm{ind}^{n+m}_{n,m} N \boxtimes L(a^m)$ is a quotient of $\mathrm{ind}^{n+m}_{n-\varepsilon,\varepsilon+m} K \boxtimes L(a^{\varepsilon+m})$. Now everything follows from Lemma 5.1.3. □

Theorem 5.1.6 *Let* $M \in \mathcal{H}_n$-mod *be irreducible and* $a \in F$*. Then, for any* $0 \leq m \leq \varepsilon_a(M)$*,* $\mathrm{soc}\, \Delta_{a^m} M$ *is an irreducible* $\mathcal{H}_{n-m,m}$*-module of the form* $L \boxtimes L(a^m)$*, with* $\varepsilon_a(L) = \varepsilon_a(M) - m$.

Proof Let $\varepsilon = \varepsilon_a(M)$. Suppose that $L \boxtimes L(a^m)$ is a constituent of $\mathrm{soc}\, \Delta_{a^m} M$. By Lemma 5.1.2, we have $\varepsilon_a(L) = \varepsilon - m$. So every such L contributes a non-trivial submodule to $\mathrm{res}^{n-\varepsilon,\varepsilon}_{n-\varepsilon,\varepsilon-m,m} \Delta_{a^\varepsilon} M$. But $\Delta_{a^\varepsilon} M$ is irreducible of the form $N \boxtimes L(a^\varepsilon)$ by Lemma 5.1.4, and so by Theorem 4.3.2(ii), the socle of $\mathrm{res}^{n-\varepsilon,\varepsilon}_{n-\varepsilon,\varepsilon-m,m} \Delta_{a^\varepsilon} M$ is $N \boxtimes L(a^{\varepsilon-m}) \boxtimes L(a^m)$. Hence $\mathrm{soc}\, \Delta_{a^m} M$ must equal $L \boxtimes L(a^m)$. □

We can also apply the theorem to study $\mathrm{res}^n_{n-1} M$, meaning the restriction of M to the subalgebra $\mathcal{H}_{n-1} \subset \mathcal{H}_n$, see (3.12). Define the functor

$$e_a := \mathrm{res}^{n-1,1}_{n-1} \circ \Delta_a : \mathcal{H}_n\text{-mod} \to \mathcal{H}_{n-1}\text{-mod}. \tag{5.6}$$

Record the following obvious equalities:

$$\varepsilon_a(M) = \max\{m \geq 0 \mid e_a^m M \neq 0\}, \tag{5.7}$$

$$\mathrm{res}^n_{n-1} M = \bigoplus_{a \in F} e_a M. \tag{5.8}$$

Also, it is clear from (5.1) that

$$\begin{gathered}\text{if} \quad \mathrm{ch}\, M = \sum_{\underline{a} \in F^n} c_{\underline{a}} [L(a_1) \boxtimes \cdots \boxtimes L(a_n)] \quad \text{then} \\ \mathrm{ch}\,(e_a M) = \sum_{\underline{a} \in F^{n-1}} c_{(a_1, \dots, a_{n-1}, a)} [L(a_1) \boxtimes \cdots \boxtimes L(a_{n-1})].\end{gathered} \tag{5.9}$$

Corollary 5.1.7 *For an irreducible $M \in \mathcal{H}_n$-mod with $\varepsilon_a(M) > 0$, the socle of $e_a M$ is irreducible, and $\varepsilon_a(\mathrm{soc}\, e_a M) = \varepsilon_a(M) - 1$.*

Proof Let L be an irreducible submodule of $e_a(M)$. The central element $z := x_1 + \cdots + x_n$ of $\mathcal{H}_n$ acts as a scalar on the whole M by Schur's Lemma, and similarly the central element $z' := x_1 + \cdots + x_{n-1}$ of $\mathcal{H}_{n-1}$ acts as a scalar on L. Hence $x_n = z - z'$ acts on L as a scalar, too. Now it follows that the scalar must be a, and that L contributes a composition factor $L \boxtimes L(a)$ to the socle of $\Delta_a(M)$. It remains to apply Theorem 5.1.6. □

Corollary 5.1.8 *For irreducible $M \in \mathcal{H}_n$-mod, the socle of $\mathrm{res}^n_{n-1} M$ is multiplicity-free.*

Proof We have $\mathrm{res}^n_{n-1} M = \bigoplus_{a \in F} e_a M$, with all but finitely many summands zero. Now, the socle of each non-zero $e_a M$ is irreducible by Corollary 5.1.7. Finally $e_a M$ and $e_b M$ are in different blocks for $a \neq b$, so their socles are definitely not isomorphic. □

5.2 Operators $\tilde{e}_a$ and $\tilde{f}_a$

Let M be an irreducible module in $\mathcal{H}_n$-mod. Define

$$\tilde{e}_a M := \mathrm{soc}\, e_a M, \quad \tilde{f}_a M := \mathrm{hd}\, \mathrm{ind}^{n+1}_{n,1} M \boxtimes L(a). \tag{5.10}$$

Note $\tilde{f}_a M$ is irreducible by Lemma 5.1.5, and $\tilde{e}_a M$ is irreducible or 0 by Corollary 5.1.7. Also from Corollary 5.1.7 we have

$$\varepsilon_a(M) = \max\{m \geq 0 \mid \tilde{e}_a^m M \neq 0\}, \tag{5.11}$$

while a special case of Lemma 5.1.5 shows that

$$\varepsilon_a(\tilde{f}_a M) = \varepsilon_a(M) + 1. \tag{5.12}$$

Lemma 5.2.1 *Let $M \in \mathcal{H}_n$-mod be irreducible, $a \in F$ and $m \geq 0$.*

(i) $\operatorname{soc} \Delta_{a^m} M \cong (\tilde{e}_a^m M) \boxtimes L(a^m)$.
(ii) $\operatorname{hd} \operatorname{ind}_{n,m}^{n+m} M \boxtimes L(a^m) \cong \tilde{f}_a^m M$.

Proof (i) If $m > \varepsilon_a(M)$, then both parts in the equality above are zero. Let $m \leq \varepsilon_a(M)$. By Corollary 5.1.7 and Theorem 5.1.6, $(\tilde{e}_a M) \boxtimes L(a)$ is a submodule of $\Delta_a M$. By applying this m times we deduce that $(\tilde{e}_a^m M) \boxtimes L(a)^{\boxtimes m}$ is a submodule of $\operatorname{res}^{n-m,m}_{n-m,1,\dots,1} \Delta_{a^m} M$, whence $(\tilde{e}_a^m M) \boxtimes L(a^m)$ is a submodule of $\Delta_{a^m} M$ by Frobenius reciprocity and Theorem 4.3.2(i). Now the result follows from Theorem 5.1.6.

(ii) By exactness of induction and Theorem 4.3.2(i), $\tilde{f}_a^m M$ is a quotient of $\operatorname{ind}_{n,m}^{n+m} M \boxtimes L(a^m)$. Now the result follows from the simplicity of the head, see Lemma 5.1.5. □

Now we refine Corollary 5.1.7.

Lemma 5.2.2 *Let $M \in \mathcal{H}_n$-mod be irreducible, $a \in F$ and $m \geq 0$. Then the socle of $e_a^m M$ is isomorphic to $(\tilde{e}_a^m M)^{\oplus m!}$.*

Proof Let $\mathcal{Z}_m$ be the center of $\mathcal{H}_m$, and $\chi = \chi_{(a,\dots,a)}$ be the character of $\mathcal{Z}_m$, which comes from the action on the Kato module $L(a^m)$. Recall that $\mathcal{H}_m$ is a free (right) module over the polynomial subalgebra $\mathcal{P}_m$ of rank $m!$, and $\mathcal{P}_m$ is a free module over the ring of symmetric functions $\mathcal{Z}_m$ of rank $m!$ (see Theorem 1.0.2). So $\mathcal{H}_m$ is a free module over $\mathcal{Z}_m$ of rank $(m!)^2$. It follows that

$$U := \operatorname{ind}_{\mathcal{Z}_m}^{\mathcal{H}_m} \chi$$

is a non-zero $\mathcal{H}_m$-module, all of whose composition factors are isomorphic to $L(a^m)$, and the multiplicity of $L(a^m)$ in U is $m!$. But by Frobenius reciprocity we have

$$\dim \operatorname{Hom}_{\mathcal{H}_m}(U, L(a^m)) = m!,$$

so by Schur's Lemma, $U \cong L(a^m)^{\oplus m!}$.

Now, let L be an irreducible submodule of $e_a^m M$. As in the proof of Corollary 5.1.7, Schur's Lemma implies that any symmetric function in the variables $x_1, \dots, x_n$ and any symmetric function in the variables $x_1, \dots, x_{n-m}$

act on L with scalars. It follows that the symmetric functions in the variables $x_{n-m+1}, \dots, x_n$ also act on L with scalars (one way to see this is to express the elementary symmetric functions in the variables $x_{n-m+1}, \dots, x_n$ in terms of those in the variables $x_1, \dots, x_n$ and the variables $x_1, \dots, x_{n-m}$). This means that the center $\mathcal{Z}_m$ of the subalgebra $\mathcal{H}_m \cong 1 \otimes \mathcal{H}_m \subset \mathcal{H}_{n-m} \otimes \mathcal{H}_m$ acts on L with scalars, and it is clear that the corresponding central character is χ. Hence we have a non-zero $H_{n-m} \otimes Z_m$-homomorphism from $L \boxtimes \chi$ to M (whose image equals L). Frobenius reciprocity now yields a non-zero homomorphism

$$L \boxtimes U \cong \operatorname{ind}_{\mathcal{H}_{n-m} \otimes \mathcal{Z}_m}^{\mathcal{H}_{n-m} \otimes \mathcal{H}_m} L \boxtimes \chi \to M,$$

whose image contains L. As $L \boxtimes U$ is a direct sum of copies of the irreducible module $L \boxtimes L(a^m)$, it follows that the $\mathcal{H}_{n-m} \otimes \mathcal{H}_m$-submodule generated by L is isomorphic to $L \boxtimes L(a^m)$. Now the result follows from Lemma 5.2.1(i). □

Lemma 5.2.3 *Let $M \in \mathcal{H}_n$-mod and $N \in \mathcal{H}_{n+1}$-mod be irreducible modules, and $a \in F$. Then $\tilde{f}_a M \cong N$ if and only if $\tilde{e}_a N \cong M$.*

Proof By Lemma 5.1.5, $\tilde{f}_a M \cong N$ is equivalent to $\operatorname{Hom}_{\mathcal{H}_{n+1}}(\operatorname{ind}_{n,1}^{n+1} M \boxtimes L(a), N) \neq 0$, which in turn is equivalent to

$$\operatorname{Hom}_{\mathcal{H}_{n,1}}(M \boxtimes L(a), \Delta_a N) \neq 0,$$

thanks to (5.4). The last property means that $M \boxtimes L(a)$ appears in the socle of $\Delta_a N$, which is equivalent to $M \cong \tilde{e}_i N$ in view of Lemma 5.2.1(i). □

From Lemma 5.2.3 we immediately deduce the following:

Corollary 5.2.4 *Let $M, N \in \mathcal{H}_n$-mod be irreducible. Then $\tilde{f}_a M \cong \tilde{f}_a N$ if and only if $M \cong N$. Similarly, providing $\varepsilon_a(M), \varepsilon_a(N) > 0$, $\tilde{e}_a M \cong \tilde{e}_a N$ if and only if $M \cong N$.*

5.3 Independence of irreducible characters

We can now prove a very useful result (we follow the argument of [V_1, Section 5.5]):

Theorem 5.3.1 *The map* ch : $K(\mathcal{H}_n\text{-mod}) \to K(\mathcal{P}_n\text{-mod})$ *is injective.*

Proof We need to show that the characters of the irreducible modules in $\mathcal{H}_n$-mod are linearly independent in $K(\mathrm{Rep}_I \mathcal{P}_n)$. Proceed by induction on n, the case $n = 0$ being trivial. Suppose $n > 0$ and there is a non-trivial $\mathbb{Z}$-linear dependence

$$\sum c_L \mathrm{ch}\, L = 0 \tag{5.13}$$

for some irreducible modules $L \in \mathcal{H}_n$-mod. Choose any $a \in F$. We will show by downward induction on $k = n, \ldots, 1$ that $c_L = 0$ for all L with $\varepsilon_a(L) = k$. Since every irreducible L has $\varepsilon_a(L) > 0$ for at least one $a \in F$, this is enough to complete the proof.

Consider first the case that $k = n$. Then $\Delta_{a^n} L = 0$ except if $L \cong L(a^n)$, by Theorem 4.3.2(i). Applying Δ_{a^n} to the equation (5.13) and using Lemma 5.1.1, we deduce that the coefficient of $\mathrm{ch}\, L(a^n)$ is zero. Thus the induction starts. Now suppose $1 \leq k < n$ and that we have shown $c_L = 0$ for all L with $\varepsilon_a(L) > k$. Apply Δ_{a^k} to the equation to deduce that

$$\sum_{L \text{ with } \varepsilon_a(L)=k} c_L \mathrm{ch}\, \Delta_{a^k} L = 0.$$

Now each such $\Delta_{a^k} L$ is isomorphic to $(\tilde{e}_a^k L) \boxtimes L(a^k)$, according to Lemmas 5.1.4 and 5.2.1(i). Moreover, for $L \not\cong L'$, $\tilde{e}_a^k L \not\cong \tilde{e}_a^k L'$ by Corollary 5.2.4. So now the induction hypothesis on n gives that all such coefficients c_L are zero, as required. □

Corollary 5.3.2 *If L is an irreducible module in $\mathcal{H}_n$-mod, then $L \cong L^\tau$.*

Proof Since $\tau(x_i) = x_i$, τ leaves characters invariant. Hence it leaves irreducibles invariant, since they are determined up to isomorphism by their character according to the theorem. □

Now we can deduce the following criterion for irreducibility:

Lemma 5.3.3 *Let $M \in \mathcal{H}_m$-mod and $N \in \mathcal{H}_n$-mod be irreducible modules. Suppose:*

(i) $\mathrm{ind}_{m,n}^{m+n} M \boxtimes N \cong \mathrm{ind}_{n,m}^{n+m} N \boxtimes M$;
(ii) *$M \boxtimes N$ appears in $\mathrm{res}_{m,n}^{m+n} \mathrm{ind}_{m,n}^{m+n} M \boxtimes N$ with multiplicity one.*

Then $\mathrm{ind}_{m,n}^{m+n} M \boxtimes N$ is irreducible.

Proof Suppose for a contradiction that $K := \mathrm{ind}_{m,n}^{m+n} M \boxtimes N$ is reducible. Then we can find a proper irreducible submodule S, and set $Q = K/S$. By Frobenius reciprocity, $M \boxtimes N$ appears in $\mathrm{res}_{m,n}^{m+n} Q$ with non-zero multiplicity.

Hence, it cannot appear in $\operatorname{res}^{m+n}_{m,n} S$ by assumption (ii). But assumption (i), Corollary 5.3.2, and Theorem 3.7.5 show that $K \cong K^\tau$. Hence, K also has a quotient isomorphic to $S^\tau \cong S$, and the Frobenius reciprocity argument implies that $M \boxtimes N$ appears in $\operatorname{res}^{m+n}_{m,n} S$, giving a contradiction. □

5.4 Labels for irreducibles

We introduce some notation to label the isomorphism classes of irreducible representations. Write $\mathbf{1}$ for the (trivial) irreducible module of $\mathcal{H}_0 \cong F$. If L is an irreducible module in $\mathcal{H}_n$-mod, we easily show using Lemma 5.2.3 repeatedly that

$$L \cong \tilde{f}_{a_n} \cdots \tilde{f}_{a_2} \tilde{f}_{a_1} \mathbf{1}$$

for at least one tuple $\underline{a} = (a_1, a_2, \ldots, a_n) \in F^n$. So if we define

$$L(\underline{a}) = L(a_1, \ldots, a_n) := \tilde{f}_{a_n} \cdots \tilde{f}_{a_2} \tilde{f}_{a_1} \mathbf{1}, \tag{5.14}$$

we obtain a labeling of all irreducibles by tuples in I^n. For example, $L(a, a, \ldots, a)$ (n times) is precisely the Kato module $L(a^n)$ introduced in (4.4). Of course, the problem with this labeling is that a given irreducible L will in general be parametrized by several *different* tuples $\underline{a} \in F^n$. But basic properties of $L(\underline{a})$ are easy to read off from the notation: for instance the central character of $L(\underline{a})$ is $\chi_{\underline{a}}$.

5.5 Alternative descriptions of ε_a

In this section we give three new interpretations of the functions ε_a.

Theorem 5.5.1 *Let $a \in F$ and M be an irreducible module in $\mathcal{H}_n$-mod. Then:*

(i) $[e_a M] = \varepsilon_a(M)[\tilde{e}_a M] + \sum c_r [N_r]$ *where the N_r are irreducible modules with* $\varepsilon_a(N_r) < \varepsilon_a(\tilde{e}_a M) = \varepsilon_a(M) - 1$;

(ii) $\varepsilon_a(M)$ *is the maximal size of a Jordan block of x_n on M with eigenvalue a;*

(iii) *The algebra* $\operatorname{End}_{\mathcal{H}_{n-1}}(e_a M)$ *is isomorphic to the algebra of truncated polynomials* $F[x]/(x^{\varepsilon_a(M)})$. *In particular,*

$$\dim \operatorname{End}_{\mathcal{H}_{n-1}}(e_a M) = \varepsilon_a(M).$$

Proof Let $\varepsilon = \varepsilon_a(M)$ and $N = \tilde{e}_a^{\varepsilon} M$.

(i) By Lemma 5.1.4 and Frobenius reciprocity, there is a short exact sequence

$$0 \longrightarrow R \longrightarrow \mathrm{ind}_{n-\varepsilon,\varepsilon}^{n} N \boxtimes L(a^{\varepsilon}) \longrightarrow M \longrightarrow 0.$$

Moreover, by Lemma 5.1.3(iii), $\varepsilon_a(L) < \varepsilon$ for all composition factors L of R.Applying the exact functor Δ_a, we obtain the exact sequence

$$0 \longrightarrow \Delta_a R \longrightarrow \Delta_a \mathrm{ind}_{n-\varepsilon,\varepsilon} N \boxtimes L(a^{\varepsilon}) \longrightarrow \Delta_a M \longrightarrow 0.$$

As $\Delta_a N = 0$, the Mackey Theorem yields

$$\Delta_a \mathrm{ind}_{n-\varepsilon,\varepsilon}^{n} N \boxtimes L(a^{\varepsilon}) \cong \mathrm{ind}_{n-\varepsilon,\varepsilon-1,1}^{n-1,1} N \boxtimes \Delta_a L(a^{\varepsilon}).$$

By considering characters, we see that

$$[\Delta_a L(a^{\varepsilon})] = \varepsilon[L(a^{\varepsilon-1}) \boxtimes L(a)].$$

Hence,

$$[\Delta_a \mathrm{ind}_{n-\varepsilon,\varepsilon}^{n} N \boxtimes L(a^{\varepsilon})] = \varepsilon[\mathrm{ind}_{n-\varepsilon,\varepsilon-1,1}^{n-1,1} N \boxtimes L(a^{\varepsilon-1}) \boxtimes L(a)]. \tag{5.15}$$

By Lemma 5.2.1(ii), the head of

$$\mathrm{ind}_{n-\varepsilon,\varepsilon-1,1}^{n-1,1} N \boxtimes L(a^{\varepsilon-1}) \boxtimes L(a)$$

is $(\tilde{f}_a^{\varepsilon-1} N) \boxtimes L(a)$ which is the same as $(\tilde{e}_a M) \boxtimes L(a)$, and all other composition factors of this module are of the form $L \boxtimes L(a)$ with $\varepsilon_a(L) < \varepsilon - 1$, thanks to Lemma 5.1.3. Moreover, all composition factors of $\Delta_a R$ are of the form $L \boxtimes L(a)$ with $\varepsilon_a(L) < \varepsilon - 1$. So we have now shown that

$$[\Delta_a M] = \varepsilon[\tilde{e}_a M \boxtimes L(a)] + \sum c_r [N_r \boxtimes L(a)]$$

for irreducibles N_r with $\varepsilon_a(N_r) < \varepsilon_a(\tilde{e}_a M)$, which implies (i).

(ii) We know that $\Delta_{a^{\varepsilon}} M \cong N \boxtimes L(a^{\varepsilon})$. So, applying the automorphism σ to Lemma 4.3.1, we deduce that the maximal size of a Jordan block of x_n on $\Delta_{a^{\varepsilon}} M$ is ε. Hence the maximal size of a Jordan block of x_n on $\Delta_a M$ is at least ε. However, the argument given above in deriving (5.15) shows that the module $\Delta_a \mathrm{ind}_{n-\varepsilon,\varepsilon}^{n} N \boxtimes L(a^{\varepsilon})$ has a filtration with ε factors, each of which is isomorphic to

$$\mathrm{ind}_{n-\varepsilon,\varepsilon-1,1}^{n-1,1} N \boxtimes L(a^{\varepsilon-1}) \boxtimes L(a).$$

Since $(x_n - a)$ annihilates $\mathrm{ind}_{n-\varepsilon,\varepsilon-1,1}^{n-1,1} N \boxtimes L(a^{\varepsilon-1}) \boxtimes L(a)$, it follows that $(x_n - a)^{\varepsilon}$ annihilates $\Delta_a \mathrm{ind}_{n-\varepsilon,\varepsilon}^{n} N \boxtimes L(a^{\varepsilon})$. So certainly $(x_n - a)^{\varepsilon}$ annihilates its quotient $\Delta_a M$. So the maximal size of a Jordan block of x_n on $\Delta_a M$ is at most ε.

(iii) The left multiplication by the element $(x_n - a)$, which centralizes the subalgebra $\mathcal{H}_{n-1}$ of $\mathcal{H}_n$, induces an $\mathcal{H}_{n-1}$-endomorphism

$$\theta : e_a M \to e_a M.$$

By (ii), $\theta^{\varepsilon-1} \neq 0$ and $\theta^{\varepsilon} = 0$. Hence, $1, \theta, \ldots, \theta^{\varepsilon-1}$ give ε linearly independent $\mathcal{H}_{n-1}$-endomorphisms of $e_a M$. They span a subalgebra of $\mathrm{End}_{\mathcal{H}_{n-1}}(e_a M)$ isomorphic to the algebra of truncated polynomials $F[x]/(x^{\varepsilon})$. However, $e_a M$ has irreducible head $\tilde{e}_a M$, and this appears in $e_a M$ with multiplicity ε by (i). So the dimension of $\mathrm{End}_{\mathcal{H}_{n-1}}(e_a M)$ is at most ε. □

The following result is quite surprising.

Corollary 5.5.2 *Let $M, N \in \mathcal{H}_n$-mod be irreducible modules with $M \not\cong N$. Then, for every $a \in F$, we have $\mathrm{Hom}_{\mathcal{H}_{n-1}}(e_a M, e_a N) = 0$.*

Proof Suppose there is a non-zero homomorphism $\theta : e_a M \to e_a N$. Then, since $e_a M$ has irreducible head $\tilde{e}_a M$, we see that $e_a N$ has $\tilde{e}_a M$ as a composition factor. Hence, by Theorem 5.5.1(i), $\varepsilon_a(\tilde{e}_a N) \geq \varepsilon_a(\tilde{e}_a M)$. However, $e_a N$ has irreducible socle $\tilde{e}_a N$, so $e_a M$ has $\tilde{e}_a N$ as a composition factor, which gives the inequality the other way round. Thus $\varepsilon_a(\tilde{e}_a N) = \varepsilon_a(\tilde{e}_a M)$. But then, $\tilde{e}_a M$ is a composition factor of $e_a N$ with $\varepsilon_a(\tilde{e}_a M) = \varepsilon_a(\tilde{e}_a N)$, hence by Theorem 5.5.1(i) again, $\tilde{e}_a M \cong \tilde{e}_a N$. But this contradicts Corollary 5.2.4. □

6
Character calculations

In the first section of this chapter we explain how to reduce the study of irreducible $\mathcal{H}_n$-modules to special blocks, namely the blocks corresponding to the orbits $S_n \cdot \underline{a}$ such that a_is appearing in $\underline{a}$ cannot be split into two subsets $B \sqcup C$ with the property $a_i - a_j \neq \pm 1$ for any $a_i \in B$ and any $a_j \in C$ (note $a_i = a_j$ is allowed). The precise statement is Theorem 6.1.4. One implication of this result is that we may concentrate on the so-called *integral* representations of $\mathcal{H}_n$, see Chapter 7. Theorem 6.1.4 is also useful for concrete calculations in the *small rank* cases, that is the cases of representations of $\mathcal{H}_n$ for $n \leq 4$.

The explicit character information for some small rank cases obtained in this chapter is exactly what we need in order to verify that the linear operators e_a on the Grothendieck group, induced by the exact functors with the same name, satisfy Serre relations of an affine Kac–Moody algebra, see Lemma 9.2.4.

We also use this small rank character information to define and investigate some generalizations $\tilde{f}_{a^r b a^s}$ of the operators $\tilde{f}_a$ on the irreducible modules. These will be used only in a couple of technical points in Section 8.4.

6.1 Some irreducible induced modules

Given $\underline{a} = (a_1, \dots, a_n) \in F^n$, let

$$\mathrm{ind}(\underline{a}) = \mathrm{ind}(a_1, \dots, a_n) := \mathrm{ind}^n_{1,\dots,1} L(a_1) \boxtimes \cdots \boxtimes L(a_n).$$

By Lemma 4.1.2, every irreducible constituent of $\mathrm{ind}(\underline{a})$ belongs to the block corresponding to the orbit $S_n \cdot \underline{a}$.

Lemma 6.1.1 *Let $\underline{a} \in F^n$. Then:*

(i) $\mathrm{res}^n_{1,\dots,1} L(\underline{a})$ *has a submodule isomorphic to* $L(a_1) \boxtimes \cdots \boxtimes L(a_n)$;
(ii) $\mathrm{ind}(\underline{a})$ *contains a copy of* $L(\underline{a})$ *in its head;*
(iii) *every irreducible module in the block corresponding to the orbit* $S_n \cdot \underline{a}$ *appears at least once as a constituent of* $\mathrm{ind}(\underline{a})$.

Proof (i) Proceed by induction on n, the case $n = 1$ being clear. Let $n > 1$. By Frobenius reciprocity, there is a non-zero, hence necessarily injective, $\mathcal{H}_{n-1,1}$-module homomorphism from $L(a_1, \dots, a_{n-1}) \boxtimes L(a_n)$ to $\mathrm{res}^n_{n-1,1} L(\underline{a})$. Hence by inductive assumption we get a copy of

$$L(a_1) \boxtimes \cdots \boxtimes L(a_{n-1}) \boxtimes L(a_n)$$

in $\mathrm{res}^n_{1,\dots,1} L(\underline{a})$.

(ii) Use (i) and Frobenius reciprocity.

(iii) By (ii), $L(\underline{a})$ appears in $\mathrm{ind}(\underline{a})$. But for any other $\underline{b}$ in the same orbit, $\mathrm{ind}(\underline{b})$ has the same character as $\mathrm{ind}(\underline{a})$, hence they have the same set of composition factors thanks to Theorem 5.3.1. Hence, $L(\underline{b})$ appears in $\mathrm{ind}(\underline{a})$. □

Lemma 6.1.2 *Let $\underline{a} \in F^m$, $\underline{b} \in F^n$ be tuples such that $a_r - b_s \neq 0, \pm 1$ for all $1 \leq r \leq m$ and $1 \leq s \leq n$. Then*

$$\mathrm{ind}^{m+n}_{m,n} L(\underline{a}) \boxtimes L(\underline{b}) \cong \mathrm{ind}^{m+n}_{n,m} L(\underline{b}) \boxtimes L(\underline{a})$$

is irreducible.

Proof By the Shuffle Lemma, $L(\underline{a}) \boxtimes L(\underline{b})$ appears in

$$\mathrm{res}^{m+n}_{m,n} \mathrm{ind}^{m+n}_{m,n} L(\underline{a}) \boxtimes L(\underline{b})$$

with multiplicity 1. So, in view of Lemma 5.3.3, it suffices to show that

$$\mathrm{ind}^{m+n}_{m,n} L(\underline{a}) \boxtimes L(\underline{b}) \cong \mathrm{ind}^{m+n}_{n,m} L(\underline{b}) \boxtimes L(\underline{a}).$$

By the Mackey Theorem and central characters argument

$$\mathrm{res}^{m+n}_{m,n} \mathrm{ind}^{m+n}_{n,m} L(\underline{b}) \boxtimes L(\underline{a})$$

contains $L(\underline{a}) \boxtimes L(\underline{b})$ as a summand with multiplicity one, all other constituents lying in different blocks. Hence by Frobenius reciprocity, there exists a non-zero homomorphism

$$f \colon \mathrm{ind}^{m+n}_{m,n} L(\underline{a}) \boxtimes L(\underline{b}) \to \mathrm{ind}^{m+n}_{n,m} L(\underline{b}) \boxtimes L(\underline{a}).$$

Every homomorphic image of $\text{ind}_{m,n}^{m+n} L(\underline{a}) \boxtimes L(\underline{b})$ contains an $\mathcal{H}_{m,n}$-submodule isomorphic to $L(\underline{a}) \boxtimes L(\underline{b})$. So, by Lemma 6.1.1(i), we see that the image of f contains a $\mathcal{P}_{m+n}$-submodule V isomorphic to

$$L(a_1) \boxtimes \cdots \boxtimes L(a_m) \boxtimes L(b_1) \boxtimes \cdots \boxtimes L(b_n).$$

Next we claim that the image of f also contains a $\mathcal{P}_{m+n}$-submodule isomorphic to

$$L(b_1) \boxtimes \cdots \boxtimes L(b_n) \boxtimes L(a_1) \boxtimes \cdots \boxtimes L(a_m). \tag{6.1}$$

Indeed, let $V = Fv$, and consider $\Phi_m v$. By (3.20) and the assumption that $a_m - b_1 \neq \pm 1$, Φ_m^2 acts on v by a non-zero scalar. So by (3.21), $\Phi_m V \neq 0$ is a $\mathcal{P}_{m+n}$-submodule isomorphic to

$$L(a_1) \boxtimes \cdots \boxtimes L(a_{m-1}) \boxtimes L(b_1) \boxtimes L(a_m) \boxtimes L(b_2) \boxtimes \cdots \boxtimes L(b_n).$$

Next apply $\Phi_{m-1}, \ldots, \Phi_1$ to move $L(b_1)$ to the first position, and continue in this way to complete the proof of the claim.

Now, by the Shuffle Lemma, all composition factors of

$$\text{res}_{1,\ldots,1}^{m+n} \text{ind}_{n,m}^{m+n} L(\underline{b}) \boxtimes L(\underline{a})$$

isomorphic to (6.1) necessarily lie in the irreducible $\mathcal{H}_{n,m}$-submodule $1 \otimes L(\underline{b}) \boxtimes L(\underline{a})$ of the induced module. Since this generates all of $\text{ind}_{n,m}^{m+n} L(\underline{b}) \boxtimes L(\underline{a})$ as an $\mathcal{H}_{m+n}$-module, this shows that f is surjective. Hence f is an isomorphism by dimension, which completes the proof. □

Remark 6.1.3 Keep the assumptions of the lemma. Set

$$\begin{aligned} \gamma &:= S_m \cdot \underline{a}, \\ \delta &:= S_n \cdot \underline{b}, \\ \gamma \cup \delta &:= S_{m+n} \cdot (\underline{a} \cup \underline{b}), \\ (\gamma, \delta) &:= S_m \times S_n \cdot (\underline{a}, \underline{b}). \end{aligned}$$

Then the argument as above shows that the functor $\text{ind}_{m,n}^{m+n}$ induces an equivalence of categories

$$\mathcal{H}_m \otimes \mathcal{H}_n\text{-mod}[(\gamma, \delta)] \simeq \mathcal{H}_{m+n}\text{-mod}[\gamma \cup \delta].$$

Theorem 6.1.4 *Let $\underline{a} \in F^m$, $\underline{b} \in F^n$ be tuples such that $a_r - b_s \neq \pm 1$ for all $1 \leq r \leq m$ and $1 \leq s \leq n$. Then*

$$\text{ind}_{m,n}^{m+n} L(\underline{a}) \boxtimes L(\underline{b}) \cong \text{ind}_{n,m}^{m+n} L(\underline{b}) \boxtimes L(\underline{a})$$

is irreducible. Moreover, every other irreducible $\mathcal{H}_{m+n}$-module lying in the same block as $\mathrm{ind}_{m,n}^{m+n} L(\underline{a}) \boxtimes L(\underline{b})$ *is of the form* $\mathrm{ind}_{m,n}^{m+n} L(\underline{a}') \boxtimes L(\underline{b}')$ *for permutations* $\underline{a}'$ *of* $\underline{a}$ *and* $\underline{b}'$ *of* $\underline{b}$.

Proof We prove the first statement by induction on $m+n$, the case $m+n=1$ being trivial. For $m+n>1$, we may assume by Lemma 6.1.2 that there exists $c \in F$ that appears in both the tuples $\underline{a}$ and $\underline{b}$. Note then that for every $r=1,\dots,m$, either $a_r = c$ or $a_r - c \neq \pm 1$, and similarly for every $s = 1,\dots,n$, either $b_s = c$ or $b_s - c \neq \pm 1$. So by the induction hypothesis, we have that

$$L(\underline{a}) \cong \mathrm{ind}_{n-r,r}^{n} L(\underline{a}') \boxtimes L(c^r) \quad \text{and} \quad L(\underline{b}) \cong \mathrm{ind}_{m-s,s}^{m} L(\underline{b}') \boxtimes L(c^s)$$

for some $r, s \geq 1$, where $\underline{a}', \underline{b}'$ are tuples with no entries equal to c. By Lemma 6.1.2,

$$\mathrm{ind}_{r,m-s}^{r+m-s} L(c^r) \boxtimes L(\underline{b}') \cong \mathrm{ind}_{m-s,r}^{m-s+r} L(\underline{b}') \boxtimes L(c^r).$$

So using Theorem 4.3.2(i) and transitivity of induction,

$$\mathrm{ind}_{m,n}^{m+n} L(\underline{a}) \boxtimes L(\underline{b}) \cong \mathrm{ind}_{n-r,m-s,r+s}^{m+n} L(\underline{a}') \boxtimes L(\underline{b}') \boxtimes L(c^{r+s}), \tag{6.2}$$

and similarly

$$\mathrm{ind}_{n,m}^{m+n} L(\underline{b}) \boxtimes L(\underline{a}) \cong \mathrm{ind}_{n-r,m-s,r+s}^{m+n} L(\underline{b}') \boxtimes L(\underline{a}') \boxtimes L(c^{r+s}), \tag{6.3}$$

Now, the right-hand sides of (6.2) and (6.3) are irreducible and isomorphic to each other by the induction hypothesis and Lemma 6.1.2. This proves the first statement of the theorem. The second statement is an easy consequence of the first one and Lemma 6.1.1(iii). □

6.2 Calculations for small rank

We will need to know the characters of certain $\mathcal{H}_n$-modules for small n.

Lemma 6.2.1 *Let $a, b \in F$ with $a - b = \pm 1$. We have:*

(i) $\mathrm{ch}\, L(a,b) = [L(a) \boxtimes L(b)]$, *and there is a non-split short exact sequence*

$$0 \longrightarrow L(a,b) \longrightarrow \mathrm{ind}(b,a) \longrightarrow L(b,a) \longrightarrow 0.$$

(ii) *If $p = 2$ then*

$$\begin{aligned}
\mathrm{ch}\, L(a,a,b) &= 2[L(a) \boxtimes L(a) \boxtimes L(b)],\\
\mathrm{ch}\, L(a,b,a) &= [L(a) \boxtimes L(b) \boxtimes L(a)],\\
\mathrm{ch}\, L(b,a,a) &= 2[L(b) \boxtimes L(a) \boxtimes L(a)],
\end{aligned}$$

and there are non-split short exact sequences

$$0 \to L(a,a,b) \to \mathrm{ind}_{2,1}^{3} L(a,b) \boxtimes L(a) \to L(a,b,a) \to 0,$$

$$0 \to L(a,b,a) \to \mathrm{ind}_{2,1}^{3} L(b,a) \boxtimes L(a) \to L(b,a,a) \to 0.$$

Proof (i) Set $M = \mathrm{ind}(b,a)$. An easy calculation shows that $N := (x_2 - a)M$ is an $\mathcal{H}_2$-submodule of M with character $[L(a) \boxtimes L(b)]$. Then $\mathrm{ch}\, M/N = [L(b) \boxtimes L(a)]$. It follows that $N \cong L(a,b)$ and $M/N \cong L(b,a)$. To prove that the extension is non-split apply Lemma 5.1.5.

(ii) Set $M := \mathrm{ind}_{2,1}^{3} L(a,b) \boxtimes L(a)$, and let $L(a,b) = Fv$ and $L(a) = Fw$. Then we have $x_1 v = av$, $x_2 v = bv$, $s_1 v = v$, and $x_3 w = aw$. Moreover,

$$\{1 \otimes v \otimes w,\ s_2 \otimes v \otimes w,\ s_1 s_2 \otimes v \otimes w\}$$

is a basis of M. Using this and relations in $\mathcal{H}_3$ we easily check that $N := (x_3 - a)M$ is an irreducible 2-dimensional $\mathcal{H}_3$-submodule of M with character $2[L(a) \boxtimes L(a) \boxtimes L(b)]$. Then M/N has character

$$[L(a) \boxtimes L(b) \boxtimes L(a)]$$

so must be isomorphic to $L(a,b,a)$. Next, we obtain the character of $L(b,a,a)$, by twisting $L(a,a,b)$ with σ, see Section 3.6. To prove that the short exact sequences do not split apply Lemma 5.1.5. □

For $a, b \in F$ with $a \neq b$ set

$$k_{ab} = \begin{cases} 0 \text{ if } a - b \neq \pm 1, \\ 1 \text{ if } a - b = \pm 1 \text{ and } p > 2, \\ 2 \text{ if } a - b = \pm 1 \text{ and } p = 2. \end{cases} \tag{6.4}$$

Lemma 6.2.2 *Let $a, b \in F$ with $a - b = \pm 1$, and set $k = k_{ab}$. Then*

$$L(a^{k+1}, b) \cong L(a^k, b, a),$$

and, for every $r, s \geq 0$ with $r + s = k$,

$$L(a^r, b, a^{s+1}) \cong \mathrm{ind}_{k+1,1}^{k+2} L(a^r, b, a^s) \boxtimes L(a)$$
$$\cong \mathrm{ind}_{1,k+1}^{k+2} L(a) \boxtimes L(a^r, b, a^s)$$

with character

$$r!(s+1)![L(a)^{\boxtimes r} \boxtimes L(b) \boxtimes L(a)^{\boxtimes (s+1)}]$$
$$+\ (r+1)!s![L(a)^{\boxtimes (r+1)} \boxtimes L(b) \boxtimes L(a)^{\boxtimes s}].$$

Proof The isomorphism

$$\mathrm{ind}_{k+1,1}^{k+2} L(a^r, b, a^s) \boxtimes L(a) \cong \mathrm{ind}_{1,k+1}^{k+2} L(a) \boxtimes L(a^r, b, a^s)$$

will follow from Corollary 5.3.2 and Theorem 3.7.5, once we prove that the first induced module is irreducible.

Let

$$M = \mathrm{ind}_{k+1,1}^{k+2} L(a^k, b) \boxtimes L(a).$$

In view of Lemma 6.2.1 and Theorem 4.3.2, we have

$$\mathrm{res}_{k+1,1}^{k+2} M \cong L(a^k, b) \boxtimes L(a) \oplus L(a^{k+1}) \boxtimes L(b). \tag{6.5}$$

So the $\mathcal{H}_{k+1,1}$-submodule $(x_{k+2} - a)M$ is isomorphic to $L(a^{k+1}) \boxtimes L(b)$. To prove that M is irreducible, it suffices to show that this submodule is *not* invariant under s_{k+1}, which is an explicit calculation. Hence, there is an irreducible $\mathcal{H}_{k+2}$-module M with character

$$(k+1)![L(a)^{\boxtimes(k+1)} \boxtimes L(b)] + k![L(a)^{\boxtimes k} \boxtimes L(b) \boxtimes L(a)].$$

Moreover, $\tilde{e}_a M \cong L(a^k, b)$ and $\tilde{e}_b M \cong L(a^{k+1})$ by (6.5). So

$$M \cong L(a^k, b, a) \cong L(a^{k+1}, b),$$

thanks to Lemma 5.2.3.

Now consider the remaining irreducibles in the block. There are at most k remaining, namely $L(a^r b a^{s+1})$ for $r \geq 0, s \geq 1$ with $r + s = k$. Twisting with the automorphism σ and using Lemma 3.6.1 gives a new irreducible module $\mathrm{ind}_{1,k+1}^{k+2} L(a) \boxtimes L(b, a^k)$ with character

$$(k+1)![L(b) \boxtimes L(a)^{\boxtimes(k+1)}] + k![L(a) \boxtimes L(b) \boxtimes L(a)^{\boxtimes k}].$$

If $p > 2$, it must be $L(b, a, a)$, and we are finished. Let $p = 2$. Consider $M := \mathrm{ind}_{3,1}^{4} L(a, b, a) \boxtimes L(a)$. By Lemma 6.2.1 and the Shuffle Lemma,

$$\mathrm{ch}\, M = 2[L(a) \boxtimes L(a) \boxtimes L(b) \boxtimes L(a)] + 2[L(a) \boxtimes L(b) \boxtimes L(a) \boxtimes L(a)].$$

Moreover the character of L must be σ-invariant, as otherwise we would produce two new modules in our block. So either $\mathrm{ch}\, L = \mathrm{ch}\, M$ or $\mathrm{ch}\, L = (1/2)\mathrm{ch}\, M$. We claim that the former happens. Indeed,

$$L(a) \boxtimes L(a) \boxtimes L(b) \boxtimes L(a) \subseteq \mathrm{res}_{\mathcal{P}_4} L$$

implies by Frobenius reciprocity and Theorem 4.3.2 that

$$L(a^2) \boxtimes L(b) \boxtimes L(a) \subseteq \mathrm{res}_{2,1,1}^{4} L.$$

Finally, we use Lemma 6.1.1(i) to identify the last module as $L(a, b, a, a)$ and the previous one as $L(b, a, a, a)$. □

6.3 Higher crystal operators

In this section we will introduce certain generalizations of the crystal operators $\tilde{f}_j$. To simplify notation, we will write simply ind in place of ind_μ^n throughout the section.

Lemma 6.3.1 *Let $a, b \in F$ with $a \neq b$. For any $r, s \geq 0$ with $r+s = k_{ab}$, and $m \geq 0$,*

$$\mathrm{ind}\, L(a^r, b, a^s) \boxtimes L(a^m) \cong \mathrm{ind}\, L(a^m) \boxtimes L(a^r, b, a^s)$$

is irreducible.

Proof We may assume that $a - b = \pm 1$, as otherwise the result is immediate from Theorem 6.1.4. Now, we claim that

$$\mathrm{ind}\, L(a^r, b, a^s) \boxtimes L(a^m) \cong \mathrm{ind}\, L(a^m) \boxtimes L(a^r, b, a^s).$$

Indeed, transitivity of induction and Lemma 6.2.2 give that

$$\begin{aligned} \mathrm{ind}\, L(a^r, b, a^s) \boxtimes L(a^m) &\cong \mathrm{ind}\, L(a^r, b, a^s) \boxtimes L(a) \boxtimes L(a^{m-1}) \\ &\cong \mathrm{ind}\, L(a) \boxtimes L(a^r, b, a^s) \boxtimes L(a^{m-1}), \end{aligned}$$

and now repeating this argument $(m-1)$ more times gives the claim.

Hence, by Corollary 5.3.2 and Theorem 3.7.5,

$$K := \mathrm{ind}\, L(a^r, b, a^s) \boxtimes L(a^m)$$

is self-dual. Now suppose for a contradiction that K is reducible. Then we can pick a proper irreducible submodule S of K, and set $Q := K/S$. Applying Lemmas 6.2.1 and 4.1.3, we see that $\mathrm{ch}\, K$ equals

$$\sum_{t=0}^{m} \binom{m}{t} (r+t)!(s+m-t)![L(a)^{\boxtimes(r+t)} \boxtimes L(b) \boxtimes L(a)^{\boxtimes(s+m-t)}].$$

By Frobenius reciprocity, Q contains an $\mathcal{H}_{r+s+1,m}$-submodule isomorphic to

$$L(a^r, b, a^s) \boxtimes L(a^m).$$

So by Lemma 6.1.1(i), the irreducible $\mathcal{P}_{r+s+m+1}$-module

$$V := L(a)^{\boxtimes r} \boxtimes L(b) \boxtimes L(a)^{\boxtimes(s+m)}$$

appears in Q as a submodule, hence in fact by Theorem 4.3.2(i) it must appear with multiplicity $r!(s+m)!$ (viewing Q as a module over $\mathcal{H}_{r,1,s+m}$). It follows that V is not a composition factor of S. But this is a contradiction, since, as K is self-dual, $S \cong S^{\tau}$ is a quotient module of K, hence it must contain V by the argument above applied to the quotient $Q = S$. □

Lemma 6.3.2 *Let $a, b \in F$ with $a \neq b$, and m be a non-negative integer. For any $r \geq 1$ and $s \geq 0$ with $r+s = k_{ab}$ and any irreducible module $M \in \mathcal{H}_n$-mod,*

$$\operatorname{hd} \operatorname{ind} M \boxtimes L(a^m) \boxtimes L(a^r, b, a^s)$$

is irreducible.

Proof By the argument in the proof of Lemma 5.1.5, it suffices to prove this in the special case that $\varepsilon_a(M) = 0$. Let $t = m+r+s+1$. Recall from the previous lemma that

$$N := \operatorname{ind} L(a^m) \boxtimes L(a^r, b, a^s)$$

is an irreducible $\mathcal{H}_t$-module. Moreover by Lemma 6.2.1,

$$\operatorname{ch} L(a^r, b, a^s) = (r!)(s!)[L(a)^{\boxtimes r} \boxtimes L(b) \boxtimes L(a)^{\boxtimes s}].$$

So, since $\varepsilon_a(M) = 0$ and $r > 0$, the Mackey Theorem and a block argument give

$$\operatorname{res}^{n+t}_{n,t} \operatorname{ind} M \boxtimes L(a^m) \boxtimes L(a^r, b, a^s) \cong (M \boxtimes N) \oplus U$$

for some $\mathcal{H}_{n,t}$-module U all of whose composition factors lie in different blocks to those of $M \boxtimes N$. Now let

$$H := \operatorname{hd} \operatorname{ind} M \boxtimes L(a^m) \boxtimes L(a^r, b, a^s).$$

It follows from above that $\operatorname{res}^{n+t}_{n,t} H \cong (M \boxtimes N) \oplus \bar{U}$, where $\bar{U}$ is some quotient module of U. Then:

$$\begin{aligned}
\operatorname{Hom}_{\mathcal{H}_{n+t}}(H, H) &\cong \operatorname{Hom}_{\mathcal{H}_{n+t}}(\operatorname{ind} M \boxtimes L(a^m) \boxtimes L(a^r, b, a^s), H) \\
&\cong \operatorname{Hom}_{\mathcal{H}_{n,t}}(M \boxtimes N, \operatorname{res}^{n+t}_{n,t} H) \\
&\cong \operatorname{Hom}_{\mathcal{H}_{n,t}}(M \boxtimes N, M \boxtimes N \oplus \bar{U}) \\
&\cong \operatorname{Hom}_{\mathcal{H}_{n,t}}(M \boxtimes N, M \boxtimes N) \cong F.
\end{aligned}$$

Since H is completely reducible, this implies that H is irreducible, as required. □

Now we can define the higher crystal operators. Let $a, b \in F$ with $a \neq b$, and $r \geq 1, s \geq 0$ satisfy $r+s = k_{ab}$. Then the special case $m = 0$ of Lemma 6.3.2 shows that

$$\tilde{f}_{a^r b a^s} M := \mathrm{hd}\, \mathrm{ind}\, M \boxtimes L(a^r, b, a^s)$$

is irreducible for every irreducible $M \in \mathcal{H}_n$-mod.

Lemma 6.3.3 *Take $a, b \in F$ with $a \neq b$ and set $k = k_{ab}$. Let $M \in \mathcal{H}_n$-mod be irreducible.*

(i) *There exists a unique integer r with $0 \leq r \leq k$ such that for every $m \geq 0$ we have*

$$\varepsilon_a(\tilde{f}_a^m \tilde{f}_b M) = m + \varepsilon_a(M) - r.$$

(ii) *Assume $m \geq k$. Then a copy of $\tilde{f}_a^m \tilde{f}_b M$ appears in the head of*

$$\mathrm{ind}\, (\tilde{f}_a^{m-k} M) \boxtimes L(a^r, b, a^{k-r}),$$

where r is as in (i). *In particular, if $r \geq 1$, then*

$$\tilde{f}_a^m \tilde{f}_b M \cong \tilde{f}_{a^r b a^{k-r}} \tilde{f}_a^{m-k} M.$$

Proof Let $\varepsilon = \varepsilon_a(M)$ and write $M = \tilde{f}_a^\varepsilon N$ for an irreducible $\mathcal{H}_{n-\varepsilon}$-module N with $\varepsilon_a(N) = 0$. It suffices to prove (i) for any fixed choice of m, the result for all other $m \geq 0$ then following immediately by (5.12). So take $m \geq k$. Note that $\tilde{f}_a^m \tilde{f}_b M = \tilde{f}_a^m \tilde{f}_b \tilde{f}_a^\varepsilon N$ is a quotient of

$$\mathrm{ind}\, N \boxtimes L(a^\varepsilon) \boxtimes L(b) \boxtimes L(a)^{\boxtimes k} \boxtimes L(a^{m-k}),$$

which by Lemma 6.2.1 has a filtration with factors isomorphic to

$$F_r := \mathrm{ind}\, N \boxtimes L(a^\varepsilon) \boxtimes L(a^r, b, a^{k-r}) \boxtimes L(a^{m-k}), \quad 0 \leq r \leq k.$$

So $\tilde{f}_a^m \tilde{f}_b M$ is a quotient of some such factor, and to prove (i) it remains to show that $\varepsilon_a(L) = \varepsilon + m - r$ for any irreducible quotient L of F_r. The inequality $\varepsilon_a(L) \leq \varepsilon + m - r$ is clear from the Shuffle Lemma. However, by transitivity of induction and Lemma 6.3.1

$$F_r \cong \mathrm{ind}\, N \boxtimes (\mathrm{ind}\, L(a^r, b, a^{k-r}) \boxtimes L(a^{\varepsilon+m-k})).$$

Moreover

$$N \boxtimes (\mathrm{ind}\, L(a^r, b, a^{k-r}) \boxtimes L(a^{\varepsilon+m-k}))$$

is irreducible by Lemma 6.3.1 again. So by Frobenius Reciprocity, it is a submodule of $\mathrm{res}_{n-\varepsilon, m+1+\varepsilon} L$. Hence $\varepsilon_a(L) \geq \varepsilon + m - r$.

For (ii), note by Lemma 6.3.1 and transitivity of induction that we also have

$$F_r \cong \operatorname{ind} N \boxtimes L(a^{m-k+\varepsilon}) \boxtimes L(a^r, b, a^{k-r}).$$

Now, by the Shuffle Lemma and Lemma 5.1.5, the only irreducible factors K of F_r with $\varepsilon_a(K) = \varepsilon + m - r$ come from its quotient

$$\operatorname{ind} \tilde{f}_a^{m-k+\varepsilon} N \boxtimes L(a^r, b, a^{k-r}) \cong \operatorname{ind} \tilde{f}_a^{m-k} M \boxtimes L(a^r, b, a^{k-r}).$$

Finally, in case $r \geq 1$, the head of the last module is precisely

$$\tilde{f}_{a^r, b, a^{k-r}} \tilde{f}_a^{m-k} M,$$

see Lemma 6.3.2. □

7

Integral representations and cyclotomic Hecke algebras

Starting from this chapter, we will consider only *integral* representations of $\mathcal{H}_n$. By this we mean those representations for which the eigenvalues of $x_1, \ldots, x_n$ are all "integral", that is belong to $I = \mathbb{Z} \cdot 1 \subset F$. In the beginning of the first section we explain why we can restrict ourselves to the category $\mathrm{Rep}_I \mathcal{H}_n$ of integral representations essentially without loss of generality. This category is very natural from another point of view. The affine algebra $\mathcal{H}_n$ has a natural family of finite dimensional quotient algebras, which are degenerate analogues of *cyclotomic Hecke algebras* of [Ch, AK, BM]. So any module over such cyclotomic quotient can be inflated to an $\mathcal{H}_n$-module. Now it turns out that the category $\mathrm{Rep}_I \mathcal{H}_n$ consists precisely of all such inflations from all cyclotomic quotients. As we now work in $\mathrm{Rep}_I \mathcal{H}_n$, the es, εs, $\tilde{f}$s, etc. will be labeled not by arbitrary $a \in F$ but only by $i \in I$. So from now on we will only have e_is, ε_is, $\tilde{f}_i$s, etc. for $i \in I$.

Note that the set I can be identified with the labeling set for simple roots of the Kac–Moody algebra $\mathfrak{g}$ of type $A^{(1)}_{p-1}$ if $p > 0$ and A_∞ if $p = 0$. Moreover the notation e_i might suggest that we want to think about these functors as the Chevalley generators of the positive part of $\mathfrak{g}$. At this stage, all the Lie theoretic notation we are going to bring in will seem completely artificial. For example, we will label the cyclotomic quotients by the dominant weights for $\mathfrak{g}$, and in Chapter 8 we will introduce a "Lie theoretic" notation for blocks! However it will gradually become clear that relations Lie Theory are not superficial at all.

Two main theorems of the chapter are Basis Theorem and Mackey Theorem for cyclotomic Hecke algebras. We also show that cyclotomic Hecke algebras are Frobenius. Finally, we draw the reader's attention to the small Section 7.4, where the irreducible $\mathcal{H}_n$-modules factoring through to the cyclotomic quotient $\mathcal{H}_n^\lambda$ are characterized in terms of the functions ε_i^*, which are the "left-hand versions" of the ε_i.

7.1 Integral representations

Recall the set I from (1.1). Let N be an irreducible $\mathcal{H}_n$-module, whose character we would like to understand. If N belongs to the block corresponding to the orbit $S_n \cdot \underline{a}$ of S_n on F^n, then, in view of Theorem 6.1.4, we may assume that $a_r - a_s \in I$ for all $1 \le r, s \le n$. Indeed, otherwise M can be decomposed as $\text{ind}^n_{m,k} M \boxtimes K$ for some irreducible $\mathcal{H}_m$-module M and some irreducible $\mathcal{H}_k$-module K, where $n = m + k$, and we are reduced to smaller ranks. Also, we may assume that $a_1 \in I$, because this can be achieved by twisting the action of $\mathcal{H}_n$ with the algebra automorphism

$$\mathcal{H}_n \to \mathcal{H}_n, \quad x_i \mapsto x_i + b, \; s_j \mapsto s_j,$$

which just shifts the formal character by a constant $b \in F$. Thus, it is sufficient to understand the irreducible $\mathcal{H}_n$-modules, which are *integral* in the following sense.

Definition 7.1.1 A $\mathcal{P}_n$-module M is called *integral* if it is finite dimensional and all eigenvalues of $x_1, \dots, x_n$ on M belong to I. An $\mathcal{H}_n$-module, or more generally an $\mathcal{H}_\mu$-module for μ a composition of n, is called *integral* if it is integral on restriction to $\mathcal{P}_n$. We write $\text{Rep}_I \mathcal{H}_n$ (resp. $\text{Rep}_I \mathcal{P}_n$, $\text{Rep}_I \mathcal{H}_\mu$) for the full subcategory of $\mathcal{H}_n$-mod (resp. $\mathcal{P}_n$-mod, $\mathcal{H}_\mu$-mod) consisting of all integral modules.

Lemma 7.1.2 *Let M be a finite dimensional $\mathcal{H}_n$-module and fix j with $1 \le j \le n$. Assume that all eigenvalues of x_j on M belong to I. Then M is integral.*

Proof It suffices to show that the eigenvalues of x_k belong to I if and only if the eigenvalues of x_{k+1} belong to I, for an arbitrary k with $1 \le k < n$. Actually, by an argument involving conjugation with the automorphism σ, it suffices just to prove the "if" part. So assume that all eigenvalues of x_{k+1} on M belong to I. Let a be an eigenvalue for the action of x_k on M. Since x_k and x_{k+1} commute, we can pick v lying in the a-eigenspace of x_k so that v is also an eigenvector for x_{k+1}, of eigenvalue b say. By assumption we have $b \in I$. Now let Φ_k be the intertwining element (3.19). By (3.21), we have $x_{k+1}\Phi_k = \Phi_k x_k$. So if $\Phi_k v \ne 0$, we get that

$$a\Phi_k v = \Phi_k x_k v = x_{k+1}\Phi_k v,$$

whence a is an eigenvalue of x_{k+1}, and so $a \in I$ by assumption. Else, $\Phi_k v = 0$ so $\Phi_k^2 v = 0$. So applying (3.20), we again get that $a \in I$. □

Lemma 7.1.3 *Let μ be a composition of n and M be an integral $\mathcal{H}_\mu$-module. Then $\mathrm{ind}_\mu^n M$ is an integral $\mathcal{H}_n$-module.*

Proof By Theorem 3.2.2, $\mathrm{ind}_\mu^n M$ is spanned by elements $w \otimes m$ for $w \in S_n$ and $m \in M$, in particular it is finite dimensional. Let

$$Y_j = \prod_{i \in I} (x_j - i), \quad 1 \le j \le n$$

(If I is infinite, take a product $\prod_{i=0}^{K} (x_j - i)$ for sufficiently large K instead). By Lemma 7.1.2, it suffices to show that Y_1^N annihilates $\mathrm{ind}_\lambda^n M$ for sufficiently large N. Consider $Y_1^N w \otimes m$ for $w \in S_n$, $m \in M$. We may write $w = u s_1 \dots s_k$ for $u \in S_{2\dots n} \cong S_{n-1}$ and $0 \le k < n$. Then Y_1^N commutes with u, so we just need to consider $Y_1^N s_1 \dots s_k \otimes m$. Now using the commutation relations, we check that $Y_1^N s_1 \dots s_k \otimes m$ can be rewritten as an $\mathcal{H}_n$-linear combination of elements of the form $1 \otimes Y_j^{N'} m$ for $1 \le j \le n$ and $N - k \le N' \le N$. Since M is integral by assumption, we can choose N sufficiently large so that each such term is zero. □

It follows that the functors ind_μ^n, res_μ^n restrict to well-defined functors

$$\mathrm{ind}_\mu^n : \mathrm{Rep}_I \mathcal{H}_\mu \to \mathrm{Rep}_I \mathcal{H}_n, \qquad \mathrm{res}_\mu^n : \mathrm{Rep}_I \mathcal{H}_n \to \mathrm{Rep}_I \mathcal{H}_\mu \tag{7.1}$$

on integral representations. Similar remarks apply to more general induction and restriction between nested parabolic subalgebras of $\mathcal{H}_n$.

7.2 Some Lie theoretic notation

We introduce some standard Lie theoretic notation, which at first will be used just as a book-keeping device, but later it will turn out to be deeply connected with the theory we are considering.

Assume first that $p > 0$. In this case we set $\ell = p - 1$ and denote by $\mathfrak{g}$ the affine Kac–Moody algebra of type $A_\ell^{(1)}$ over $\mathbb{C}$, see [Kc, Ch. 4, Table Aff 1]. In particular, we label the Dynkin diagram by the index set $I = \{0, 1, \dots, \ell\}$ as follows:

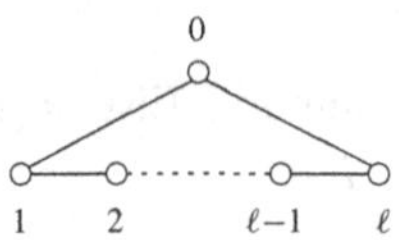

if $\ell \ge 2$, and

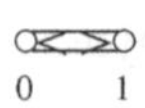

if $\ell = 1$.

The weight lattice is denoted P, the simple roots are $\{\alpha_i \mid i \in I\} \subset P$ and the corresponding simple coroots are $\{h_i \mid i \in I\} \subset P^*$. The Cartan matrix $\big(\langle h_i, \alpha_j\rangle\big)_{0\le i,j\le \ell}$ is

$$\begin{pmatrix} 2 & -1 & 0 & \cdots & 0 & 0 & -1 \\ -1 & 2 & -1 & \cdots & 0 & 0 & 0 \\ 0 & -1 & 2 & \cdots & 0 & 0 & 0 \\ & & & \ddots & & & \\ 0 & 0 & 0 & \ldots & 2 & -1 & 0 \\ 0 & 0 & 0 & \ldots & -1 & 2 & -1 \\ -1 & 0 & 0 & \ldots & 0 & -1 & 2 \end{pmatrix} \quad \text{if } \ell \ge 2,$$

and

$$\begin{pmatrix} 2 & -2 \\ -2 & 2 \end{pmatrix} \quad \text{if } \ell = 1.$$

Let $\{\Lambda_i \mid i \in I\} \subset P$ denote fundamental dominant weights, so that $\langle h_i, \Lambda_j\rangle = \delta_{i,j}$, and let $P_+ \subset P$ denote the set of all dominant integral weights. Set

$$c = \sum_{i=0}^{\ell} h_i, \qquad \delta = \sum_{i=0}^{\ell} \alpha_i. \tag{7.2}$$

Then the $\Lambda_0, \ldots, \Lambda_\ell, \delta$ form a $\mathbb{Z}$-basis for P, and $\langle c, \alpha_i\rangle = \langle h_i, \delta\rangle = 0$ for all $i \in I$.

In the case $p = 0$, we make the following changes to these definitions. First, we let $\ell = \infty$, and $\mathfrak{g}$ denotes the Kac–Moody algebra of type A_∞, see [Kc, Section 7.11]. So $I = \mathbb{Z}$, corresponding to the nodes of the Dynkin diagram

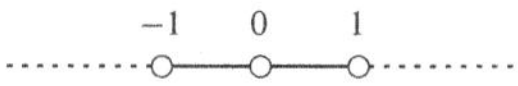

Note certain notions, for example the element c from (7.2), only make sense if we pass to the completed algebra a_∞, see [Kc, Section 7.12], though the intended meaning whenever we make use of them should be obvious regardless.

Now, for either $\ell < \infty$ or $\ell = \infty$, we let $U_{\mathbb{Q}}$ denote the $\mathbb{Q}$-subalgebra of the universal enveloping algebra of $\mathfrak{g}$ generated by the Chevalley generators e_i, f_i, h_i $(i \in I)$. Recall these are subject only to the relations

$$[h_i, h_j] = 0, \qquad [e_i, f_j] = \delta_{i,j} h_i, \tag{7.3}$$

$$[h_i, e_j] = \langle h_i, \alpha_j\rangle e_j, \qquad [h_i, f_j] = -\langle h_i, \alpha_j\rangle f_j, \tag{7.4}$$

$$(\mathrm{ad}\, e_i)^{1-\langle h_i, \alpha_k\rangle} e_k = 0, \qquad (\mathrm{ad}\, f_i)^{1-\langle h_i, \alpha_k\rangle} f_k = 0 \tag{7.5}$$

for all $i, j, k \in I$ with $i \neq k$. We let $U_{\mathbb{Z}}$ denote the $\mathbb{Z}$-form of $U_{\mathbb{Q}}$ generated by the divided powers

$$e_i^{(n)} = e_i^n/n! \quad \text{and} \quad f_i^{(n)} = f_i^n/n!.$$

Then $U_{\mathbb{Z}}$ has the usual triangular decomposition

$$U_{\mathbb{Z}} = U_{\mathbb{Z}}^- U_{\mathbb{Z}}^0 U_{\mathbb{Z}}^+.$$

We are particularly concerned here with the plus part $U_{\mathbb{Z}}^+$, generated by all $e_i^{(n)}$. It is a graded Hopf algebra over $\mathbb{Z}$ via the *principal grading* where $\deg(e_i^{(n)}) = n$ for all $i \in I, n \geq 0$.

7.3 Degenerate cyclotomic Hecke algebras

For $\lambda \in P_+$ we set

$$f_\lambda := \prod_{i \in I} (x_1 - i)^{\langle h_i, \lambda \rangle} \in \mathcal{H}_n. \tag{7.6}$$

(Note the product is finite even if $\ell = \infty$). Let $\mathcal{J}_\lambda$ denote the two-sided ideal of $\mathcal{H}_n$ generated by f_λ, and define the *(degenerate) cyclotomic Hecke algebra* to be the quotient

$$\mathcal{H}_n^\lambda := \mathcal{H}_n / \mathcal{J}_\lambda.$$

Also, *by agreement*, $\mathcal{H}_0^\lambda \cong F$. The algebra $\mathcal{H}_n^\lambda$ defined for $\lambda \in P_+$ should not be confused with the parabolic subalgebra $\mathcal{H}_\mu \subset \mathcal{H}_n$ defined earlier for μ a composition of n.

The following result explains the relation between integral modules and cyclotomic Hecke algebras. It shows that an $\mathcal{H}_n$-module is integral if and only if it is an inflation of a finite dimensional $\mathcal{H}_n^\lambda$-module for some "sufficiently large" λ.

Lemma 7.3.1 *Let M be a finite dimensional $\mathcal{H}_n$-module. Then M is integral if and only if $\mathcal{J}_\lambda M = 0$ for some $\lambda \in P_+$.*

Proof If $\mathcal{J}_\lambda M = 0$, then the eigenvalues of x_1 on M are all in I, by definition of $\mathcal{J}_\lambda$. Hence M is integral in view of Lemma 7.1.2. Conversely, suppose that M is integral. Then the minimal polynomial of x_1 on M is of the form $\prod_{i \in I} (t - i)^{\lambda_i}$ for some $\lambda_i \geq 0$. So if we set

$$\lambda = \lambda_0 \Lambda_0 + \lambda_1 \Lambda_1 + \cdots + \lambda_\ell \Lambda_\ell \in P_+,$$

we certainly have that $\mathcal{J}_\lambda M = 0$. □

Lemma 7.3.1 allows us to introduce the functors

$$\mathrm{pr}^\lambda : \mathrm{Rep}_I \mathcal{H}_n \to \mathcal{H}_n^\lambda\text{-mod}, \qquad \mathrm{infl}^\lambda : \mathcal{H}_n^\lambda\text{-mod} \to \mathrm{Rep}_I \mathcal{H}_n. \tag{7.7}$$

Here, infl^λ is simply inflation along the canonical epimorphism $\mathcal{H}_n \to \mathcal{H}_n^\lambda$, while on a module M, $\mathrm{pr}^\lambda M = M/\mathcal{J}_\lambda M$ with the induced action of $\mathcal{H}_n^\lambda$. The functor infl^λ is right adjoint to pr^λ, that is there is a functorial isomorphism

$$\mathrm{Hom}_{\mathcal{H}_n^\lambda}(\mathrm{pr}^\lambda M, N) \cong \mathrm{Hom}_{\mathcal{H}_n}(M, \mathrm{infl}^\lambda N). \tag{7.8}$$

Note we will generally be sloppy and omit the functor infl^λ in our notation. In other words, we generally identify $\mathcal{H}_n^\lambda$-mod with the full subcategory of $\mathrm{Rep}_I \mathcal{H}_n$ consisting of all modules M with $\mathcal{J}_\lambda M = 0$.

7.4 The $*$-operation

Suppose M is an irreducible module in $\mathrm{Rep}_I \mathcal{H}_n$ and $0 \le m \le n$. Using Lemma 3.6.1 for the second equality in (7.10), define

$$\tilde{e}_i^* M = (\tilde{e}_i(M^\sigma))^\sigma, \tag{7.9}$$

$$\tilde{f}_i^* M = (\tilde{f}_i(M^\sigma))^\sigma = \mathrm{hd}\,\mathrm{ind}_{1,n}^{n+1} L(i) \boxtimes M, \tag{7.10}$$

$$\varepsilon_i^*(M) = \varepsilon_i(M^\sigma) = \max\{m \ge 0 \mid (\tilde{e}_i^*)^m M \neq 0\}. \tag{7.11}$$

We may think of the "starred" notions as left-hand versions of the original notions, which are right-hand versions. For example, $\varepsilon_i^*(M)$ can be worked out just from knowledge of the character of M as the maximal k such that $[L(i)^{\boxtimes k} \boxtimes \dots]$ appears in $\mathrm{ch}\, M$, while for $\varepsilon_i(M)$ we would take $[\cdots \boxtimes L(i)^{\boxtimes k}]$ here.

Recalling the definition of the ideal $\mathcal{J}_\lambda$ generated by the element (7.6), Theorem 5.5.1(ii) has the following important corollary.

Corollary 7.4.1 *Let M be an irreducible module in $\mathrm{Rep}_I \mathcal{H}_n$ and $\lambda \in P_+$. Then $\mathrm{pr}^\lambda M = M$ if and only if $\varepsilon_i^*(M) \le \langle h_i, \lambda\rangle$ for all $i \in I$.*

Proof In view of Theorem 5.5.1(ii), $\varepsilon_i^*(M)$ is the maximal size of a Jordan block of x_1 on M with eigenvalue i. The result follows immediately. □

7.5 Basis Theorem for cyclotomic Hecke algebras

The goal in this section is to describe the Ariki–Koike explicit basis for $\mathcal{H}_n^\lambda$, see [AK]. Our method does not use representation theory of H_n^λ and so it is very different from that of [AK].

Let $d = \langle c, \lambda \rangle$. Then f_λ is a monic polynomial of degree d. Write

$$f_\lambda = x_1^d + a_{d-1}x_1^{d-1} + \cdots + a_1 x_1 + a_0.$$

Set $f_1 = f_\lambda$ and for $i = 2, \ldots, n$, define inductively

$$f_i = s_{i-1} f_{i-1} s_{i-1}.$$

The first lemma follows easily by induction using (3.9).

Lemma 7.5.1 *For $i = 1, \ldots, n$, we have*

$$f_i = x_i^d + (\text{terms lying in } \mathcal{P}_{i-1} x_i^e \mathcal{G}_i \text{ for } 0 \le e < d).$$

Given $Z = \{z_1 < \cdots < z_u\} \subseteq \{1, \ldots, n\}$, let

$$f_Z = f_{z_1} f_{z_2} \cdots f_{z_u} \in \mathcal{H}_n.$$

Also, define

$$\Pi_n = \{(\alpha, Z) \mid Z \subseteq \{1, \ldots, n\}, \alpha \in \mathbb{Z}_+^n \text{ with } \alpha_i < d \text{ whenever } i \notin Z\},$$
$$\Pi_n^+ = \{(\alpha, Z) \in \Pi_n \mid Z \neq \varnothing\}.$$

Lemma 7.5.2 *$\mathcal{H}_n$ is a free right $\mathcal{G}_n$-module on basis*

$$\{x^\alpha f_Z \mid (\alpha, Z) \in \Pi_n\}.$$

Proof Define a lexicographic ordering on $\mathbb{Z}_+^n$: $\alpha \prec \alpha'$ if and only if

$$\alpha_n = \alpha'_n, \ \ldots, \ \alpha_{k+1} = \alpha'_{k+1}, \ \alpha_k < \alpha'_k$$

for some $k = 1, \ldots, n$. Define a function

$$\gamma : \Pi_n \to \mathbb{Z}_+^n$$

by $\gamma(\alpha, Z) := (\gamma_1, \ldots, \gamma_n)$, where

$$\gamma_i = \begin{cases} \alpha_i & \text{if } i \notin Z, \\ \alpha_i + d & \text{if } i \in Z. \end{cases}$$

Using induction on n and Lemma 7.5.1, we prove for $(\alpha, Z) \in \Pi_n$

$$x^\alpha f_Z = x^{\gamma(\alpha,Z)} + (\text{terms lying in } x^\beta \mathcal{G}_n \text{ for } \beta \prec \gamma(\alpha, Z)). \tag{7.12}$$

Since $\gamma : \Pi_n \to \mathbb{Z}_+^n$ is a bijection and we already know that the

$$\{x^\alpha \mid \alpha \in \mathbb{Z}_+^n\}$$

form a basis for $\mathcal{H}_n$ viewed as a right $\mathcal{G}_n$-module by Theorem 3.2.2, (7.12) implies the lemma. □

Lemma 7.5.3 *For $n > 1$ we have $\mathcal{G}_{n-1} f_n \mathcal{G}_n = f_n \mathcal{G}_n$.*

Proof It suffices to show that the left multiplication by the elements $s_1, \dots, s_{n-2}$ leaves the space $f_n \mathcal{G}_n$ invariant. But this follows from the definition of f_n and braid relations in S_n. □

Lemma 7.5.4 *We have $\mathcal{J}_\lambda = \sum_{i=1}^n \mathcal{P}_n f_i \mathcal{G}_n$.*

Proof We have

$$\begin{aligned}
\mathcal{J}_\lambda &= \mathcal{H}_n f_1 \mathcal{H}_n = \mathcal{H}_n f_1 \mathcal{P}_n \mathcal{G}_n = \mathcal{H}_n f_1 \mathcal{G}_n \\
&= \mathcal{P}_n \mathcal{G}_n f_1 \mathcal{G}_n = \sum_{i=1}^n \sum_{u \in S_{2\dots n}} \mathcal{P}_n s_{i-1} \dots s_1 u f_1 \mathcal{G}_n \\
&= \sum_{i=1}^n \mathcal{P}_n s_{i-1} \dots s_1 f_1 \mathcal{G}_n = \sum_{i=1}^n \mathcal{P}_n f_i \mathcal{G}_n,
\end{aligned}$$

as required. □

Lemma 7.5.5 *For $d > 0$ we have $\mathcal{J}_\lambda = \sum_{(\alpha, Z) \in \Pi_n^+} x^\alpha f_Z \mathcal{G}_n$.*

Proof Proceed by induction on n, the case $n = 1$ being obvious. Let $n > 1$. Denote $\mathcal{J}'_\lambda := \mathcal{H}_{n-1} f_\lambda \mathcal{H}_{n-1}$, so

$$\mathcal{J}'_\lambda = \sum_{(\alpha', Z') \in \Pi_{n-1}^+} x^{\alpha'} f_{Z'} \mathcal{G}_{n-1} \tag{7.13}$$

by the induction hypothesis. Let

$$\mathcal{I} = \sum_{(\alpha, Z) \in \Pi_n^+} x^\alpha f_Z \mathcal{G}_n.$$

Obviously $\mathcal{I} \subseteq \mathcal{J}_\lambda$. So in view of Lemma 7.5.4, it suffices to show that $x^\alpha f_i \mathcal{G}_n \subseteq \mathcal{I}$ for each $\alpha \in \mathbb{Z}_+^n$ and each $i = 1, \dots, n$.

Consider first $x^\alpha f_n \mathcal{G}_n$. Write $x^\alpha = x_n^{\alpha_n} x^\beta$ for $\beta \in \mathbb{Z}_+^{n-1}$. Expanding x^β in terms of the basis of $\mathcal{H}_{n-1}$ from Lemma 7.5.2, we see that

$$x^\alpha f_n \mathcal{G}_n \subseteq \sum_{(\alpha', Z') \in \Pi_{n-1}} x_n^{\alpha_n} x^{\alpha'} f_{Z'} \mathcal{G}_{n-1} f_n \mathcal{G}_n,$$

which is contained in $\mathcal{J}$ thanks to Lemma 7.5.3.

Finally, consider $x^\alpha f_i \mathcal{G}_n$ with $i < n$. Write $x^\alpha = x_n^{\alpha_n} x^\beta$ for $\beta \in \mathbb{Z}_+^{n-1}$. By the induction hypothesis,

$$x^\alpha f_i \mathcal{G}_n = x_n^{\alpha_n} x^\beta f_i \mathcal{G}_n \subseteq \sum_{(\alpha', Z') \in \Pi_{n-1}^+} x_n^{\alpha_n} x^{\alpha'} f_{Z'} \mathcal{G}_n.$$

Now we show by induction on α_n that $x_n^{\alpha_n} x^{\alpha'} f_{Z'} \mathcal{G}_n \subseteq \mathcal{J}$ for each $(\alpha', Z') \in \Pi_{n-1}^+$. This is immediate if $\alpha_n < d$, so take $\alpha_n \geq d$ and consider the induction step. Expanding f_n using Lemma 7.5.1, the set

$$x_n^{\alpha_n - d} x^{\alpha'} f_{Z'} f_n \mathcal{G}_n \subseteq \mathcal{J}$$

looks like the desired $x_n^{\alpha_n} x^{\alpha'} f_{Z'} \mathcal{G}_n$ plus a sum of terms belonging to $x_n^{\alpha_n - d + e} \mathcal{J}'_\lambda \mathcal{G}_n$ with $0 \leq e < d$. It now suffices to show that each such $x_n^{\alpha_n - d + e} \mathcal{J}'_\lambda \mathcal{G}_n \subseteq \mathcal{J}$. But by (7.13),

$$x_n^{\alpha_n - d + e} \mathcal{J}'_\lambda \mathcal{G}_n \subseteq \sum_{(\alpha', Z') \in \Pi_{n-1}^+} x_n^{\alpha_n - d + e} x^{\alpha'} f_{Z'} \mathcal{G}_n$$

and each such term lies in $\mathcal{J}$ by induction, since $0 \leq \alpha_n - d + e < \alpha_n$. □

Theorem 7.5.6 *The canonical images of the elements*

$$\{x^\alpha w \mid \alpha \in \mathbb{Z}_+^n \text{ with } \alpha_1, \ldots, \alpha_n < d, \ w \in S_n\}$$

form a basis for $\mathcal{H}_n^\lambda$.

Proof By Lemmas 7.5.2 and 7.5.5, the elements $\{x^\alpha f_Z \mid (\alpha, Z) \in \Pi_n^+\}$ form a basis for $\mathcal{J}_\lambda$ viewed as a right $\mathcal{G}_n$-module. Hence Lemma 7.5.2 implies that the elements

$$\{x^\alpha \mid \alpha \in \mathbb{Z}_+^n \text{ with } \alpha_1, \ldots, \alpha_n < d\}$$

form a basis for a complement to $\mathcal{J}_\lambda$ in $\mathcal{H}_n$ viewed as a right $\mathcal{G}_n$-module. The theorem follows at once. □

Remark 7.5.7 It follows from Theorem 7.5.6 that $\mathcal{H}_n^{\Lambda_0}$ is isomorphic to the group algebra FS_n.

7.6 Cyclotomic Mackey Theorem

We will need a special case of a Mackey Theorem for $\mathcal{H}_n^\lambda$. Given any $y \in \mathcal{H}_n$, we will denote its canonical image in $\mathcal{H}_n^\lambda$ by the same symbol. Thus, Theorem 7.5.6 says that

$$B_n := \{x^\alpha w \mid \alpha \in \mathbb{Z}_+^n \text{ with } \alpha_1, \dots, \alpha_n < d,\ w \in S_n\} \tag{7.14}$$

is a basis for $\mathcal{H}_n^\lambda$. Also Theorem 7.5.6 implies that the subalgebra of $\mathcal{H}_{n+1}^\lambda$ generated by $x_1, \dots, x_n$ and w, for $w \in S_n$, is isomorphic to $\mathcal{H}_n^\lambda$. We will write $\mathrm{ind}_{\mathcal{H}_n^\lambda}^{\mathcal{H}_{n+1}^\lambda}$ and $\mathrm{res}_{\mathcal{H}_n^\lambda}^{\mathcal{H}_{n+1}^\lambda}$ for the induction and restriction functors between $\mathcal{H}_n^\lambda$ and $\mathcal{H}_{n+1}^\lambda$, to avoid confusion with the affine analogue from (3.11) and (3.12). So,

$$\mathrm{ind}_{\mathcal{H}_n^\lambda}^{\mathcal{H}_{n+1}^\lambda} M = \mathcal{H}_{n+1}^\lambda \otimes_{\mathcal{H}_n^\lambda} M.$$

Lemma 7.6.1

(i) *$\mathcal{H}_{n+1}^\lambda$ is a free right $\mathcal{H}_n^\lambda$-module on basis*

$$\{x_j^a s_j \dots s_n \mid 0 \le a < d,\ 1 \le j \le n+1\}.$$

(ii) *As $(\mathcal{H}_n^\lambda, \mathcal{H}_n^\lambda)$-bimodules*

$$\mathcal{H}_{n+1}^\lambda = \mathcal{H}_n^\lambda s_n \mathcal{H}_n^\lambda \oplus \bigoplus_{0 \le a < d} x_{n+1}^a \mathcal{H}_n^\lambda.$$

(iii) *For $0 \le a < d$, there are isomorphisms*

$$\mathcal{H}_n^\lambda s_n \mathcal{H}_n^\lambda \cong \mathcal{H}_n^\lambda \otimes_{\mathcal{H}_{n-1}^\lambda} \mathcal{H}_n^\lambda \qquad \text{and} \qquad x_{n+1}^a \mathcal{H}_n^\lambda \cong \mathcal{H}_n^\lambda$$

of $(\mathcal{H}_n^\lambda, \mathcal{H}_n^\lambda)$-bimodules.

Proof (i) By Theorem 7.5.6 and dimension considerations, we just need to check that $\mathcal{H}_{n+1}^\lambda$ is generated as a right $\mathcal{H}_n^\lambda$-module by the given elements. This follows using (3.9).

(ii) It suffices to notice, using (i) and (3.9), that

$$\{x_j^a s_j \dots s_n \mid 0 \le a < d,\ 1 \le j \le n\}$$

is a basis of $\mathcal{H}_n^\lambda s_n \mathcal{H}_n^\lambda$ as a free right $\mathcal{H}_n^\lambda$-module.

(iii) The isomorphism $x_{n+1}^a \mathcal{H}_n^\lambda \cong \mathcal{H}_n^\lambda$ is clear from (i). Furthermore, the map

$$\mathcal{H}_n^\lambda \times \mathcal{H}_n^\lambda \to \mathcal{H}_n^\lambda s_n \mathcal{H}_n^\lambda,\ (u, v) \mapsto u s_n v$$

is $\mathcal{H}_{n-1}^\lambda$-balanced, so it induces a homomorphism

$$\Phi : \mathcal{H}_n^\lambda \otimes_{\mathcal{H}_{n-1}^\lambda} \mathcal{H}_n^\lambda \to \mathcal{H}_n^\lambda s_n \mathcal{H}_n^\lambda$$

of $(\mathcal{H}_n^\lambda, \mathcal{H}_n^\lambda)$-bimodules. By (i), $\mathcal{H}_n^\lambda \otimes_{\mathcal{H}_{n-1}^\lambda} \mathcal{H}_n^\lambda$ is a free right $\mathcal{H}_n^\lambda$-module on a basis

$$\{x_j^a s_j \dots s_{n-1} \otimes 1 \mid 1 \leq j \leq n,\ 0 \leq a < d\}.$$

But Φ maps these elements to a basis for $\mathcal{H}_n^\lambda s_n \mathcal{H}_n^\lambda$ as a free right $\mathcal{H}_n^\lambda$-module, using a fact observed in the proof of (ii). This shows that Φ is an isomorphism. □

We have now decomposed $\mathcal{H}_{n+1}^\lambda$ as an $(\mathcal{H}_n^\lambda, \mathcal{H}_n^\lambda)$-bimodule. So the same argument as for Theorem 3.5.2 gives:

Theorem 7.6.2 *Let M be an $\mathcal{H}_n^\lambda$-module. Then there is a natural isomorphism*

$$\operatorname{res}_{\mathcal{H}_n^\lambda}^{\mathcal{H}_{n+1}^\lambda} \operatorname{ind}_{\mathcal{H}_n^\lambda}^{\mathcal{H}_{n+1}^\lambda} M \cong M^{\oplus d} \oplus \operatorname{ind}_{\mathcal{H}_{n-1}^\lambda}^{\mathcal{H}_n^\lambda} \operatorname{res}_{\mathcal{H}_{n-1}^\lambda}^{\mathcal{H}_n^\lambda} M$$

of $\mathcal{H}_n^\lambda$-modules.

7.7 Duality for cyclotomic algebras

We wish next to prove that the induction functor $\operatorname{ind}_{\mathcal{H}_n^\lambda}^{\mathcal{H}_{n+1}^\lambda}$ commutes with the τ-duality. We need a little preliminary work.

Lemma 7.7.1 *For $1 \leq i \leq n$ and $a \geq 0$, we have*

$$s_n \dots s_i x_i^a s_i \dots s_n = x_{n+1}^a + \left(\text{terms lying in } \mathcal{H}_n^\lambda s_n \mathcal{H}_n^\lambda + \sum_{k=0}^{a-2} x_{n+1}^k \mathcal{H}_n^\lambda\right).$$

Proof We apply induction on $n = i, i+1, \dots$. In case $n = i$, the result follows from a calculation using (3.9). The induction step is similar, noting that s_n centralizes $\mathcal{H}_{n-1}^\lambda$. □

Lemma 7.7.2 *There exists an $(\mathcal{H}_n^\lambda, \mathcal{H}_n^\lambda)$-bimodule homomorphism $\theta : \mathcal{H}_{n+1}^\lambda \to \mathcal{H}_n^\lambda$ such that* $\ker \theta$ *contains no non-zero left ideals of $\mathcal{H}_{n+1}^\lambda$.*

Proof By Lemma 7.6.1(ii), we know that

$$\mathcal{H}_{n+1}^\lambda = x_{n+1}^{d-1} \mathcal{H}_n^\lambda \oplus \bigoplus_{a=0}^{d-2} x_{n+1}^a \mathcal{H}_n^\lambda \oplus \mathcal{H}_n^\lambda s_n \mathcal{H}_n^\lambda$$

as an $(\mathcal{H}_n^\lambda, \mathcal{H}_n^\lambda)$-bimodule. Let $\theta : \mathcal{H}_{n+1}^\lambda \to x_{n+1}^{d-1} \mathcal{H}_n^\lambda$ be the projection on to the first summand of this bimodule decomposition, which by Lemma 7.6.1(iii) is isomorphic to $\mathcal{H}_n^\lambda$ as an $(\mathcal{H}_n^\lambda, \mathcal{H}_n^\lambda)$-bimodule. So we just need to show that if $y \in \mathcal{H}_{n+1}^\lambda$ has the property that $\theta(hy) = 0$ for all $h \in \mathcal{H}_{n+1}^\lambda$, then $y = 0$. Using Lemma 7.6.1(i), we may write

$$y = \sum_{a=0}^{d-1} x_{n+1}^a t_a + \sum_{a=0}^{d-1} \sum_{j=1}^{n} x_j^a s_j \dots s_n u_{a,j}$$

for some elements $t_a, u_{a,j} \in \mathcal{H}_n^\lambda$. As $\theta(y) = 0$, we must have $t_{d-1} = 0$. Now $\theta(x_{n+1}y) = 0$ implies $t_{d-2} = 0$. Similarly $t_{d-3} = t_{d-4} = \dots = 0$. Next, considering $\theta(s_n y), \theta(x_{n+1} s_n y), \theta(x_{n+1}^2 s_n y), \dots$, and using Lemma 7.7.1, we get $u_{d-1,n} = u_{d-2,n} = \dots = 0$. Now repeat the argument again, this time considering $\theta(s_n s_{n-1} y), \theta(x_{n+1} s_n s_{n-1} y), \dots$, to get that all $u_{a,n-1} = 0$. Continuing in this way we eventually arrive at the desired conclusion that $y = 0$. □

Now we are ready to prove the main result of the section:

Theorem 7.7.3 *There is a natural isomorphism*

$$\mathcal{H}_{n+1}^\lambda \otimes_{\mathcal{H}_n^\lambda} M \cong \mathrm{Hom}_{\mathcal{H}_n^\lambda}(\mathcal{H}_{n+1}^\lambda, M)$$

for all $\mathcal{H}_n^\lambda$-modules M.

Proof We show that there exists an isomorphism

$$\varphi : \mathcal{H}_{n+1}^\lambda \to \mathrm{Hom}_{\mathcal{H}_n^\lambda}(\mathcal{H}_{n+1}^\lambda, \mathcal{H}_n^\lambda)$$

of $(\mathcal{H}_{n+1}^\lambda, \mathcal{H}_n^\lambda)$-bimodules. Then, applying the functor $? \otimes_{\mathcal{H}_n^\lambda} M$, we obtain natural isomorphisms

$$\mathcal{H}_{n+1}^\lambda \otimes_{\mathcal{H}_n^\lambda} M \xrightarrow{\varphi \otimes \mathrm{id}} \mathrm{Hom}_{\mathcal{H}_n^\lambda}(\mathcal{H}_{n+1}^\lambda, \mathcal{H}_n^\lambda) \otimes_{\mathcal{H}_n^\lambda} M \cong \mathrm{Hom}_{\mathcal{H}_n^\lambda}(\mathcal{H}_{n+1}^\lambda, M),$$

as required. Note the existence of the second isomorphism here uses the fact that $\mathcal{H}_{n+1}^\lambda$ is a projective left $\mathcal{H}_n^\lambda$-module and [AF, 20.10].

To construct φ, let θ be as in Lemma 7.7.2, and define $\varphi(h)$ to be the map $h\theta$, for each $h \in \mathcal{H}_{n+1}^\lambda$, where

$$h\theta : \mathcal{H}_{n+1}^\lambda \to \mathcal{H}_n^\lambda, \; h' \mapsto \theta(h'h).$$

We can easily check that $\varphi : \mathcal{H}_{n+1}^\lambda \to \mathrm{Hom}_{\mathcal{H}_n^\lambda}(\mathcal{H}_{n+1}^\lambda, \mathcal{H}_n^\lambda)$ is then a well-defined homomorphism of $(\mathcal{H}_{n+1}^\lambda, \mathcal{H}_n^\lambda)$-bimodules. To see that it is an isomorphism, it suffices by dimensions to check it is injective. If $\varphi(h) = 0$ for some $h \in \mathcal{H}_{n+1}^\lambda$ then for every $x \in \mathcal{H}_{n+1}^\lambda$, $\theta(xh) = 0$, that is the left ideal $\mathcal{H}_{n+1}^\lambda h$ is contained in $\ker \theta$. So Lemma 7.7.2 implies $h = 0$. □

Corollary 7.7.4 *$\mathcal{H}_n^\lambda$ is a Frobenius algebra, that is there is an isomorphism of left $\mathcal{H}_n^\lambda$-modules $\mathcal{H}_n^\lambda \cong \mathrm{Hom}_F(\mathcal{H}_n^\lambda, F)$ between the left regular module and the F-linear dual of the right regular module.*

Proof Proceed by induction on n. For the induction step,

$$\begin{aligned}\mathcal{H}_n^\lambda &\cong \mathcal{H}_n^\lambda \otimes_{\mathcal{H}_{n-1}^\lambda} \mathcal{H}_{n-1}^\lambda \cong \mathcal{H}_n^\lambda \otimes_{\mathcal{H}_{n-1}^\lambda} \mathrm{Hom}_F(\mathcal{H}_{n-1}^\lambda, F)\\ &\cong \mathrm{Hom}_{\mathcal{H}_{n-1}^\lambda}(\mathcal{H}_n^\lambda, \mathrm{Hom}_F(\mathcal{H}_{n-1}^\lambda, F)) \cong \mathrm{Hom}_F(\mathcal{H}_{n-1}^\lambda \otimes_{\mathcal{H}_{n-1}^\lambda} \mathcal{H}_n^\lambda, F)\\ &\cong \mathrm{Hom}_F(\mathcal{H}_n^\lambda, F),\end{aligned}$$

applying Theorem 7.7.3 and adjointness of $\otimes$ and Hom. □

For the next corollary, recall the duality induced by τ (3.17) on finite dimensional $\mathcal{H}_n$-modules. Since τ leaves the two-sided ideal $\mathcal{J}_\lambda$ invariant, it induces a duality also denoted τ on finite dimensional $\mathcal{H}_n^\lambda$-modules.

Corollary 7.7.5 *The exact functor $\mathrm{ind}_{\mathcal{H}_n^\lambda}^{\mathcal{H}_{n+1}^\lambda}$ is both left and right adjoint to $\mathrm{res}_{\mathcal{H}_n^\lambda}^{\mathcal{H}_{n+1}^\lambda}$. Moreover, it commutes with duality in the sense that there is a natural isomorphism*

$$\mathrm{ind}_{\mathcal{H}_n^\lambda}^{\mathcal{H}_{n+1}^\lambda}(M^\tau) \cong (\mathrm{ind}_{\mathcal{H}_n^\lambda}^{\mathcal{H}_{n+1}^\lambda} M)^\tau$$

for all finite dimensional $\mathcal{H}_n^\lambda$-modules M.

Proof The fact that $\mathrm{ind}_{\mathcal{H}_n^\lambda}^{\mathcal{H}_{n+1}^\lambda} = \mathcal{H}_{n+1}^\lambda \otimes_{\mathcal{H}_n^\lambda} ?$ is right adjoint to $\mathrm{res}_{\mathcal{H}_n^\lambda}^{\mathcal{H}_{n+1}^\lambda}$ is immediate from Theorem 7.7.3, since $\mathrm{Hom}_{\mathcal{H}_n^\lambda}(\mathcal{H}_{n+1}^\lambda, ?)$ is right adjoint to restriction by adjointness of $\otimes$ and Hom. But on finite dimensional modules, a standard check shows that the functor $\tau \circ \mathrm{ind}_{\mathcal{H}_n^\lambda}^{\mathcal{H}_{n+1}^\lambda} \circ \tau$ is also right adjoint to restriction. Now the remaining part of the corollary follows by uniqueness of adjoint functors. □

Let $r \geq 1$. By Theorem 7.5.6, the subalgebra

$$\mathcal{G}'_r := \langle s_{n-r+1}, s_{n-r+2}, \dots, s_{n-1} \rangle \subseteq \mathcal{H}_{n+r}^\lambda$$

is isomorphic to $\mathcal{G}_r$. This subalgebra commutes with the subalgebra $\mathcal{H}_n^\lambda$, and $\mathcal{H}_n^\lambda \mathcal{G}'_r$ is isomorphic to $\mathcal{H}_n^\lambda \otimes \mathcal{G}_r$. From now on we will consider $\mathcal{H}_n^\lambda \otimes \mathcal{G}_r$ as a subalgebra of $\mathcal{H}_{n+r}^\lambda$ in this way. Our goal is to generalize Theorem 7.7.3 and Corollary 7.7.5 from subalgebra $\mathcal{H}_n^\lambda \subset \mathcal{H}_{n+1}^\lambda$ to subalgebras $\mathcal{H}_n^\lambda \mathcal{G}'_r = \mathcal{H}_n^\lambda \otimes$

$\mathcal{G}_r \subset \mathcal{H}^\lambda_{n+r}$. First, we need to decompose $\mathcal{H}^\lambda_{n+r}$ as an $(\mathcal{H}^\lambda_n \mathcal{G}'_r, \mathcal{H}^\lambda_n \mathcal{G}'_r)$-bimodule. To this end, note that the set of distinguished double coset representatives

$$D_{(n,r),(n,r)} = \{1 = \sigma_0, \sigma_1, \dots, \sigma_m\},$$

where $m = \min(n, r)$, and

$$\sigma_i := \prod_{j=0}^{i-1} (n-j, n+i-j) \qquad (0 \le i \le m).$$

Denote the corresponding double cosets by

$$C_i := (S_n \times S_r)\sigma_i(S_n \times S_r) \qquad (0 \le i \le m).$$

Recall the basis B_{n+r} of $\mathcal{H}^\lambda_{n+r}$ from (7.14).

Theorem 7.7.6 *Set*

$$X := \mathcal{H}^\lambda_n \mathcal{G}'_r x^{d-1}_{n+1} \dots x^{d-1}_{n+r} \mathcal{H}^\lambda_n \mathcal{G}'_r$$

and

$$Y := \sum \mathcal{H}^\lambda_n \mathcal{G}'_r x^{a_{i+1}}_{n+i+1} \cdots x^{a_r}_{n+r} \sigma_i \mathcal{H}^\lambda_n \mathcal{G}'_r,$$

the sum running over all $0 \le i \le m$ and $0 \le a_{i+1} \le \cdots \le a_r < d$ such that $a_1 + \cdots + a_r < r(d-1)$ if $i = 0$. Then, as $(\mathcal{H}^\lambda_n \mathcal{G}'_r, \mathcal{H}^\lambda_n \mathcal{G}'_r)$-bimodules, $X \cong \mathcal{H}^\lambda_n \mathcal{G}'_r$, and

$$\mathcal{H}^\lambda_{n+r} = X \oplus Y.$$

Moreover, X has basis

$$B_X := \{x_1^{b_1} \dots x_{n+r}^{b_{n+r}} w \in B_{n+r} \mid w \in S_n \times S_r,$$
$$b_{n+1} = \cdots = b_{n+r} = d-1\},$$

and Y has basis

$$B_Y := B_{n+r} \setminus B_X.$$

Proof Introduce a partial order on B_{n+r} by putting

$$h_1 = x_1^{b_1} \dots x_{n+r}^{b_{n+r}} w < h_2 = x_1^{c_1} \dots x_{n+r}^{c_{n+r}} u$$

if and only if either $w \in C_i$ and $u \in C_j$ with $i < j$, or $i = j$ and $b_{n+1} + \cdots + b_{n+r} < c_{n+1} + \cdots + c_{n+r}$.

We now prove that $\mathcal{H}^\lambda_{n+r} = X + Y$. It suffices to prove that every element of B_{n+r} belongs to $X + Y$. Suppose this is false. Let

$$h := x_1^{b_1} \dots x_{n+r}^{b_{n+r}} w \in B_{n+r}$$

be a minimal element with respect to our partial order such that $h \notin X+Y$. Assume that $w \in C_i$. Let

$$V = \mathrm{span}(h' \in B_{n+r} \mid h' < h) \subseteq X+Y,$$

and write $\equiv$ for equality modulo V. Let $w = g_1\sigma_i g_2$ for $g_1, g_2 \in S_n \times S_r$. Using Lemma 3.2.1, we get

$$\begin{aligned} h &= x_1^{b_1} \dots x_n^{b_n} x_{n+1}^{b_{n+1}} \dots x_{n+r}^{b_{n+r}} g_1\sigma_i g_2 \\ &\equiv x_1^{b_1} \dots x_n^{b_n} g_1 x_{g_1^{-1}(n+1)}^{b_{n+1}} \dots x_{g_1^{-1}(n+r)}^{b_{n+r}} \sigma_i g_2. \end{aligned}$$

As sets, $\{g_1^{-1}(n+1), \dots, g_1^{-1}(n+r)\} = \{n+1, \dots, n+r\}$. So, renaming bs with cs, we can write

$$h \equiv x_1^{b_1} \dots x_n^{b_n} g_1 x_{n+1}^{c_{n+1}} \dots x_{n+r}^{c_{n+r}} \sigma_i g_2,$$

Moreover, if $g \in S_n \times S_r$ fixes all numbers, except possibly $n+i+1, \dots, n+r$, then g commutes with σ_i, and so $w = g_1\sigma_i g_2 = g_1 g\sigma_i g^{-1} g_2$. So by changing g_1 to $g_1 g$ and g_2 to $g^{-1}g_2$, if necessary, we can achieve $c_{n+i+1} \leq \dots \leq c_{n+r}$. Finally, using Lemma 3.2.1 again, we have

$$\begin{aligned} & x_1^{b_1} \dots x_n^{b_n} g_1 x_{n+1}^{c_{n+1}} \dots x_{n+r}^{c_{n+r}} \sigma_i g_2 \\ &= x_1^{b_1} \dots x_n^{b_n} g_1 (x_{n+i+1}^{c_{n+i+1}} \dots x_{n+r}^{c_{n+r}})(x_{n+1}^{c_{n+1}} \dots x_{n+i}^{c_{n+i}})\sigma_i g_2 \\ &\equiv x_1^{b_1} \dots x_n^{b_n} g_1 (x_{n+i+1}^{c_{n+i+1}} \dots x_{n+r}^{c_{n+r}})\sigma_i (x_{n-i+1}^{c_{n+1}} \dots x_n^{c_{n+i}}) g_2. \end{aligned}$$

The last element is in $X+Y$, because $x_1^{b_1} \dots x_n^{b_n} g_1$ and $x_{n-i+1}^{c_{n+1}} \dots x_n^{c_{n+i}} g_2$ belong to $\mathcal{H}_n^\lambda \mathcal{G}_r'$. Hence $h \in X+Y$, giving a contradiction. Thus $\mathcal{H}_{n+r}^\lambda = X+Y$.

Next, note by Theorem 3.3.1 that the element $x_{n+1}^{d-1} \dots x_{n+r}^{d-1}$ commutes with $\mathcal{H}_n^\lambda \mathcal{G}_r'$, so the statement about the basis of X follows immediately. It also follows that $X \cong \mathcal{H}_n^\lambda \mathcal{G}_r'$. Now, it suffices to show that elements of Y are linear combinations of elements from B_Y. But Y is spanned by elements of the form

$$G := x_1^{b_1} \dots x_n^{b_n} g_1 x_{n+i+1}^{a_{i+1}} \dots x_{n+r}^{a_r} \sigma_i x_1^{c_1} \dots x_n^{c_n} g_2,$$

with $g_1, g_2 \in S_n \times S_r$, $0 \leq a_k, b_k, c_k < d$, $0 \leq a_{i+1} \leq \dots \leq a_r < d$, and $a_1 + \dots + a_r < r(d-1)$ if $i = 0$. So it suffices to show that G is a linear combination of elements from B_Y. Write $g_1 = yz$ for $y \in S_n$ and $z \in S_r'$. Then

$$G = x_1^{b_1} \dots x_n^{b_n} y x_1^{c_1} \dots x_{n-i}^{c_{n-i}} z x_{n+i+1}^{a_{i+1}} \dots x_{n+r}^{a_r} \sigma_i x_{n-i+1}^{c_{n-i+1}} \dots x_n^{c_n} g_2.$$

Note $yx_1^{c_1}\dots x_{n-i}^{c_{n-i}} \in \mathcal{H}_n^\lambda$, so it can be written as a linear combination of the standard basis elements of the form $x_1^{b_1'}\dots x_n^{b_n'}y'$ for $y' \in S_n$. So we may assume that

$$G = x_1^{b_1}\dots x_n^{b_n} yzx_{n+i+1}^{a_{i+1}}\dots x_{n+r}^{a_r}\sigma_i x_{n-i+1}^{c_{n-i+1}}\dots x_n^{c_n} g_2.$$

Assume first that $i = 0$. Then

$$G = x_1^{b_1}\dots x_n^{b_n} yzx_{n+1}^{a_1}\dots x_{n+r}^{a_r} g_2 = x_1^{b_1}\dots x_n^{b_n} zx_{n+1}^{a_1}\dots x_{n+r}^{a_r} yg_2.$$

In view of Lemma 3.2.1 applied to $zx_{n+1}^{a_1}\dots x_{n+r}^{a_r}$ we conclude that G is a linear combination of elements in B_Y.

Now, let $i > 0$. Applying Lemma 3.2.1 to $\sigma_i x_{n-i+1}^{c_{n-i+1}}\dots x_n^{c_n}$ we see that this can be written as a linear combination of elements of the form

$$x_{n-i+1}^{d_{n-i+1}}\dots x_n^{d_n} x_{n+1}^{d_{n+1}}\dots x_{n+i}^{d_{n+i}}\sigma$$

where either $\sigma = \sigma_i$ or the degree

$$d_{n-i+1} + \dots + d_n + d_{n+1} + \dots + d_{n+i} < i(d-1). \tag{7.15}$$

So we may assume that

$$G = x_1^{b_1}\dots x_n^{b_n} yzx_{n+i+1}^{a_{i+1}}\dots x_{n+r}^{a_r} x_{n-i+1}^{d_{n-i+1}}\dots x_n^{d_n} x_{n+1}^{d_{n+1}}\dots x_{n+i}^{d_{n+i}}\sigma g_2$$

and (7.15) holds. Using Lemma 3.2.1 to commute z past the element $x_{n+i+1}^{a_{i+1}}\dots x_{n+r}^{a_r} x_{n+1}^{d_{n+1}}\dots x_{n+i}^{d_{n+i}}$ and y past $x_{n-i+1}^{d_{n-i+1}}\dots x_n^{d_n}$, we conclude again that G is a linear combination of elements from B_Y, as the maximal possible degree $r(d-1)$ on $x_{n+1}, \dots, x_{n+r}$ cannot be achieved. □

Corollary 7.7.7 *There exists an $(\mathcal{H}_n^\lambda \mathcal{G}_r', \mathcal{H}_n^\lambda \mathcal{G}_r')$-bimodule homomorphism $\theta : \mathcal{H}_{n+r}^\lambda \to \mathcal{H}_n^\lambda \mathcal{G}_r'$ such that* $\ker\theta$ *contains no non-zero left ideals of $\mathcal{H}_{n+r}^\lambda$.*

Proof Using the notation of Theorem 7.7.6, let θ be the projection to X along Y. So we just need to show that if $y \in \mathcal{H}_{n+r}^\lambda$ has the property that $hy \in Y$ for all $h \in \mathcal{H}_{n+r}^\lambda$, then $y = 0$. Assume for a contradiction that there is a non-zero such y. By applying the antiautomorphism τ to the bases B_X and B_Y we see that X also has a basis

$$B_X' = \{wx_1^{b_1}\dots x_n^{b_n} x_{n+1}^{d-1}\dots x_{n+r}^{d-1} \mid w \in S_n \times S_r,\ 0 \le b_1, \dots, b_n < d\},$$

and Y also has basis

$$B_Y' = \{wx_1^{b_1}\dots x_{n+r}^{b_{n+r}} \mid w \notin S_n \times S_r \text{ or } b_{n+1} + \dots + b_{n+r} < r(d-1)\}.$$

Write y as a linar combination of basis elements from B_Y':

$$y = \sum c_{\underline{b},w} wx_1^{b_1}\dots x_{n+r}^{b_{n+r}}.$$

Multiplying on the left with elements from S_{n+r}, we may assume that some terms with $w = w_0$, the longest element of S_{n+r}, appear. Order them lexicographically so that

$$w_0 x_1^{b_1} \dots x_{n+r}^{b_{n+r}} < w_0 x_1^{c_1} \dots c_{n+r}^{c_{n+r}}$$

if $b_k < c_k$ and $c_l = b_l$ for any $l > k$. Let $w_0 x_1^{c_1} \dots x_{n+r}^{c_{n+r}}$ be the top term. Then, in view of Lemma 3.2.1, the product

$$y' := x_{n+1}^{d-1-c_{n+1}} \dots x_{n+r}^{d-1-c_{n+r}} y$$

contains the term

$$w_0 x_1^{c_1} \dots x_n^{c_n} x_{n+1}^{d-1} \dots x_{n+r}^{d-1}$$

(when decomposed in terms of the basis $B'_X \cup B'_Y$). But then $w_0^{-1} y'$ contains

$$x_1^{c_1} \dots x_n^{c_n} x_{n+1}^{d-1} \dots x_{n+r}^{d-1} \in B'_X,$$

giving a contradiction. □

Corollary 7.7.8 *Let $r \geq 1$. Then:*

(i) *There is a natural isomorphism*

$$\mathcal{H}_{n+r}^\lambda \otimes_{\mathcal{H}_n^\lambda \otimes \mathcal{G}_r} M \cong \mathrm{Hom}_{\mathcal{H}_n^\lambda \otimes \mathcal{G}_r}(\mathcal{H}_{n+r}^\lambda, M)$$

for all $\mathcal{H}_n^\lambda \otimes \mathcal{G}_r$-modules M.

(ii) *The exact functor $\mathrm{ind}_{\mathcal{H}_n^\lambda \otimes \mathcal{G}_r}^{\mathcal{H}_{n+r}^\lambda}$ is both left and right adjoint to the functor $\mathrm{res}_{\mathcal{H}_n^\lambda \otimes \mathcal{G}_r}^{\mathcal{H}_{n+r}^\lambda}$. Moreover, it commutes with duality in the sense that there is a natural isomorphism*

$$\mathrm{ind}_{\mathcal{H}_n^\lambda \otimes \mathcal{G}_r}^{\mathcal{H}_{n+r}^\lambda}(M^\tau) \cong (\mathrm{ind}_{\mathcal{H}_n^\lambda \otimes \mathcal{G}_r}^{\mathcal{H}_{n+r}^\lambda} M)^\tau$$

for all finite dimensional $\mathcal{H}_n^\lambda \otimes \mathcal{G}_r$-modules M.

Proof The proof is the same as that of Theorem 7.7.3 and Corollary 7.7.5, except that it uses Corollary 7.7.7 instead of Lemma 7.7.2. □

7.8 Presentation for degenerate cyclotomic Hecke algebras

Cyclotomic Hecke algebras are usually given by generators and relations, and in this section we describe such a presentation for $\mathcal{H}_n^\lambda$. Note this result will not be used anywhere in this book.

Proposition 7.8.1 *The algebra* $\mathcal{H}_n^\lambda$ *is generated by its elements* x_1 *and* $s_1, \dots, s_{n-1}$, *subject only to the following relations:*

$$\prod_{i\in I}(x_1 - i)^{\langle h_i,\lambda\rangle} = 0, \tag{7.16}$$

$$x_1 s_l = s_l x_1 \quad (2 \le l < n), \tag{7.17}$$

$$x_1(s_1x_1s_1 + s_1) = (s_1x_1s_1 + s_1)x_1, \tag{7.18}$$

$$s_k^2 = 1,\ s_ks_m = s_ms_k,\ s_ks_{k+1}s_k = s_{k+1}s_ks_{k+1}. \tag{7.19}$$

for all admissible k *and* m, *with* $|k-m| > 1$.

Proof Let $\mathcal{A}$ be the algebra given by generators y_1 and $t_1, \dots, t_{n-1}$ and relations (7.16)–(7.19) with y_1 instead of x_1 and t_ks instead of s_ks. It suffices to show that there exist algebra homomorphisms

$$\alpha : \mathcal{A} \to \mathcal{H}_n^\lambda,\quad y_1 \mapsto x_1,\ t_k \mapsto s_k \quad (1 \le k < n),$$

$$\beta : \mathcal{H}_n^\lambda \to \mathcal{A},\quad x_1 \mapsto y_1,\ s_k \mapsto t_k \quad (1 \le k < n).$$

The existence of α follows from the easily checked fact that elements s_k and x_1 in $\mathcal{H}_n^\lambda$ satisfy the same relations as the elements t_k and y_1 in $\mathcal{A}$. To construct β it suffices to construct an algebra homomorphism

$$\hat{\beta} : \mathcal{H}_n \to \mathcal{A},\quad x_1 \mapsto y_1,\ s_k \mapsto t_k \quad (1 \le k < n), \tag{7.20}$$

such that $\hat{\beta}(f_\lambda) = 0$, see (7.6). Well, let us define $\hat{\beta}$ using (7.20) and the recurrent formula

$$\hat{\beta} : x_k \mapsto t_{k-1}\hat{\beta}(x_{k-1})t_{k-1} + t_{k-1} \quad (2 \le k \le n).$$

All we have to check is that this makes sense, that is that $\hat{\beta}(s_k)$ and $\hat{\beta}(x_m)$ satisfy the defining relations of $\mathcal{H}_n$, of which only the relations (3.1) and (3.6) are not immediate. Now we verify all the relations of the form (3.1) and (3.6) involving $\hat{\beta}(x_k)$ using induction on $k = 1, 2, \dots, n$ and relations in $\mathcal{A}$. □

8
Functors e_i^λ and f_i^λ

In this chapter we will finally define the "induction analogues" f_i^λ, $\tilde{f}_i^\lambda$, and φ_i^λ of e_i, $\tilde{e}_i$, and ε_i, respectively. These will depend crucially on λ, that is on the cyclotomic algebra $\mathcal{H}_n^\lambda$ we are working with. In this respect they will differ from e_i^λ, $\tilde{e}_i^\lambda$, and ε_i^λ, which are the same as e_i, $\tilde{e}_i$, and ε_i, if we consider a module over $\mathcal{H}_n^\lambda$ as a module over $\mathcal{H}_n$ via inflation. This crucial role of the cyclotomic Hecke algebras in the definition of f_i^λ, $\tilde{f}_i^\lambda$, and φ_i^λ explains why we could not define these notions in Chapter 5.

One natural definition of $\tilde{f}_i^\lambda$ is easy to come up with if we recall that for an $\mathcal{H}_n$-module M, e_iM is roughly speaking restriction from $\mathcal{H}_n$ to $\mathcal{H}_{n-1}$ followed by a projection to a block. If M is integral, which we always assume, then it is also a module over $\mathcal{H}_n^\lambda$ for some sufficiently large λ, and we can think of e_i as the restriction from $\mathcal{H}_n^\lambda$ to $\mathcal{H}_{n-1}^\lambda$ followed by a projection to a block. It is crucial here that if an $\mathcal{H}_n$-module factors through $\mathcal{H}_n^\lambda$, then its restriction to $\mathcal{H}_{n-1}$ factors through $\mathcal{H}_{n-1}^\lambda$. Now, we can define f_i^λ as induction from $\mathcal{H}_n^\lambda$ to $\mathcal{H}_{n+1}^\lambda$ followed by a projection to a block. We can see immediately that this definition depends crucially on λ.

Although the definition of f_i^λ in terms of induction of cyclotomic Hecke algebras is quite natural, we will need a different description of this functor, see Lemma 8.2.3. This "inverse limit" description is one of the key observations of Grojnowski. It allows us to describe the function φ_i^λ, defined originally in (8.17), as the "stabilization" point of the limit, which should be thought of as the analogue of the description of ε_i given in Theorem 5.5.1(ii). The tricky definition of f_i^λ is responsible for the fact that the analogue of Theorem 5.5.1 is quite difficult to get. This is achieved by the end of this chapter, see Theorem 8.5.9.

Finally, we mention that important "divided power" generalizations of the functors e_i, e_i^λ, and f_i^λ are studied in Section 8.3.

8.1 New notation for blocks

We will use a new notation for blocks in $\text{Rep}_I \mathcal{H}_n$. In view of Section 4.2 the blocks (or central characters) in the category $\text{Rep}_I \mathcal{H}_n$ are labeled by the S_n-orbits on I^n. If $\underline{i} \in I^n$, define its *content* $\text{cont}(\underline{i}) \in P$ by

$$\text{cont}(\underline{i}) = \sum_{i \in I} \gamma_i \alpha_i, \qquad \text{where} \qquad \gamma_i = \sharp\{j = 1, \dots, n \mid i_j = i\}. \tag{8.1}$$

So $\text{cont}(\underline{i})$ is an element of the set Γ_n of non-negative integral linear combinations $\gamma = \sum_{i \in I} \gamma_i \alpha_i$ of the simple roots such that $\sum_{i \in I} \gamma_i = n$. Obviously, the S_n-orbit of $\underline{i}$ is uniquely determined by the content of $\underline{i}$, so we obtain a labeling of the orbits of S_n on I^n by the elements of Γ_n. We will also use the notation χ_γ for the central character $\chi_{\underline{i}}$, where $\underline{i}$ is any element of I^n with $\text{cont}(\underline{i}) = \gamma$. We will write as usual $\text{Rep}_I \mathcal{H}_n[\gamma]$ for the corresponding block, $M[\gamma]$ for the corresponding block component of the module M, etc.

We can extend some of these notions to $\mathcal{H}_n^\lambda$-modules, for $\lambda \in P_+$. In particular, if $M \in \mathcal{H}_n^\lambda$-mod, we also write $M[\gamma]$ for the summand $M[\gamma]$ of M defined by first viewing M as an $\mathcal{H}_n$-module by inflation. Also write $\mathcal{H}_n^\lambda$-mod$[\gamma]$ for the full subcategory of $\mathcal{H}_n^\lambda$-mod consisting of the modules M with $M = M[\gamma]$. Then we have a decomposition

$$\mathcal{H}_n^\lambda\text{-mod} \cong \bigoplus_{\gamma \in \Gamma_n} \mathcal{H}_n^\lambda\text{-mod}[\gamma]. \tag{8.2}$$

Note though that we should not yet refer to $\mathcal{H}_n^\lambda$-mod$[\gamma]$ as a block of $\text{Rep}\, \mathcal{H}_n^\lambda$: the center of $\mathcal{H}_n^\lambda$ may be larger than the image of the center of $\mathcal{H}_n$, so we cannot yet assert that $Z(\mathcal{H}_n^\lambda)$ acts on $M[\gamma]$ by a single central character. Also we no longer know precisely which $\gamma \in \Gamma_n$ have the property that $\mathcal{H}_n^\lambda$-mod$[\gamma]$ is non-trivial. These questions will be settled in Section 9.6.

8.2 Definitions

Fix $\lambda \in P_+$ throughout the section. Recall from (5.6) the functors e_i for $i \in I$. If M is an $\mathcal{H}_n^\lambda$-module then $e_i M$ is automatically an $\mathcal{H}_{n-1}^\lambda$-module (recall that we always identify the category of $\mathcal{H}_m^\lambda$-modules with a full subcategory of $\mathcal{H}_m$-modules via inflation). So the restriction of the functor e_i gives a functor

$$e_i^\lambda : \mathcal{H}_n^\lambda\text{-mod} \to \mathcal{H}_{n-1}^\lambda\text{-mod}. \tag{8.3}$$

There is an alternative definition of this functor: if M is a module in $\mathcal{H}_n^\lambda$-mod$[\gamma]$ for some fixed $\gamma = \sum_{j\in I} \gamma_j \alpha_j \in \Gamma_n$ then

$$e_i^\lambda M = \begin{cases} (\mathrm{res}^{\mathcal{H}_n^\lambda}_{\mathcal{H}_{n-1}^\lambda} M)[\gamma - \alpha_i] & \text{if } \gamma_i > 0, \\ 0 & \text{if } \gamma_i = 0. \end{cases} \tag{8.4}$$

This description suggests how to define an analogous (additive) functor

$$f_i^\lambda : \mathcal{H}_n^\lambda\text{-mod} \to \mathcal{H}_{n+1}^\lambda\text{-mod}. \tag{8.5}$$

Using (8.2) and additivity, it suffices to define this on an object M belonging to $\mathcal{H}_n^\lambda$-mod$[\gamma]$ for fixed $\gamma = \sum_{j\in I} \gamma_j \alpha_j \in \Gamma_n$. Then we set

$$f_i^\lambda M = (\mathrm{ind}^{\mathcal{H}_{n+1}^\lambda}_{\mathcal{H}_n^\lambda} M)[\gamma + \alpha_i]. \tag{8.6}$$

To complete the definition of the functor f_i^λ, it is defined on a morphism θ simply by restriction of the corresponding morphism $\mathrm{ind}^{\mathcal{H}_{n+1}^\lambda}_{\mathcal{H}_n^\lambda} \theta$.

Remark 8.2.1 Note that the functor f_i^λ *depends fundamentally on* λ, unlike e_i^λ, which is just the restriction of its affine counterpart e_i.

Lemma 8.2.2 *For* $\lambda \in P_+$ *and each* $i \in I$:

(i) e_i^λ *and* f_i^λ *are both left and right adjoint to each other, hence they are exact and send projectives to projectives;*

(ii) e_i^λ *and* f_i^λ *commute with duality, that is there is a natural isomorphism*

$$e_i^\lambda(M^\tau) \cong (e_i^\lambda M)^\tau, \qquad f_i^\lambda(M^\tau) \cong (f_i^\lambda M)^\tau$$

for each finite dimensional $\mathcal{H}_n^\lambda$*-module* M.

Proof We know that $e_i^\lambda M$ and $f_i^\lambda M$ are summands of $\mathrm{res}^{\mathcal{H}_n^\lambda}_{\mathcal{H}_{n-1}^\lambda} M$ and $\mathrm{ind}^{\mathcal{H}_{n+1}^\lambda}_{\mathcal{H}_n^\lambda} M$ respectively. Moreover, τ-duality leaves central characters invariant because $\tau(x_j) = x_j$ for each j. Now everything follows easily, applying Corollary 7.7.5. □

From (5.8) and Lemma 8.2.2(i), we have

$$\mathrm{res}^{\mathcal{H}_n^\lambda}_{\mathcal{H}_{n-1}^\lambda} M = \bigoplus_{i\in I} e_i^\lambda M, \qquad \mathrm{ind}^{\mathcal{H}_{n+1}^\lambda}_{\mathcal{H}_n^\lambda} M = \bigoplus_{i\in I} f_i^\lambda M. \tag{8.7}$$

We will use an alternative description of e_i^λ and f_i^λ. Recall the definition of the $\mathcal{H}_1$-modules $L_m(i)$ for $i \in I$, $m \geq 0$ from Section 4.4. The limits in the next lemma are taken with respect to the systems induced by the maps (4.7).

Lemma 8.2.3 *For every $M \in \mathcal{H}_n^\lambda$-mod and $i \in I$, there are natural isomorphisms*

$$e_i^\lambda M \cong \varinjlim \mathrm{pr}^\lambda \mathrm{Hom}_{\mathcal{H}_1'}(L_m(i), M),$$

$$f_i^\lambda M \cong \varprojlim \mathrm{pr}^\lambda \mathrm{ind}_{n,1}^{n+1} M \boxtimes L_m(i),$$

where $\mathcal{H}_1'$ denotes the subalgebra of $\mathcal{H}_{n-1,1}$ generated by x_n and the $\mathcal{H}_{n-1}$-module structure on $\mathrm{Hom}_{\mathcal{H}_1'}(L_m(i), M)$ is defined by

$$(hf)(v) = h(f(v))$$

for $f \in \mathrm{Hom}_{\mathcal{H}_1'}(L_m(i), M)$, $h \in \mathcal{H}_{n-1}$, and $v \in L_m(i)$.

Proof For e_i^λ, it suffices to consider the effect on $M \in \mathcal{H}_n^\lambda$-mod$[\gamma]$ for $\gamma = \sum_{j\in I} \gamma_j \alpha_j \in \Gamma_n$ with $\gamma_i > 0$, both sides of what we are trying to prove clearly being zero if $\gamma_i = 0$. Then, for all sufficiently large m, Lemma 4.4.2 (in the special case $n = 1$) implies that

$$\mathrm{Hom}_{\mathcal{H}_1'}(L_m(i), M) \cong (\mathrm{res}^{\mathcal{H}_n^\lambda}_{\mathcal{H}_{n-1}^\lambda} M)[\gamma - \alpha_i].$$

Hence

$$\varinjlim \mathrm{pr}^\lambda \mathrm{Hom}_{\mathcal{H}_1'}(L_m(i), M) \cong (\mathrm{res}^{\mathcal{H}_n^\lambda}_{\mathcal{H}_{n-1}^\lambda} M)[\gamma - \alpha_i] = e_i^\lambda M.$$

To deduce the statement about f_i^λ, it now suffices by uniqueness of adjoint functors to show that $\varprojlim \mathrm{pr}^\lambda \mathrm{ind}_{n,1}^{n+1} ? \boxtimes L_m(i)$ is left adjoint to $\varinjlim \mathrm{pr}^\lambda \mathrm{Hom}_{\mathcal{H}_1'}(L_m(i), ?)$. Let $N \in \mathcal{H}_{n-1}^\lambda$-mod and $M \in \mathcal{H}_n^\lambda$-mod. As explained in the previous paragraph, the direct system

$$\mathrm{pr}^\lambda \mathrm{Hom}_{\mathcal{H}_1'}(L_1(i), M) \hookrightarrow \mathrm{pr}^\lambda \mathrm{Hom}_{\mathcal{H}_1'}(L_2(i), M) \hookrightarrow \dots$$

stabilizes after finitely many terms. We claim that the inverse system

$$\mathrm{pr}^\lambda \mathrm{ind}_{n,1}^{n+1} N \boxtimes L_1(i) \twoheadleftarrow \mathrm{pr}^\lambda \mathrm{ind}_{n,1}^{n+1} N \boxtimes L_2(i) \twoheadleftarrow \dots$$

also stabilizes after finitely many terms. To see this, it suffices to show that the dimension of $\mathrm{pr}^\lambda \mathrm{ind}_{n,1}^{n+1} N \boxtimes L_m(i)$ is bounded above independently of m. For, let w be a vector which generates the "Jordan block" $L_m(i)$ as an $\mathcal{H}_1'$-module, and $W = Fw$ be the 1-dimensional subspace of $L_m(i)$ spanned by w. Then $\mathrm{ind}_{n,1}^{n+1} N \boxtimes L_m(i)$ is generated as an $\mathcal{H}_{n+1}$-module by the subspace $W' = 1 \otimes (N \otimes W)$, of dimension independent of m. Finally, we need to observe that $\mathrm{pr}^\lambda \mathrm{ind}_{n,1}^{n+1} N \boxtimes L_m(i)$ is a quotient of the vector space $\mathcal{H}_{n+1}^\lambda \otimes_F W'$, whose dimension is independent of m.

Now we can complete the proof of adjointness. Using the fact that the direct and inverse systems stabilize after finitely many terms, we have natural isomorphisms

$$\begin{aligned}
&\mathrm{Hom}_{\mathcal{H}_n^\lambda}(\varprojlim \mathrm{pr}^\lambda \mathrm{ind}_{n-1,1}^n N \boxtimes L_m(i), M)\\
&\cong \varinjlim \mathrm{Hom}_{\mathcal{H}_n^\lambda}(\mathrm{pr}^\lambda \mathrm{ind}_{n-1,1}^n N \boxtimes L_m(i), M)\\
&\cong \varinjlim \mathrm{Hom}_{\mathcal{H}_n}(\mathrm{ind}_{n-1,1}^n N \boxtimes L_m(i), M)\\
&\cong \varinjlim \mathrm{Hom}_{\mathcal{H}_{n-1,1}}(N \boxtimes L_m(i), \mathrm{res}_{n-1,1}^n M)\\
&\cong \varinjlim \mathrm{Hom}_{\mathcal{H}_{n-1}}(N, \mathrm{Hom}_{\mathcal{H}_1'}(L_m(i), M))\\
&\cong \varinjlim \mathrm{Hom}_{\mathcal{H}_{n-1}^\lambda}(N, \mathrm{pr}^\lambda \mathrm{Hom}_{\mathcal{H}_1'}(L_m(i), M))\\
&\cong \mathrm{Hom}_{\mathcal{H}_{n-1}^\lambda}(N, \varinjlim \mathrm{pr}^\lambda \mathrm{Hom}_{\mathcal{H}_1'}(L_m(i), M)).
\end{aligned}$$

This completes the argument. □

Now we define the *cyclotomic crystal operators* on irreducible modules

$$\tilde{e}_i^\lambda = \mathrm{pr}^\lambda \circ \tilde{e}_i \circ \mathrm{infl}^\lambda, \tag{8.8}$$

$$\tilde{f}_i^\lambda = \mathrm{pr}^\lambda \circ \tilde{f}_i \circ \mathrm{infl}^\lambda \tag{8.9}$$

for each $i \in I$ and $\lambda \in P_+$ (cf. Section 5.2). Let $B(\infty)$ and $B(\lambda)$ denote the set of isomorphism classes of irreducible modules in $\mathrm{Rep}_I\, \mathcal{H}_n$ and $\mathcal{H}_n^\lambda$-mod for all $n \geq 0$, respectively. Note that we may consider $B(\lambda)$ as a subset of $B(\infty)$ via inflation. We have maps

$$\begin{aligned}
\tilde{e}_i &: B(\infty) \to B(\infty) \cup \{0\},\\
\tilde{f}_i &: B(\infty) \to B(\infty),\\
\tilde{e}_i^\lambda, \tilde{f}_i^\lambda &: B(\lambda) \to B(\lambda) \cup \{0\}.
\end{aligned}$$

Remark 8.2.4 As with e_i^λ and f_i^λ, there is an important difference between $\tilde{e}_i^\lambda$ and $\tilde{f}_i^\lambda$. The operator $\tilde{e}_i^\lambda$ is just the restriction of $\tilde{e}_i$ from $B(\infty)$ to $B(\lambda)$, that is

$$\mathrm{pr}^\lambda \circ \tilde{e}_i \circ \mathrm{infl}^\lambda M = \tilde{e}_i \circ \mathrm{infl}^\lambda M$$

for all irreducible $\mathcal{H}_n^\lambda$-modules M. This is certainly *not* the case for $\tilde{f}_i$: even though $\tilde{f}_i^\lambda$ also "tries" to be the restriction of $\tilde{f}_i$, but the problem is that $\tilde{f}_i$ does not leave $B(\lambda)$ invariant, so it will often be the case that $\tilde{f}_i^\lambda M = 0$, even though $\tilde{f}_i M$ is never zero.

Theorem 8.2.5 *Let $\lambda \in P_+$ and $i \in I$. Then for any irreducible $\mathcal{H}_n^\lambda$-module M we have:*

(i) *$e_i^\lambda M$ is non-zero if and only if $\tilde{e}_i^\lambda M \neq 0$, in which case $e_i^\lambda M$ is a self-dual indecomposable module with irreducible socle and head isomorphic to $\tilde{e}_i^\lambda M$;*

(ii) *$f_i^\lambda M$ is non-zero if and only if $\tilde{f}_i^\lambda M \neq 0$, in which case $f_i^\lambda M$ is a self-dual indecomposable module with irreducible socle and head isomorphic to $\tilde{f}_i^\lambda M$.*

Proof (i) By Corollary 5.1.7, $e_i^\lambda M$ has irreducible socle $\tilde{e}_i^\lambda M$ whenever $e_i^\lambda M$ is non-zero. The remaining facts follow since M is self-dual by Lemma 5.3.2, and e_i^λ commutes with duality by Lemma 8.2.2(ii).

(ii) Let $M \in \mathcal{H}_n^\lambda$-mod, $N \in \mathcal{H}_{n+1}^\lambda$-mod be irreducible modules. Then, by Lemma 8.2.2(i), $\mathrm{Hom}_{\mathcal{H}_{n+1}^\lambda}(f_i^\lambda M, N) = \mathrm{Hom}_{\mathcal{H}_n^\lambda}(M, e_i^\lambda N)$. By (i), the latter is zero unless $M = \tilde{e}_i^\lambda N$, or equivalently $N = \tilde{f}_i M$ by Lemma 5.2.3, in which case the Hom-space is 1-dimensional. By Schur's lemma we now have that $\mathrm{hd}\, f_i^\lambda M \cong \tilde{f}_i^\lambda M$. Finally, note $f_i^\lambda M$ is self-dual by Lemma 8.2.2(ii) so everything else follows. □

8.3 Divided powers

Continue with $\lambda \in P_+$ and fix $i \in I$. We can generalize the definitions of $e_i, e_i^\lambda, f_i^\lambda$ to define functors denoted $e_i^{(r)}, (e_i^\lambda)^{(r)}, (f_i^\lambda)^{(r)}$. It will be the case that $e_i^{(1)} = e_i$, $(e_i^\lambda)^{(1)} = e_i$, $(f_i^\lambda)^{(1)} = f_i^\lambda$. For the definitions, we make use of the covering modules $L_m(i^r)$ from Section 4.4.

Let M be a module in $\mathrm{Rep}_I\, \mathcal{H}_n$. If $r > n$, we set $e_i^{(r)} M = 0$. Otherwise, let $\mathcal{H}_r'$ denote the subalgebra of $\mathcal{H}_n$ generated by

$$x_{n-r+1}, \ldots, x_n, s_{n-r+1}, \ldots, s_{n-1}.$$

We have a direct system

$$\mathrm{Hom}_{\mathcal{H}_r'}(L_1(i^r), M) \hookrightarrow \mathrm{Hom}_{\mathcal{H}_r'}(L_2(i^r), M) \hookrightarrow \ldots$$

induced by the inverse system (4.7). Now define

$$e_i^{(r)} M = \varinjlim \mathrm{Hom}_{\mathcal{H}_r'}(L_m(i^r), M). \tag{8.10}$$

As in the case $r = 1$, if M is an $\mathcal{H}_n^\lambda$-module, then $e_i^{(r)} M$ is an $\mathcal{H}_{n-r}^\lambda$-module, so, on restriction, $e_i^{(r)}$ gives a functor

$$(e_i^\lambda)^{(r)} : \mathcal{H}_n^\lambda\text{-mod} \to \mathcal{H}_{n-r}^\lambda\text{-mod}.$$

Lemma 4.4.3 yields another description of $(e_i^\lambda)^{(r)}$. Let M be a module in $\mathcal{H}_n^\lambda$-mod$[\gamma]$ for some fixed $\gamma \in \Gamma_n$. Then we have a functorial isomorphism

$$(e_i^\lambda)^{(r)} M \cong (\mathrm{res}^{\mathcal{H}_n^\lambda}_{\mathcal{H}_{n-r}^\lambda} M[\gamma - r\alpha_i])^{S'_r}, \tag{8.11}$$

where $-^{S'_r}$ stands for the invariants of the symmetric group

$$S'_r = \langle s_{n-r+1}, s_{n-r+2}, \dots, s_{n-1} \rangle, \tag{8.12}$$

which is a subgroup of the multiplicative group of $\mathcal{H}_n^\lambda$, commuting with $\mathcal{H}_{n-r}^\lambda$. Equivalently, instead of considering $\mathrm{res}^{\mathcal{H}_n^\lambda}_{\mathcal{H}_{n-r}^\lambda} M[\gamma - r\alpha_i]$, we could take the simultaneous generalized i-eigenspace of the last r polynomial generators $x_{n-r+1}, \dots, x_n$ on M, which is invariant with respect to $\mathcal{H}_{n-r}^\lambda$. We could also take coinvariants in the definition (8.11), in view of the following functorial isomorphism:

$$(e_i^\lambda)^{(r)} M \cong (\mathrm{res}^{\mathcal{H}_n^\lambda}_{\mathcal{H}_{n-r}^\lambda} M[\gamma - r\alpha_i])_{S'_r}. \tag{8.13}$$

This follows from the following lemma.

Lemma 8.3.1 *Let G be a finite group, A be an associative F-algebra, and $\mathfrak{C}$ be the full subcategory of $A \otimes FG$-*mod *which consists of the modules whose restrictions to FG are free. Then the functors $(-)^G$ and $(-)_G$ from $\mathfrak{C}$ to A-*mod *are isomorphic.*

Proof Let $e := \sum_{g \in G} g \in FG$ and Aug be the augmentation ideal in FG. Take $M \in \mathfrak{C}$. Then $M^G = eM$ and $M_G = M/\mathrm{Aug} \cdot M$. Consider the map

$$\varphi_M : M_G \to M^G, \quad m + \mathrm{Aug} \cdot M \mapsto em.$$

This is well-defined because $e \cdot \mathrm{Aug} = 0$. Since FG commutes with A, φ_M is an A-homomorphism, and, by assumption, it is an isomorphism of A-modules. Moreover, this isomorphism is clearly functorial in M. □

Finally, in view of Lemma 4.4.3, we could also take the largest submodule on which S'_r acts as the sign representation instead of taking S'_r-invariants in (8.11) (or the largest quotient module on which S'_r acts as the sign representation instead of taking S'_r-coinvariants in (8.13)).

To define $(f_i^\lambda)^{(r)}$, which as usual only makes sense in the cyclotomic case, let M be an $\mathcal{H}_n^\lambda$-module. We have an inverse system

$$M \boxtimes L_1(i^r) \twoheadleftarrow M \boxtimes L_2(i^r) \twoheadleftarrow \dots$$

of $\mathcal{H}_{n,r}$-modules induced by the maps from (4.7). Define

$$(f_i^\lambda)^{(r)} M = \varprojlim \mathrm{pr}^\lambda \mathrm{ind}_{n,r}^{n+r} M \boxtimes L_m(i^r). \tag{8.14}$$

As in the case $r = 1$, we verify that the limit stabilizes after finitely many steps and so $(f_i^\lambda)^{(r)}$ is a functor $\mathcal{H}_n^\lambda\text{-mod} \to \mathcal{H}_{n+r}^\lambda\text{-mod}$. As in the proof of Lemma 8.2.3 we show that $(f_i^\lambda)^{(r)}$ is right adjoint to $(e_i^\lambda)^{(r)}$. Now we use the uniqueness of adjoints to get another description of $(f_i^\lambda)^{(r)}$. Let M be a module in $\mathcal{H}_n^\lambda\text{-mod}[\gamma]$ for some fixed $\gamma \in \Gamma_n$, and let $\mathbf{1}$ and $\mathbf{sgn}$ stand for the trivial and sign $\mathcal{G}_r$-modules, respectively. Then

$$\begin{aligned}(f_i^\lambda)^{(r)} M &\cong (\mathrm{ind}_{\mathcal{H}_n^\lambda \otimes \mathcal{G}_r}^{\mathcal{H}_{n+r}^\lambda} M \boxtimes \mathbf{1})[\gamma + r\alpha_i] \\ &\cong (\mathrm{ind}_{\mathcal{H}_n^\lambda \otimes \mathcal{G}_r}^{\mathcal{H}_{n+r}^\lambda} M \boxtimes \mathbf{sgn})[\gamma + r\alpha_i],\end{aligned} \tag{8.15}$$

the isomorphisms being functorial.

We collect the main properties of $(e_i^\lambda)^{(r)}$ and $(f_i^\lambda)^{(r)}$ in the following theorem.

Theorem 8.3.2 *Let $\lambda \in P_+$, $i \in I$, $r \geq 1$, and M be an irreducible $\mathcal{H}_n^\lambda$-module.*

(i) *$(e_i^\lambda)^{(r)}$ and $(f_i^\lambda)^{(r)}$ are both left and right adjoint to each other; in particular, they are exact and send projectives to projectives.*

(ii) *There exist isomorphisms of functors*

$$e_i^r \cong (e_i^{(r)})^{\oplus r!}, \quad (e_i^\lambda)^r \cong ((e_i^\lambda)^{(r)})^{\oplus r!}, \quad (f_i^\lambda)^r \cong ((f_i^\lambda)^{(r)})^{\oplus r!}.$$

(iii) *$e_i^{(r)}$, $(e_i^\lambda)^{(r)}$, and $(f_i^\lambda)^{(r)}$ commute with duality.*

(iv) *$(e_i^\lambda)^{(r)} M$ is non-zero if and only if $(\tilde{e}_i^\lambda)^r M \neq 0$, in which case $(e_i^\lambda)^{(r)} M$ is a self-dual indecomposable module with irreducible socle and head isomorphic to $(\tilde{e}_i^\lambda)^r M$.*

(v) *$(f_i^\lambda)^{(r)} M$ is non-zero if and only if $(\tilde{f}_i^\lambda)^r M \neq 0$, in which case $(f_i^\lambda)^{(r)} M$ is a self-dual indecomposable module with irreducible socle and head isomorphic to $(\tilde{f}_i^\lambda)^r M$.*

Proof (i) To see that the functors are both right and left adjoint to each other, use Lemma 8.3.1, Corollary 7.7.8(ii) and the alternative descriptions of the functors obtained in (8.11), (8.15).

(ii) Lemmas 4.4.1 and 4.4.2 show that there is an isomorphism of functors $(e_i^\lambda)^r \cong ((e_i^\lambda)^{(r)})^{\oplus r!}$. So, by (i), both $((f_i^\lambda)^{(r)})^{\oplus r!}$ and f_i^r are adjoint to $(e_i^\lambda)^r$. Therefore they are isomorphic by uniqueness of adjoint functors.

(iii) is proved as Lemma 8.2.2(ii) but using Corollary 7.7.8(ii) instead of Corollary 7.7.5.

(iv) In view of (ii), $(e_i^\lambda)^{(r)} M \neq 0$ if and only if $(e_i^\lambda)^r M \neq 0$, which, in view of (5.7) and (5.11), is equivalent to $(\tilde{e}_i^\lambda)^r M \neq 0$. The self-duality of $(e_i^\lambda)^{(r)} M$

follows from (iii) and self-duality of M. The result on the socle is contained in Lemma 5.2.2, and the result on the head follows from self-duality.

(v) Arguement as in the proof of Theorem 8.2.5(ii). □

Since we have defined the exact functors above on module categories we get induced linear maps denoted by the same symbols at the level of Grothendieck groups, namely,

$$e_i^{(r)} : K(\mathrm{Rep}_I\,\mathcal{H}_n) \to K(\mathrm{Rep}_I\,\mathcal{H}_{n-r}),$$
$$(e_i^\lambda)^{(r)} : K(\mathcal{H}_n^\lambda\text{-mod}) \to K(\mathcal{H}_{n-r}^\lambda\text{-mod}),$$
$$(f_i^\lambda)^{(r)} : K(\mathcal{H}_n^\lambda\text{-mod}) \to K(\mathcal{H}_{n+r}^\lambda\text{-mod}).$$

Also, in view of Theorem 8.3.2(i), we get linear maps on Grothendieck groups of *projective modules*:

$$(e_i^\lambda)^{(r)} : K(\mathcal{H}_n^\lambda\text{-proj}) \to K(\mathcal{H}_{n-r}^\lambda\text{-proj}),$$
$$(f_i^\lambda)^{(r)} : K(\mathcal{H}_n^\lambda\text{-proj}) \to K(\mathcal{H}_{n+r}^\lambda\text{-proj}).$$

We record a corollary of Theorem 8.3.2(ii):

Lemma 8.3.3 *As operators on the corresponding Grothendieck groups,*

$$e_i^r = (r!)e_i^{(r)},\ (e_i^\lambda)^r = (r!)(e_i^\lambda)^{(r)},\ (f_i^\lambda)^r = (r!)f_i^{(r)}.$$

8.4 Functions φ_i^λ

Fix $\lambda \in P_+$ throughout this section. Let M be an irreducible $\mathcal{H}_n^\lambda$-module. Define

$$\varepsilon_i^\lambda(M) = \max\{m \geq 0 \mid (\tilde{e}_i^\lambda)^m M \neq 0\}, \tag{8.16}$$
$$\varphi_i^\lambda(M) = \max\{m \geq 0 \mid (\tilde{f}_i^\lambda)^m M \neq 0\}. \tag{8.17}$$

As $\tilde{e}_i^\lambda$ is simply the restriction of $\tilde{e}_i$, we have $\varepsilon_i^\lambda(M) = \varepsilon_i(M)$, see Remark 8.2.4 and equation (5.11). On the other hand, φ_i^λ depends crucially on λ. We will see shortly (Corollary 8.4.4) that $\varphi_i^\lambda(M) < \infty$ always, so that the definition makes sense.

Recall that we interpret $\mathcal{H}_0^\lambda$ as F. Let $\mathbf{1}_\lambda \cong F$ denote the trivial irreducible $\mathcal{H}_0^\lambda$-module.

Lemma 8.4.1 *For any $i \in I$ we have $\varepsilon_i^\lambda(\mathbf{1}_\lambda) = 0$ and $\varphi_i^\lambda(\mathbf{1}_\lambda) = \langle h_i, \lambda\rangle$.*

Proof The statement involving ε_i^λ is obvious. For φ_i^λ, note that $\tilde{f}_i^m \mathbf{1}_\lambda = L(i^m)$ and

$$\varepsilon_i^*(L(i^m)) = m, \qquad \varepsilon_j^*(L(i^m)) = 0$$

for every $j \neq i$. Hence by Corollary 7.4.1, $\mathrm{pr}^\lambda L(i^m) \neq 0$ if and only if $m \leq \langle h_i, \lambda \rangle$. This implies the required result. □

Lemma 8.4.2 *Let $i, j \in I$ with $i \neq j$ and M be an irreducible module in* $\mathrm{Rep}_I \mathcal{H}_n$. *Then $\varepsilon_j^*(\tilde{f}_i^m M) \leq \varepsilon_j^*(M)$ for every $m \geq 0$.*

Proof Follows from the Shuffle Lemma. □

Lemma 8.4.3 *Let $i, j \in I$ with $i \neq j$. Suppose that M is an irreducible $\mathcal{H}_n^\lambda$-module such that $\varphi_j^\lambda(M) > 0$. Then*

$$\varphi_i^\lambda(\tilde{f}_j M) - \varepsilon_i^\lambda(\tilde{f}_j M) \leq \varphi_i^\lambda(M) - \varepsilon_i^\lambda(M) - \langle h_i, \alpha_j \rangle.$$

Proof Set

$$\varepsilon := \varepsilon_i^\lambda(M), \ \ \varphi := \varphi_i^\lambda(M), \ \ k := -\langle h_i, \alpha_j \rangle.$$

By Lemma 6.3.3, there exist unique $r, s \geq 0$ with $r+s = k$ such that $\varepsilon_i(\tilde{f}_j M) = \varepsilon - r$. We need to show that $\varphi_i^\lambda(\tilde{f}_j M) \leq \varphi + s$, which follows if we can show that $\mathrm{pr}^\lambda \tilde{f}_i^m \tilde{f}_j M = 0$ for all $m > \varphi + s$. It suffices to prove that

$$\varepsilon_i^*(\tilde{f}_i^{m+s} \tilde{f}_j M) \geq \varepsilon_i^*(\tilde{f}_i^m M) \tag{8.18}$$

for all $m \geq 0$. Indeed, by the definition of φ, we have $\mathrm{pr}^\lambda \tilde{f}_i^m M = 0$ for any $m > \varphi$. In view of Corollary 7.4.1, this means that $\varepsilon_j^*(\tilde{f}_i^m M) > \langle h_j, \lambda \rangle$ for some $j \in I$. But by Lemma 8.4.2, such j can only equal i. Thus, $\varepsilon_i^*(\tilde{f}_i^m M) > \langle h_i, \lambda \rangle$ for all $m > \varphi$. So (8.18) implies that $\varepsilon_i^*(\tilde{f}_i^m \tilde{f}_j M) > \langle h_i, \lambda \rangle$ for all $m > \varphi + s$, hence by Corollary 7.4.1 once more, $\mathrm{pr}^\lambda \tilde{f}_i^m \tilde{f}_j M = 0$ as required.

To prove (8.18), note that $r \leq \varepsilon$, so $s + \varepsilon \geq k$. Hence, Lemma 6.3.3(ii) shows that there is a surjection

$$\mathrm{ind}_{n,m-r,r+s+1}^{n+m+s+1} M \boxtimes L(i^{m-r}) \boxtimes L(i^r, j, i^s) \twoheadrightarrow \tilde{f}_i^{m+s} \tilde{f}_j M.$$

By Lemma 6.2.1, $\mathrm{res}_{r,s+1}^{r+s+1} L(i^r, j, i^s) \cong L(i^r) \boxtimes L(j, i^s)$. Hence by Frobenius reciprocity, there is a surjection

$$\mathrm{ind}_{r,s+1}^{r+s+1} L(i^r) \boxtimes L(j, i^s) \twoheadrightarrow L(i^r, j, i^s).$$

Combining, we have proved existence of a surjection

$$\mathrm{ind}_{n,m,s+1}^{n+m+s+1} M \boxtimes L(i^m) \boxtimes L(j, i^s) \twoheadrightarrow \tilde{f}_i^{m+s} \tilde{f}_j M.$$

Hence by Frobenius reciprocity there is a non-zero map

$$(\mathrm{ind}_{n,m}^{n+m} M \boxtimes L(i^m)) \boxtimes L(j, i^s) \to \mathrm{res}_{n+m,s+1}^{n+m+s+1} \tilde{f}_i^{m+s} \tilde{f}_j M.$$

Since the left-hand module has irreducible head $\tilde{f}_i^m M \boxtimes L(j, i^s)$, we deduce that $\tilde{f}_i^{m+s} \tilde{f}_j M$ has a constituent isomorphic to $\tilde{f}_i^m M$ on restriction to the subalgebra $\mathcal{H}_{n+m} \subseteq \mathcal{H}_{n+m+s+1}$. This implies the claim. □

Corollary 8.4.4 *Let $\lambda \in P_+$ and M be an irreducible $\mathcal{H}_n^\lambda$-module with central character χ_γ for some $\gamma \in \Gamma_n$. Then*

$$\varphi_i^\lambda(M) - \varepsilon_i^\lambda(M) \leq \langle h_i, \lambda - \gamma \rangle.$$

Proof Proceed by induction on n, the case $n = 0$ being immediate by Lemma 8.4.1. For $n > 0$, we may write $M = \tilde{f}_j N$ for some irreducible $\mathcal{H}_{n-1}^\lambda$-module N with $\varphi_j^\lambda(N) > 0$. By induction,

$$\varphi_i^\lambda(N) - \varepsilon_i^\lambda(N) \leq \langle h_i, \lambda - \gamma + \alpha_j \rangle.$$

The conclusion follows from Lemma 8.4.3. □

8.5 Alternative descriptions of φ_i^λ

Now we wish to prove the analogue of Theorem 5.5.1 for the function φ_i^λ. This is considerably more difficult to do. Let M be an irreducible $\mathcal{H}_n^\lambda$-module. Recall that

$$f_i^\lambda M = \varprojlim \mathrm{pr}^\lambda \mathrm{ind}_{n,1}^{n+1} M \boxtimes L_m(i)$$

and that the inverse limit stabilizes after finitely many terms. Define $\tilde{\varphi}_i^\lambda(M)$ to be the stabilization point of the limit, that is the least $m \geq 0$ such that $f_i^\lambda M = \mathrm{pr}^\lambda \mathrm{ind}_{n,1}^{n+1} M \boxtimes L_m(i)$. Later it will turn out that $\tilde{\varphi}_i^\lambda = \varphi_i^\lambda$, see Corollary 8.5.7.

Lemma 8.5.1 *Let M be an irreducible $\mathcal{H}_n^\lambda$-module and $i \in I$. Then:*

(i) $[f_i^\lambda M] = \tilde{\varphi}_i^\lambda(M)[\tilde{f}_i^\lambda M] + \sum c_r [N_r]$ *where the N_r are irreducible $\mathcal{H}_{n+1}^\lambda$-modules with $\varepsilon_i^\lambda(N_r) < \varepsilon_i^\lambda(\tilde{f}_i^\lambda M)$.*

(ii) *The algebra $\mathrm{End}_{\mathcal{H}_{n+1}^\lambda}(f_i^\lambda M)$ is isomorphic to the algebra of truncated polynomials $k[x]/(x^{\tilde{\varphi}_i^\lambda(M)})$.*

Proof For any $m \geq 1$ denote

$$V_m := \mathrm{ind}_{n,1}^{n+1} M \boxtimes L_m(i).$$

(i) Since pr^λ is right exact, the natural surjection $L_m(i) \twoheadrightarrow L_{m-1}(i)$ and the natural embedding $L_m(i) \hookrightarrow L_{m+1}(i)$ of the "Jordan blocks" (see Section 4.4) induce a commutative diagram

$$\begin{array}{ccccccc} \mathrm{pr}^\lambda V_1 & \xrightarrow{\alpha_m} & \mathrm{pr}^\lambda V_m & \xrightarrow{\beta_m} & \mathrm{pr}^\lambda V_{m-1} & \to & 0 \\ \| & & \downarrow & & & & \\ \mathrm{pr}^\lambda V_1 & \xrightarrow{\alpha_{m+1}} & \mathrm{pr}^\lambda V_{m+1} & \xrightarrow{\beta_{m+1}} & \mathrm{pr}^\lambda V_m & \to & 0 \end{array}$$

where the rows are exact. Note if $\alpha_m = 0$ then $\alpha_{m+1} = 0$. It follows that if β_m is an isomorphism so is $\beta_{m'}$ for every $m' \geq m$. So by definition of $\tilde{\varphi}_i^\lambda(M)$, the maps $\beta_1, \beta_2, \dots, \beta_{\tilde{\varphi}_i^\lambda(M)}$ are not isomorphisms but all other $\beta_{m'}$, $m' > \tilde{\varphi}_i^\lambda(M)$, are isomorphisms.

Now to prove (i), we show by induction on $m = 0, 1, \dots, \tilde{\varphi}_i^\lambda(M)$ that

$$[f_i^\lambda M] = [\mathrm{pr}^\lambda V_m] = m[\tilde{f}_i^\lambda M] + \text{lower terms},$$

where the lower terms are irreducible $\mathcal{H}_{n+1}^\lambda$-modules N with $\varepsilon_i^\lambda(N) < \varepsilon_i^\lambda(\tilde{f}_i^\lambda M)$. This is vacuous if $m = 0$. For $m > 0$, β_m is not an isomorphism, so $\alpha_m \neq 0$. Hence, by Lemma 5.1.5, the image of α_m contains a copy of $\tilde{f}_i^\lambda M$ plus lower terms. Now the induction step is immediate.

(ii) Take $m = \tilde{\varphi}_i^\lambda(M)$. We easily show, using the explicit construction of $L_m(i)$ in Section 4.4, that there is an endomorphism

$$\theta : L_m(i) \to L_m(i)$$

of $\mathcal{H}_1$-modules, such that the image of θ^k is isomorphic to $L_{m-k}(i)$ for each $0 \leq k \leq m$. Functoriality yields algebra homomorphisms

$$\mathrm{End}_{\mathcal{H}_1}(L_m(i)) \hookrightarrow \mathrm{End}_{\mathcal{H}_{n,1}}(M \boxtimes L_m(i)) \hookrightarrow \mathrm{End}_{\mathcal{H}_{n+1}}(V_m).$$

So θ induces an $\mathcal{H}_{n+1}$-endomorphism $\tilde{\theta}$ of V_m, such that the image of $\tilde{\theta}^k$ is isomorphic to V_{m-k} for $0 \leq k \leq m$. Now apply the right exact functor pr^λ to get an $\mathcal{H}_{n+1}^\lambda$-endomorphism

$$\hat{\theta} : f_i^\lambda M \to f_i^\lambda M$$

induced by $\tilde{\theta}$. Note $\hat{\theta}^m = 0$ and $\hat{\theta}^{m-1} \neq 0$ because its image coincides with the image of the non-zero map α_m in the proof of (i). Hence, $1, \hat{\theta}, \dots, \hat{\theta}^{m-1}$ are linearly independent, endomorphisms of $f_i^\lambda M$. Now the proof of (ii) is completed in the same way as in the proof of Theorem 5.5.1(iii). □

Corollary 8.5.2 *Let M, N be irreducible $\mathcal{H}_n^\lambda$-modules with $M \not\cong N$. Then for every $i \in I$ we have* $\mathrm{Hom}_{\mathcal{H}_{n+1}^\lambda}(f_i^\lambda M, f_i^\lambda N) = 0$.

Proof Repeat the argument in the proof of Corollary 5.5.2, but using $\tilde{\varphi}_i^\lambda$ and Lemma 8.5.1(i) in place of ε_i and Theorem 5.5.1(i). □

We now start proving that $\tilde{\varphi}_i^\lambda = \varphi_i^\lambda$. Note right away from the definitions that for any irreducible $\mathcal{H}_n^\lambda$-module M we have $\tilde{\varphi}_i^\lambda(M) = 0$ if and only if $\varphi_i^\lambda(M) = 0$.

Lemma 8.5.3 *If M is an irreducible $\mathcal{H}_n^\lambda$-module then*

$$\sum_{i=0}^{\ell}(\tilde{\varphi}_i^\lambda(M) - \varepsilon_i^\lambda(M)) = \langle c, \lambda\rangle.$$

Proof Using Lemma 8.5.1(ii), decompositions (8.7), Frobenius reciprocity, Theorem 7.6.2, Schur's Lemma, and Theorem 5.5.1(iii), we have that

$$\begin{aligned}
\sum_{i=0}^{\ell}\tilde{\varphi}_i^\lambda(M) &= \sum_{i=0}^{\ell}\dim \mathrm{End}_{\mathcal{H}_{n+1}^\lambda}(f_i^\lambda M)\\
&= \dim \mathrm{End}_{\mathcal{H}_{n+1}^\lambda}(\mathrm{ind}_{\mathcal{H}_n^\lambda}^{\mathcal{H}_{n+1}^\lambda} M)\\
&= \dim \mathrm{Hom}_{\mathcal{H}_n^\lambda}(M, \mathrm{res}_{\mathcal{H}_n^\lambda}^{\mathcal{H}_{n+1}^\lambda}\mathrm{ind}_{\mathcal{H}_n^\lambda}^{\mathcal{H}_{n+1}^\lambda} M)\\
&= \dim \mathrm{End}_{\mathcal{H}_n^\lambda}(M)^{\oplus\langle c,\lambda\rangle}\\
&\quad + \dim \mathrm{Hom}_{\mathcal{H}_n^\lambda}(M, \mathrm{ind}_{\mathcal{H}_{n-1}^\lambda}^{\mathcal{H}_n^\lambda}\mathrm{res}_{\mathcal{H}_{n-1}^\lambda}^{\mathcal{H}_n^\lambda} M)\\
&= \langle c, \lambda\rangle + \dim \mathrm{End}_{\mathcal{H}_{n-1}^\lambda}(\mathrm{res}_{\mathcal{H}_{n-1}^\lambda}^{\mathcal{H}_n^\lambda} M)\\
&= \langle c, \lambda\rangle + \sum_{i=0}^{\ell}\varepsilon_i^\lambda(M),
\end{aligned}$$

and the conclusion follows. □

Lemma 8.5.4 *Let M be an irreducible $\mathcal{H}_n^\lambda$-module and $i \in I$. Then:*

(i) $[e_i^\lambda f_i^\lambda M : M] = \varepsilon_i^\lambda(\tilde{f}_i^\lambda M)\tilde{\varphi}_i^\lambda(M)$ *and*
$[f_i^\lambda e_i^\lambda M : M] = \varepsilon_i^\lambda(M)\tilde{\varphi}_i^\lambda(\tilde{e}_i^\lambda M)$;

(ii) $\mathrm{soc}\, e_i^\lambda f_i^\lambda M \cong M^{\oplus\tilde{\varphi}_i^\lambda(M)}$ *and* $\mathrm{soc}\, f_i^\lambda e_i^\lambda M \cong M^{\oplus\varepsilon_i^\lambda(M)}$.

Proof (i) follows from Theorem 5.5.1(i) and Lemma 8.5.1(i).

(ii) Let N be an irreducible $\mathcal{H}_n^\lambda$-module. Then by Lemma 8.2.2(i),

$$\mathrm{Hom}_{\mathcal{H}_n^\lambda}(N, e_i^\lambda f_i^\lambda M) \cong \mathrm{Hom}_{\mathcal{H}_{n+1}^\lambda}(f_i^\lambda N, f_i^\lambda M),$$

and the first part of (ii) follows from Corollary 8.5.2 and Lemma 8.5.1(ii). The second part is similar. □

The proof of the next lemma is based on [V$_2$, Lemma 6.1].

Lemma 8.5.5 *Let M be an irreducible $\mathcal{H}_n^\lambda$-module and $i \in I$. There is an $\mathcal{H}_n^\lambda$-module homomorphism $\psi : f_i^\lambda e_i^\lambda M \to e_i^\lambda f_i^\lambda M$ such that the following composition is surjective:*

$$f_i^\lambda e_i^\lambda M \xrightarrow{\psi} e_i^\lambda f_i^\lambda M \xrightarrow{\text{can}} e_i^\lambda f_i^\lambda M / \operatorname{soc} e_i^\lambda f_i^\lambda M,$$

Proof Let $k = \tilde{\varphi}_i^\lambda(M)$ and

$$\pi : \operatorname{ind}_{n,1}^{n+1} M \boxtimes L_k(i) \twoheadrightarrow \operatorname{pr}^\lambda \operatorname{ind}_{n,1}^{n+1} M \boxtimes L_k(i) = f_i^\lambda M$$

be the quotient map. Let $\mathcal{H}_1'$ denote the subalgebra of $\mathcal{H}_n$ generated by x_n. Recall from Section 4.4 that viewed as an $\mathcal{H}_1'$-module, we have that $L_k(i) \cong \mathcal{H}_1'/((x_n - i)^k)$. In particular, $L_k(i)$ is a cyclic module generated by the image $\tilde{1}$ of $1 \in \mathcal{H}_1'$.

We first observe that for any $m \geq \varepsilon_i^\lambda(M) + k$, the element $(x_n - i)^m$ annihilates the vector

$$s_n \otimes (u \otimes v) \in \operatorname{ind}_{n,1}^{n+1} M \boxtimes L_k(i)$$

for any $u \in M$, $v \in L_k(i)$. This is obtained using the relations (3.10) in $\mathcal{H}_{n+1}$: We ultimately appeal to the facts that $(x_n - i)^{\varepsilon_i^\lambda(M)}$ annihilates u (see Theorem 5.5.1(ii)) and $(x_{n+1} - i)^k$ annihilates v.

Therefore, for any $m \geq \varepsilon_i^\lambda(M) + k$, the following equality holds in $f_i^\lambda M$:

$$(x_n - i)^m \pi\big(s_n \otimes (u \otimes v)\big) = 0. \tag{8.19}$$

Next, it is not difficult to check that there exists a unique $\mathcal{H}_{n-1,1}$-homomorphism

$$(e_i M) \boxtimes \mathcal{H}_1' \to \operatorname{res}_{n-1,1}^n e_i f_i^\lambda M, \quad u \otimes 1 \mapsto \pi\big(s_n \otimes (u \otimes \tilde{1})\big),$$

for each $u \in e_i M \subseteq M$. It follows from (8.19) that this homomorphism factors to induce a well-defined $\mathcal{H}_{n-1,1}$-module homomorphism $(e_i M) \boxtimes L_m(i) \to \operatorname{res}_{n-1,1}^n e_i f_i^\lambda M$. We then get from Frobenius reciprocity an induced map

$$\psi_m : \operatorname{ind}_{n-1,1}^n (e_i M) \boxtimes L_m(i) \to e_i f_i^\lambda M \tag{8.20}$$

for each $m \geq \varepsilon_i^\lambda(M) + k$. Each ψ_m factors through the quotient

$$\operatorname{pr}^\lambda \operatorname{ind}_{n-1,1}^n (e_i M) \boxtimes L_m(i),$$

so we get an induced map

$$\psi : f_i^\lambda e_i^\lambda M = \varprojlim \operatorname{pr}^\lambda \operatorname{ind}_{n-1,1}^n (e_i M) \boxtimes L_m(i) \to e_i f_i^\lambda M = e_i^\lambda f_i^\lambda M.$$

It remains to show that the composite of ψ with the canonical epimorphism from $e_i^\lambda f_i^\lambda M$ to $e_i^\lambda f_i^\lambda M/\operatorname{soc} e_i^\lambda f_i^\lambda M$ is surjective.

By Mackey Theorem there exists an exact sequence

$$0 \to M \boxtimes L_k(i) \to \operatorname{res}^{n+1}_{n,1}\big(\operatorname{ind}^{n+1}_{n,1} M \boxtimes L_k(i)\big)$$
$$\to \operatorname{ind}^{n,1}_{n-1,1,1} {}^{s_n}\big((\operatorname{res}^{n}_{n-1,1} M) \boxtimes L_k(i)\big) \to 0.$$

In other words, there is an $\mathcal{H}_{n,1}$-isomorphism from

$$\operatorname{ind}^{n,1}_{n-1,1,1} {}^{s_n}\big((\operatorname{res}^{n}_{n-1,1} M) \boxtimes L_k(i)\big)$$

to

$$\operatorname{res}^{n+1}_{n,1}\big(\operatorname{ind}^{n+1}_{n,1} M \boxtimes L_k(i)\big)/\big(M \boxtimes L_k(i)\big),$$

with

$$h \otimes (u \otimes v) \mapsto h s_n \otimes u \otimes v + M \boxtimes L_k(i)$$

for $h \in \mathcal{H}_n, u \in M, v \in L_k(i)$, where $M \boxtimes L_k(i)$ is embedded into $\operatorname{res}^{n+1}_{n,1}(\operatorname{ind}^{n+1}_{n,1} M \boxtimes L_k(i))$ as $1 \otimes M \otimes L_k(i)$. Since $\dim L_k(i) = k$, Lemma 8.5.4 (ii) gives

$$\operatorname{res}^{n,1}_{n} M \boxtimes L_k(i) \cong M^{\oplus k} \cong \operatorname{soc} e_i^\lambda f_i^\lambda M.$$

So, applying the exact functor e_i to the isomorphism above, we get an isomorphism

$$\operatorname{ind}^{n}_{n-1,1} e_i M \boxtimes L_k(i) \xrightarrow{\sim} e_i\big(\operatorname{ind}^{n+1}_{n,1} M \boxtimes L_k(i)\big)/\operatorname{soc} e_i^\lambda f_i^\lambda M,$$
$$h \otimes u \otimes v \mapsto h s_n \otimes u \otimes v + \operatorname{soc} e_i^\lambda f_i^\lambda M.$$

(Note that '$\operatorname{soc} e_i^\lambda f_i^\lambda M$' is *not* the socle of $e_i\big(\operatorname{ind}^{n+1}_{n,1} M \boxtimes L_k(i)\big)$, so '$\operatorname{soc} e_i^\lambda f_i^\lambda M$' just means *some* submodule isomorphic to $\operatorname{soc} e_i^\lambda f_i^\lambda M$.) It follows that there is a surjection

$$\theta : \operatorname{ind}^{n}_{n-1,1} e_i M \boxtimes L_k(i) \twoheadrightarrow e_i^\lambda f_i^\lambda M/\operatorname{soc} e_i^\lambda f_i^\lambda M,$$

such that the diagram

$$\begin{array}{ccc}
\operatorname{ind}^{n}_{n-1,1}(e_i M) \boxtimes L_m(i) & \xrightarrow{\psi_m} & e_i^\lambda f_i^\lambda M \\
\downarrow & & \downarrow{\scriptstyle \mathrm{can}} \\
\operatorname{ind}^{n}_{n-1,1}(e_i M) \boxtimes L_k(i) & \xrightarrow{\theta} & e_i f_i^\lambda M/\operatorname{soc} e_i^\lambda f_i^\lambda M
\end{array}$$

commutes for all $m \geq \varepsilon_i^\lambda(M) + k$, where ψ_m is the map from (8.20) and the left-hand arrow is the natural surjection. Now surjectivity of θ implies surjectivity of $\mathrm{can} \circ \psi_m$ and of $\mathrm{can} \circ \psi$. □

Lemma 8.5.6 *Let M be an irreducible $\mathcal{H}_n^\lambda$-module with $\varepsilon_i^\lambda(M) > 0$. Then*

$$\tilde{\varphi}_i^\lambda(\tilde{e}_i^\lambda M) = \tilde{\varphi}_i^\lambda(M) + 1.$$

Proof Let us first show that

$$\tilde{\varphi}_i^\lambda(\tilde{e}_i^\lambda M) \geq \tilde{\varphi}_i^\lambda(M) + 1. \tag{8.21}$$

Recall that $\varphi_i^\lambda(M) = 0$ if and only if $\tilde{\varphi}_i^\lambda(M) = 0$. Suppose first that $\varphi_i^\lambda(M) = 0$. Then $\varphi_i^\lambda(\tilde{e}_i^\lambda M) \neq 0$ in view of Lemma 5.2.3. Now, $\tilde{\varphi}_i^\lambda(M) = 0$ and $\tilde{\varphi}_i^\lambda(\tilde{e}_i^\lambda M) \neq 0$, so the conclusion certainly holds in this case. Next, assume that $\varphi_i^\lambda(M) > 0$, hence $\tilde{\varphi}_i^\lambda(M) > 0$. Note by Lemma 8.5.4,

$$\begin{aligned}[e_i^\lambda f_i^\lambda M/\operatorname{soc} e_i^\lambda f_i^\lambda M : M] &= \varepsilon_i^\lambda(\tilde{f}_i^\lambda M)\tilde{\varphi}_i^\lambda(M) - \tilde{\varphi}_i^\lambda(M) \\ &= \varepsilon_i^\lambda(M)\tilde{\varphi}_i^\lambda(M) \neq 0.\end{aligned}$$

In particular, the map ψ in Lemma 8.5.5 is non-zero. Now Lemma 8.5.5 implies that the multiplicity of M as a composition factor of $\operatorname{im}\psi$ is *strictly* greater than $\varepsilon_i^\lambda(M)\tilde{\varphi}_i^\lambda(M)$, since at least one composition factor of $\operatorname{soc}\operatorname{im}\psi \subseteq \operatorname{soc} e_i^\lambda f_i^\lambda M$ must be sent to zero on composing with the second map can. Using another part of Lemma 8.5.4, this shows that $\varepsilon_i^\lambda(M)\tilde{\varphi}_i^\lambda(\tilde{e}_i^\lambda M) > \varepsilon_i^\lambda(M)\tilde{\varphi}_i^\lambda(M)$ and (8.21) follows.

Now using (8.21) and Lemma 8.5.4, we see that in the Grothendieck group,

$$[e_i^\lambda f_i^\lambda M - f_i^\lambda e_i^\lambda M : M] \leq (\tilde{\varphi}_i^\lambda(M) - \varepsilon_i^\lambda(M)),$$

with equality if and only if equality holds in (8.21). By central character considerations, for $i \neq j$, we have $[e_i^\lambda f_j^\lambda M : M] = [f_j^\lambda e_i^\lambda M : M] = 0$. So using (8.7) we deduce that

$$[\operatorname{res}^{\mathcal{H}_{n+1}^\lambda}_{\mathcal{H}_n^\lambda} \operatorname{ind}^{\mathcal{H}_{n+1}^\lambda}_{\mathcal{H}_n^\lambda} M - \operatorname{ind}^{\mathcal{H}_n^\lambda}_{\mathcal{H}_{n-1}^\lambda} \operatorname{res}^{\mathcal{H}_n^\lambda}_{\mathcal{H}_{n-1}^\lambda} M : M] \leq \sum_{i=0}^{\ell} (\tilde{\varphi}_i^\lambda(M) - \varepsilon_i^\lambda(M))$$

with equality if and only if equality holds in (8.21) for all $i \in I$. Now Lemma 8.5.3 shows that the right-hand side equals $\langle c, \lambda \rangle$, which does indeed equal the left-hand side thanks to Theorem 7.6.2. □

Corollary 8.5.7 *For any irreducible $\mathcal{H}_n^\lambda$-module M we have*

$$\varphi_i^\lambda(M) = \tilde{\varphi}_i^\lambda(M).$$

Proof We proceed by induction on $\varphi_i^\lambda(M)$, the conclusion being known already in the case $\varphi_i^\lambda(M) = 0$. For the induction step, take an irreducible $\mathcal{H}_n^\lambda$-module N with $\varphi_i^\lambda(N) > 0$, so $N = \tilde{e}_i^\lambda M$, where $M = \tilde{f}_i^\lambda N$ is an irreducible

$\mathcal{H}_{n+1}^\lambda$-module with $\varepsilon_i^\lambda(M) > 0$, $\varphi_i^\lambda(M) < \varphi_i^\lambda(N)$. Then by Lemma 8.5.6 and the induction hypothesis,

$$\tilde{\varphi}_i^\lambda(N) = \tilde{\varphi}_i^\lambda(\tilde{e}_i^\lambda M) = \tilde{\varphi}_i^\lambda(M) + 1 = \varphi_i^\lambda(M) + 1 = \varphi_i^\lambda(\tilde{e}_i^\lambda M) = \varphi_i^\lambda(N).$$

This completes the induction step. □

As a first consequence, we can improve Corollary 8.4.4:

Lemma 8.5.8 *Let M be an irreducible $\mathcal{H}_n^\lambda$-module of central character χ_γ for $\gamma \in \Gamma_n$. Then*

$$\varphi_i^\lambda(M) - \varepsilon_i^\lambda(M) = \langle h_i, \lambda - \gamma \rangle.$$

Proof In view of Corollary 8.4.4, it suffices to show that

$$\sum_{i=0}^{\ell} (\varphi_i^\lambda(M) - \varepsilon_i^\lambda(M)) = \langle c, \lambda \rangle.$$

But this is immediate from Lemma 8.5.3 and Corollary 8.5.7. □

We are finally ready to assemble the results of the section to obtain the full analogue of Theorem 5.5.1 for φ_i^λ:

Theorem 8.5.9 *Let $i \in I$ and M be an irreducible $\mathcal{H}_n^\lambda$-module. Then:*

(i) $[f_i^\lambda M] = \varphi_i^\lambda(M)[\tilde{f}_i^\lambda M] + \sum c_r [N_r]$ *where the N_r are irreducibles with* $\varphi_i^\lambda(N_r) < \varphi_i^\lambda(\tilde{f}_i^\lambda M) = \varphi_i^\lambda(M) - 1$;

(ii) $\varphi_i^\lambda(M)$ *is the least $m \geq 0$ such that* $f_i^\lambda M = \mathrm{pr}^\lambda \mathrm{ind}_{n,1}^{n+1} M \boxtimes L_m(i)$;

(iii) *the algebra* $\mathrm{End}_{\mathcal{H}_{n+1}^\lambda}(f_i^\lambda M)$ *is isomorphic to the algebra of truncated polynomials* $k[x]/(x^{\varphi_i^\lambda(M)})$. *In particular,*

$$\dim \mathrm{End}_{\mathcal{H}_{n+1}^\lambda}(f_i^\lambda M) = \varphi_i^\lambda(M).$$

Proof (i) Since $\varphi_i^\lambda(M) = \tilde{\varphi}_i^\lambda(M)$ by Corollary 8.5.7, we know by Lemma 8.5.1(i) that

$$[f_i^\lambda M] = \varphi_i^\lambda(M)[\tilde{f}_i^\lambda M] + \sum c_r [N_r]$$

where the N_r are irreducibles with $\varepsilon_i^\lambda(N_r) < \varepsilon_i^\lambda(\tilde{f}_i^\lambda M)$. Suppose that $M \in \mathrm{Rep}_\gamma \mathcal{H}_n^\lambda$, for $\gamma \in \Gamma_n$. Then $\tilde{f}_i^\lambda M$ and each N_r have central character $\chi_{\gamma+\alpha_i}$, since they are all composition factors of $f_i^\lambda M$. So by Lemma 8.5.8,

$$\varphi_i^\lambda(N_r) = \langle h_i, \lambda - \gamma - \alpha_i \rangle + \varepsilon_i^\lambda(N_r),$$
$$\varphi_i^\lambda(\tilde{f}_i^\lambda M) = \langle h_i, \lambda - \gamma - \alpha_i \rangle + \varepsilon_i^\lambda(\tilde{f}_i^\lambda M).$$

It follows that $\varphi_i^\lambda(N_r) < \varphi_i^\lambda(\tilde{f}_i^\lambda M)$:

(ii) This is just the definition of $\tilde{\varphi}_i^\lambda(M)$ combined with Corollary 8.5.7.

(iii) This follows from Lemma 8.5.1(ii) and Corollary 8.5.7. □

Now we can generalize some results from e_i^λ and f_i^λ to $(e_i^\lambda)^{(r)}$ and $(f_i^\lambda)^{(r)}$, respectively.

Proposition 8.5.10 *Let $i \in I$, M, N be irreducible $\mathcal{H}_n^\lambda$-modules with $M \ncong N$, $\varepsilon = \varepsilon_i^\lambda(M)$, and $\varphi = \varphi_i^\lambda(M)$. Then:*

(i) $[(e_i^\lambda)^{(r)}M] = \binom{\varepsilon}{r}[(\tilde{e}_i^\lambda)^r M] + \sum c_r[N_r]$, *where the N_r are irreducibles with $\varepsilon_i^\lambda(N_r) < \varepsilon_i^\lambda((\tilde{e}_i^\lambda)^r M) = \varepsilon - r$;*

(i′) $[(f_i^\lambda)^{(r)}M] = \binom{\varphi}{r}[(\tilde{f}_i^\lambda)^r M] + \sum c_r[N_r]$, *where the N_r are irreducibles with $\varphi_i^\lambda(N_r) < \varphi_i^\lambda((\tilde{f}_i^\lambda)^r M) = \varphi - r$;*

(ii) $\mathrm{Hom}_{\mathcal{H}_{n-r}^\lambda}((e_i^\lambda)^{(r)}M, (e_i^\lambda)^{(r)}N) = 0$;

(ii′) $\mathrm{Hom}_{\mathcal{H}_{n+r}^\lambda}((f_i^\lambda)^{(r)}M, (f_i^\lambda)^{(r)}N) = 0$.

Proof Parts (i) and (i′) follow by applying Theorems 5.5.1, 8.5.9 r times and using Lemma 8.3.3. Parts (ii) and (ii′) follow from (i) and (i′), see the proofs of Corollaries 5.5.2 and 8.5.2. □

8.6 More on endomorphism algebras

The results of this section will only be referred to in Section 11.2.

We want to study the endomorphism algebras $\mathrm{End}_{H_{n-r}^\lambda}((e_i^\lambda)^{(r)}M)$ and $\mathrm{End}_{H_{n+r}^\lambda}((f_i^\lambda)^{(r)}M)$ for an irreducible $\mathcal{H}_n^\lambda$-module M and arbitrary r. For $r = 1$ the result is Theorems 5.5.1(iii) and 8.5.9(iii).

First, we slightly refine Theorem 5.5.1(iii).

Lemma 8.6.1 *Let $i \in I$, M be an irreducible module in $\mathrm{Rep}_I \mathcal{H}_n$, and $\varepsilon = \varepsilon_i(M)$. Let θ be the endomorphism of e_iM given by multiplying with $(x_n - i)$. Then $\mathrm{id} = \theta^0, \theta, \ldots, \theta^{\varepsilon-1}$ is a basis of $\mathrm{End}_{\mathcal{H}_{n-1}}(e_iM)$, $\theta^\varepsilon = 0$, and $\mathrm{im}\ \theta^{\varepsilon-1} = \mathrm{soc}\, e_iM \cong \tilde{e}_iM$.*

Proof The only thing not proved in the proof of Theorem 5.5.1(iii) is the fact that the image of $\theta^{\varepsilon-1}$ is the socle of e_iM, which we know is isomorphic to $\tilde{e}_iM$. Assume the image is strictly bigger. As the head of e_iM is also $\tilde{e}_iM$, we see that the module $\mathrm{im}\ \theta^{\varepsilon-1}$ is not irreducible and has irreducible head and socle both isomorphic to $\tilde{e}_iM$. Therefore there exists a non-trivial nilpotent endomorphism ψ of $\mathrm{im}\ \theta^{\varepsilon-1}$ with $\mathrm{im}\,\psi = \mathrm{soc}\,\mathrm{im}\ \theta^{\varepsilon-1}$. It follows that $\psi \circ \theta^{\varepsilon-1}$

is an element of $\mathrm{End}_{\mathcal{H}_{n-1}}(e_iM)$ such that $\theta^0, \theta^1, \dots, \theta^{\varepsilon-1}, \psi\circ\theta^{\varepsilon-1}$ are linearly independent. This contradiction completes the proof of the lemma. □

Proposition 8.6.2 *Let $i \in I$, $r \geq 1$, M be an irreducible $\mathcal{H}_n^\lambda$-module, and $\varepsilon = \varepsilon_i^\lambda(M)$. Then*

$$\dim \mathrm{End}_{H_{n-r}^\lambda}((e_i^\lambda)^{(r)}M) = \binom{\varepsilon}{r}.$$

Proof We can work with the affine Hecke algebra $\mathcal{H}_n$ instead of $\mathcal{H}_n^\lambda$. If $r > \varepsilon$, the equality obviously holds, so let $r \leq \varepsilon$. As $e_i^{(r)}M$ is indecomposable with irreducible head $\tilde{e}_i^rM$, and the multiplicity of $\tilde{e}_i^rM$ in $e_i^{(r)}M$ is $\binom{\varepsilon}{r}$, it follows that

$$\dim \mathrm{End}_{H_{n-r}}(e_i^{(r)}M) \leq \binom{\varepsilon}{r}.$$

For the converse inequality, recall the subgroup S_r' from (8.12) and consider the set of elements:

$$Y = \{y(w; a_{n-r+1}, \dots, a_n) := w(x_{n-r+1} - i)^{a_{n-r+1}} \dots (x_n - i)^{a_n}\},$$

where w runs over all elements of S_r', and the a_ms run over all integers satisfying $0 \leq a_{n-k} < \varepsilon - k$ for $0 \leq k < r$. The elements of Y commute with $\mathcal{H}_{n-r}$, so multiplying e_i^rM with $y \in Y$ gives an endomorphisms f_y of e_i^rM. Now, $e_i^rM = (e_i^{(r)}M)^{\oplus r!}$, so

$$\dim \mathrm{End}_{\mathcal{H}_{n-r}}(e_i^rM) = (r!)^2 \dim \mathrm{End}_{H_{n-r}}(e_i^{(r)}M).$$

As $|Y| = r!\binom{\varepsilon}{r}$, it suffices to prove that $\{f_y \mid y \in Y\}$ are linearly independent. Assume there is a non-trivial linear combination

$$\sum_{y\in Y} c_y f_y = 0 \tag{8.22}$$

in $\mathrm{End}_{\mathcal{H}_{n-r}}(e_i^rM)$. Let $X = \{y \in Y \mid c_y \neq 0\}$. Let X_n be the set of the elements $y = y(w; a_{n-r+1}, \dots a_n) \in X$, which have minimal possible value of a_n. Say, this value is b_n. Now, (8.22) implies

$$\sum_{y\in X} c_y f_y (x_n - i)^{\varepsilon-1-b_n} M = 0.$$

By Lemma 8.6.1, if $y \in X \setminus X_n$ then $f_y(x_n - i)^{\varepsilon-1-b_n}M = 0$, and if $y = y(w; a_{n-r+1}, \dots, a_{n-1}, b_n) \in X_n$ then

$$\begin{aligned} &f_y(x_n - i)^{\varepsilon-1-b_n}M \\ &= w(x_{n-r+1} - i)^{a_{n-r+1}} \dots (x_{n-1} - i)^{a_{n-1}}(x_n - i)^{\varepsilon-1}M \\ &= w(x_{n-r+1} - i)^{a_{n-r+1}} \dots (x_{n-1} - i)^{a_{n-1}}\tilde{e}_iM, \end{aligned}$$

where $\tilde{e}_i M = \operatorname{soc} e_i M$. Thus, we have

$$\sum_{y \in X_n} c_y w(x_{n-r+1} - i)^{a_{n-r+1}} \dots (x_{n-1} - i)^{a_{n-1}} \tilde{e}_i M = 0,$$

where the summation is over all $y = y(w; a_{n-r+1}, \dots, b_n) \in X_n$. Next, let X_{n-1} be the set of the elements $y(w; a_{n-r+1}, \dots, a_{n-1}, b_n) \in X_n$ which have minimal possible value of a_{n-1}. Say, this value is b_{n-1}. Then, arguing as above, we get

$$\sum_{y \in X_{n-1}} c_y w(x_{n-r+1} - i)^{a_{n-r+1}} \dots (x_{n-2} - i)^{a_{n-2}} \tilde{e}_i^2 M = 0,$$

where the summation is over all $y = y(w; a_{n-r+1}, \dots, b_{n-1}, b_n) \in X_{n-1}$, and $\tilde{e}_i^2 M$ is considered as a submodule of $e_i^2 M$ by considering it as the socle of $e_i \tilde{e}_i M$, where $\tilde{e}_i M$ is the socle of $e_i M$. Continuing in this manner $r-2$ more times, we get a non-empty subset X_{n-r+1} of Y such that $c_y \neq 0$ for any $y \in X_{n-r+1}$, and such that

$$\sum_{y \in X_{n-r+1}} c_y w \tilde{e}_i^r M = 0, \tag{8.23}$$

where the summation is over all $y = y(w; b_{n-r+1}, \dots, b_n) \in X_{n-r+1}$, and $\tilde{e}_i^r M$ is a submodule of $e_i^r M$.

By Theorem 8.3.2(ii),(iv), $\operatorname{soc} e_i^r M = \operatorname{soc} (e_i^{(r)} M)^{\oplus r!} = (\tilde{e}_i^r m)^{\oplus r!}$. However, $\operatorname{soc} \Delta_{i^r} M \cong (\tilde{e}_i^r M) \boxtimes L(i^r)$, and the dimension of the Kato module $L(i^r)$ is $r!$, so any irreducible submodule $\tilde{e}_i^r M$ of $e_i^r M$ is an $\mathcal{H}_{n-r}$-submodule of the $\mathcal{H}_{n-r,r}$-module $(\tilde{e}_i^r M) \boxtimes L(i^r) = \operatorname{soc} \Delta_{i^r} M$. As $L(i^r)$ over S_r' is just the regular module, (8.23) implies that $c_y = 0$ for all $y \in X_{n-r+1}$, which is a contradiction. □

Remark 8.6.3 Recently, Chuang and Rouquier [CR] have pushed this result a little further and actually described $\operatorname{End}_{\mathcal{H}_{n-r}}(e_i^{(r)} M)$ as an algebra. Namely,

$$\operatorname{End}_{\mathcal{H}_{n-r}}(e_i^{(r)} M) \cong \mathcal{Z}_r / (\mathcal{Z}_r \cap \mathcal{P}_\varepsilon \mathcal{Z}_\varepsilon^+), \tag{8.24}$$

where $\mathcal{P}_r$ (and $\mathcal{Z}_r$) are embedded into $\mathcal{P}_\varepsilon$ in the last r variables, and we use notation of Theorem 1.0.2. This result can be recovered by a more careful analysis of the proof above as follows. Note that the elements of the subalgebra $\mathcal{H}_r' \subset \mathcal{H}_n$ act naturally on the restriction $\operatorname{res}^{\mathcal{H}_n}_{\mathcal{H}_{n-r}} M$ and hence on $e_i^r M$ as endomorphisms. In fact, all $\mathcal{H}_{n-r}$-endomorphisms of $e_i^r M$ come from $\mathcal{H}_r'$, since the endomorphisms from the spanning set Y certainly do. Therefore $\operatorname{End}_{\mathcal{H}_{n-r}}(e_i^r M) \cong \mathcal{H}_r / J_r$, where J_r is the annihilator in $\mathcal{H}_r'$ of $e_i^r M$.

We also know that $\operatorname{End}_{\mathcal{H}_{n-r}}(e_i^r M)$ is the $r! \times r!$ matrix algebra over $\operatorname{End}_{\mathcal{H}_{n-r}}(e_i^{(r)} M)$, and $\dim \operatorname{End}_{\mathcal{H}_{n-r}}(e_i^{(r)} M) = \binom{\varepsilon}{r}$. So, if we can exhibit a central subalgebra A in $\mathrm{End}_{\mathcal{H}_{n-r}}(e_i^r M)$ of dimension $\binom{\varepsilon}{r}$, it will follow that $A \cong \mathrm{End}_{\mathcal{H}_{n-r}}(e_i^{(r)} M)$. In fact, it is even sufficient to exhibit $\binom{\varepsilon}{r}$ linearly independent central elements of $\mathrm{End}_{\mathcal{H}_{n-r}}(e_i^r M)$ – they will automatically span a central subalgebra of the right dimension. Such elements are easy to come up with. Consider the monomial symmetric polynomials in $(x_{n-r+1} - a), \dots, (x_n - a)$ of degree at most $\varepsilon - r$ in each variable. These are central even in $\mathcal{H}'_r$, and, in view of the linear independence of elements of Y, produce linearly independent endomorphisms.

It remains to understand the subalgebra A of $\mathcal{H}'_r/J$ spanned by our special symmetric polynomials. It is clear from above that this subalgebra is exactly the image of the center of $\mathcal{H}'_r$, which consists of *all* symmetric polynomials in $(x_{n-r+1} - a), \dots, (x_n - a)$. Now, note that $\mathcal{H}'_r/J_r$ embeds into $\mathcal{H}'_\varepsilon/J_\varepsilon$. Also, in view of Lemma 5.1.4 and (4.8), J_ε is the ideal generated by the symmetric polynomials in $(x_{n-\varepsilon+1} - a), \dots, (x_n - a)$ without the free term. So $A \cong \mathcal{Z}_r/(\mathcal{Z}_r \cap \mathcal{H}_\varepsilon \mathcal{Z}_\varepsilon^+) \cong \mathcal{Z}_r/(\mathcal{Z}_r \cap \mathcal{P}_\varepsilon \mathcal{Z}_\varepsilon^+)$.

Remark 8.6.4 Proposition 8.6.2, Lemma 8.5.8, and Remark 9.4.5 below imply for $i \in I$, $r \geq 1$, and an irreducible $\mathcal{H}_n^\lambda$-module M:

$$\dim \mathrm{End}_{H_{n+r}^\lambda}((f_i^\lambda)^{(r)} M) = \binom{\varphi_i^\lambda(M)}{r}. \tag{8.25}$$

We do not know how to get a description of $\mathrm{End}_{H_{n+r}^\lambda}((f_i^\lambda)^{(r)} M)$ as an algebra in the spirit of (8.24).

9

Construction of $U_{\mathbb{Z}}^{+}$ and irreducible modules

In this chapter we obtain some major results connecting representation theory of affine and cyclotomic Hecke algebras with affine Kac–Moody algebras. First, we form the Grothendieck group $K(\infty)$ of integral representations of all affine algebras $\mathcal{H}_n$, $n \geq 0$, and show that the operations of parabolic restriction and induction define on $K(\infty)$ a structure of a commutative (but not cocommutative) graded Hopf algebra over $\mathbb{Z}$. We will construct an explicit isomorphism of graded Hopf algebras between the Kostant–Tits $\mathbb{Z}$-form $U_{\mathbb{Z}}^{+}$ of the positive part of $\mathfrak{g}$ and the graded dual $K(\infty)^*$, see Theorem 9.5.3.

Now fix λ and consider the Grothendieck group $K(\lambda)$ of finite dimensional representations of all cyclotomic algebras $\mathcal{H}_n^{\lambda}$, $n \geq 0$. In this case the graded dual $K(\lambda)^*$ can be identified with the Grothendieck group of the projective modules over all $\mathcal{H}_n^{\lambda}$, $n \geq 0$, and there is a natural decomposition map $\omega : K(\lambda)^* \to K(\lambda)$, sending a projective module to the linear combination of its composition factors with multiplicities. We will prove that ω is injective (this does not follow from general nonsense!). This will allow us to identify $K(\lambda)^*$ as a sublattice of $K(\lambda)$. Further, $K(\lambda)$ is embedded into $K(\infty)$ via inflation, and is in fact a right subcomodule of $K(\infty)$, see (9.12). So it naturally becomes a left module over $K(\infty)^* \cong U_{\mathbb{Z}}^{+}$. Using the explicit nature of the last isomorphism, we can check that the action of the generator $e_i^r/r!$ of $U_{\mathbb{Z}}^{+}$ (on the Grothendieck group $K(\lambda)$) comes from the exact functor $(e_i^{\lambda})^{(r)}$ (on the module category). Now we define the action of the whole $U_{\mathbb{Z}}$ on $K(\lambda)$ with the action of $f_i^r/r!$ coming, of course, from the exact functors $(f_i^{\lambda})^{(r)}$. The action of h_i is given in terms of λ and block data, see (9.19). It turns out that Chevalley relations are satisfied and so we do get an action of the $\mathbb{Z}$-form $U_{\mathbb{Z}}$ on $K(\lambda)$. Moreover, $K(\lambda)^* \subset K(\lambda)$ is invariant with respect to this action.

The next step is to extend the scalars to $\mathbb{Q}$ to get the action of $U_{\mathbb{Q}}$ on $K(\lambda)_{\mathbb{Q}}$. The natural pairing between $K(\lambda)^*$ and $K(\lambda)$ yields a non-degenerate symmetric bilinear form $(.,.)$ on $K(\lambda)_{\mathbb{Q}}$, under which the two sublattices $K(\lambda), K(\lambda)^* \subset K(\lambda)$ are dual to each other. The important Theorem 9.5.1 identifies the $U_{\mathbb{Q}}$-module $K(\lambda)_{\mathbb{Q}}$, *constructed purely in terms of representation theory of affine and cyclotomic Hecke algebras*, as the irreducible integrable $U_{\mathbb{Q}}$-module $V(\lambda)$ of highest weight λ. Moreover, under this identification:

- the highest weight vector v_+ corresponds to the class $[\mathbf{1}_\lambda]$ of the trivial $\mathcal{H}_0^\lambda$-module in the Grothendieck group;
- the form $(.,.)$ is nothing but the Shapovalov–Jantzen contravariant form on $V(\lambda)$, satisfying $(v_+, v_+) = 1$;
- blocks of the algebras $\mathcal{H}_n^\lambda$ correspond to the weight spaces of $V(\lambda)$ (this will follow only after Corollary 9.6.2 is proved);
- the sublattice $K(\lambda)^* \subset K(\lambda)_{\mathbb{Q}}$ corresponds to the smallest sublattice $U_{\mathbb{Z}}^- v$, of $V(\lambda)$ containing v_+ and invariant with respect to the Kostant–Tits $\mathbb{Z}$-form $U_{\mathbb{Z}}$.

In the end of this chapter we will prove Theorem 9.6.1, which claims that two modules over a cyclotomic Hecke algebra $\mathcal{H}_n^\lambda$ are in the same block if and only if their inflations are in the same block of the affine Hecke algebra $\mathcal{H}_n$. This, of course, would be immediate if we knew that under the natural projection $\mathcal{H}_n \twoheadrightarrow \mathcal{H}_n^\lambda$ the center of $\mathcal{H}_n$ mapped *on to* the center of $\mathcal{H}_n^\lambda$. Even though this is probably true, we have no idea how to prove it. The following fact is proved instead: if, in the category $\operatorname{Rep}_I \mathcal{H}_n$, we have an extension X of an irreducible module M by an irreducible module N, and both M and N factor through $\mathcal{H}_n^\lambda$, then X also factors through $\mathcal{H}_n^\lambda$. The proof of this fact relies on a nice property of the functors e_i proved in Corollary 5.5.2. Note that Theorem 9.6.1 provides us with a convenient classification of blocks for cyclotomic Hecke algebras, as the blocks of affine Hecke algebras are easy to understand, see Section 4.2.

9.1 Grothendieck groups

Let us write

$$K(\infty) = \bigoplus_{n \geq 0} K(\operatorname{Rep}_I \mathcal{H}_n) \tag{9.1}$$

for the sum over all n of the Grothendieck groups of the categories $\operatorname{Rep}_I \mathcal{H}_n$. Also, write

$$K(\infty)_{\mathbb{Q}} = \mathbb{Q} \otimes_{\mathbb{Z}} K(\infty),$$

extending scalars. Thus $K(\infty)$ is a free $\mathbb{Z}$-module with canonical basis given by $B(\infty)$, the isomorphism classes of irreducible modules, and $K(\infty)_{\mathbb{Q}}$ is the $\mathbb{Q}$-vector space on basis $B(\infty)$. We will always view $K(\infty)$ as a lattice in $K(\infty)_{\mathbb{Q}}$.

We let $K(\infty)^*$ denote the *restricted dual* of $K(\infty)$, namely the set of functions $f: K(\infty) \to \mathbb{Z}$ such that f vanishes on all but finitely many elements of $B(\infty)$. Equivalently, $K(\infty)^*$ is the *graded dual* of $K(\infty)$. Thus $K(\infty)^*$ is also a free $\mathbb{Z}$-module, with canonical basis

$$\{\delta_M \mid [M] \in B(\infty)\}$$

dual to the basis $B(\infty)$ of $K(\infty)$, that is $\delta_M([M]) = 1$, $\delta_M([N]) = 0$ for $[N] \in B(\infty)$ with $N \not\cong M$. Note for an arbitrary $N \in \mathrm{Rep}_I \mathcal{H}_n$, $\delta_M([N])$ simply computes the multiplicity $[N : M]$ of the irreducible module M as a composition factor of N. Finally, we write

$$B(\infty)^*_{\mathbb{Q}} := \mathbb{Q} \otimes_{\mathbb{Z}} B(\infty)^*,$$

which can be identified with the restricted (or graded) dual of $B(\infty)_{\mathbb{Q}}$.

Entirely similar definitions can be made for each $\lambda \in P_+$. Set

$$K(\lambda) = \bigoplus_{n \geq 0} K(\mathcal{H}_n^\lambda\text{-mod}). \tag{9.2}$$

Again, $K(\lambda)$ is a free $\mathbb{Z}$-module on the basis $B(\lambda)$ of isomorphism classes of irreducible $\mathcal{H}_n^\lambda$-modules. Moreover, infl^λ induces canonical embeddings

$$\mathrm{infl}^\lambda : K(\lambda) \hookrightarrow K(\infty), \quad \mathrm{infl}^\lambda : B(\lambda) \hookrightarrow B(\infty). \tag{9.3}$$

We will generally identify $K(\lambda)$ and $B(\lambda)$ with their images under these embeddings. We also define $K(\lambda)^*$ and $K(\lambda)_{\mathbb{Q}} = \mathbb{Q} \otimes_{\mathbb{Z}} K(\lambda)$ as above. Note that $K(\infty)$, $K(\infty)_{\mathbb{Q}}$, $K(\infty)^*$, $K(\lambda)$, etc. are all naturally graded.

Recall the functors e_i and more generally the divided power functors $e_i^{(r)}$ for $r \geq 1$, defined in Chapter 8. These induce linear maps

$$e_i^{(r)} : K(\infty) \to K(\infty) \tag{9.4}$$

for each $r \geq 1$. Similarly, $(e_i^\lambda)^{(r)}$ and $(f_i^\lambda)^{(r)}$ from Chapter 8 induce maps

$$(e_i^\lambda)^{(r)}, (f_i^\lambda)^{(r)} : K(\lambda) \to K(\lambda). \tag{9.5}$$

Note by Lemma 8.3.3 that

$$e_i^r = (r!)e_i^{(r)}, \qquad (e_i^\lambda)^r = (r!)(e_i^\lambda)^{(r)}, \qquad (f_i^\lambda)^r = (r!)(f_i^\lambda)^{(r)}. \tag{9.6}$$

Extending scalars, the maps $e_i^{(r)}$, $(e_i^\lambda)^{(r)}$, $(f_i^\lambda)^{(r)}$ induce linear maps on $K(\infty)_{\mathbb{Q}}$ and $K(\lambda)_{\mathbb{Q}}$, too.

9.2 Hopf algebra structure

Now we wish to give $K(\infty)$ the structure of a graded Hopf algebra over $\mathbb{Z}$. Note that $\boxtimes$ induces the canonical isomorphism

$$K(\operatorname{Rep}_I \mathcal{H}_m) \otimes_{\mathbb{Z}} K(\operatorname{Rep}_I \mathcal{H}_n) \to K(\operatorname{Rep}_I \mathcal{H}_{m,n}) \tag{9.7}$$

for each $m, n \geq 0$. The exact functor $\operatorname{ind}_{m,n}^{m+n}$ induces a well-defined map

$$\operatorname{ind}_{m,n}^{m+n} : K(\operatorname{Rep}_I \mathcal{H}_{m,n}) \to K(\operatorname{Rep}_I \mathcal{H}_{m+n}).$$

Composing with the isomorphism (9.7) and taking the direct sum over all $m, n \geq 0$, we obtain a homogeneous map

$$\diamond : K(\infty) \otimes_{\mathbb{Z}} K(\infty) \to K(\infty). \tag{9.8}$$

By transitivity of induction, this makes $K(\infty)$ into an associative graded $\mathbb{Z}$-algebra. By Corollary 5.3.2, the duality τ induces the identity map at the level of Grothendieck groups, so Theorem 3.7.5 implies that the multiplication $\diamond$ is commutative. Moreover, there is a unit

$$\iota : \mathbb{Z} \to K(\infty) \tag{9.9}$$

mapping 1 to the class of the trivial module $[\mathbf{1}] \in K(\operatorname{Rep}_I \mathcal{H}_0) \subset K(\infty)$.

Now we define the comultiplication. The exact functor $\operatorname{res}_{m,n-m}^n$ induces a map

$$\operatorname{res}_{m,n-m}^n : K(\operatorname{Rep}_I \mathcal{H}_n) \to K(\operatorname{Rep}_I \mathcal{H}_{m,n-m}).$$

On composing with the isomorphism (9.7), we obtain maps

$$\Delta_{m,n-m}^n : K(\operatorname{Rep}_I \mathcal{H}_n) \to K(\operatorname{Rep}_I \mathcal{H}_m) \otimes_{\mathbb{Z}} K(\operatorname{Rep}_I H_{n-m}),$$

$$\Delta^n = \sum_{n_1+n_2=n} \Delta_{n_1,n_2}^n : K(\operatorname{Rep}_I \mathcal{H}_n) \to \bigoplus_{n_1+n_2=n} K(\operatorname{Rep}_I \mathcal{H}_{n_1}) \otimes_{\mathbb{Z}} K(\operatorname{Rep}_I \mathcal{H}_{n_2}).$$

Now taking the direct sum over all $n \geq 0$ gives a homogeneous map

$$\Delta : K(\infty) \to K(\infty) \otimes_{\mathbb{Z}} K(\infty). \tag{9.10}$$

Transitivity of restriction implies that Δ is coassociative, while the homogeneous projection on to $K(\operatorname{Rep}_I \mathcal{H}_0) \cong \mathbb{Z}$ gives a counit

$$\varepsilon : K(\infty) \to \mathbb{Z}. \tag{9.11}$$

Thus $K(\infty)$ is also a graded coalgebra over $\mathbb{Z}$. Now finally:

Theorem 9.2.1 *$(K(\infty), \diamond, \Delta, \iota, \varepsilon)$ is a commutative, graded Hopf algebra over $\mathbb{Z}$.*

Proof It just remains to check that Δ is an algebra homomorphism, which follows using the Mackey Theorem (Theorem 3.5.2). □

Remark 9.2.2 For $\lambda \in P_+$, there is in general no natural way to give $K(\lambda)$ the structure of a Hopf algebra, unlike $K(\infty)$. An exception arises in the special case $\lambda = \Lambda_0$ where $\mathcal{H}_n^{\Lambda_0} \cong \mathcal{G}_n$. In this case, the parabolic subalgebras $\mathcal{G}_\mu \subseteq \mathcal{G}_n$ can be used to make

$$K(\Lambda_0) = \bigoplus_{n \geq 0} K(\mathcal{G}_n\text{-mod})$$

into a graded Hopf algebra in exactly the same way as $K(\infty)$ above. We will not make use of this.

The following lemma, explaining how to compute the action of e_i on $K(\infty)$ explicitly in terms of Δ, follows from the definitions:

Lemma 9.2.3 *Let M be a module in $\mathrm{Rep}_I\, \mathcal{H}_n$. Write*

$$\Delta^n_{n-1,1}[M] = \sum_r [M_r] \otimes [N_r]$$

for irreducible $\mathcal{H}_{n-1}$-modules M_r and irreducible $\mathcal{H}_1$-modules N_r. Then

$$e_i[M] = \sum_{r \text{ with } N_r \cong L(i)} [M_r].$$

Lemma 9.2.4 *The operators $e_i : K(\infty) \to K(\infty)$ satisfy the Serre relations, that is*

$$e_i e_j = e_j e_i \qquad \text{if } |i-j| > 1,$$
$$e_i^2 e_j + e_j e_i^2 = 2 e_i e_j e_i \qquad \text{if } |i-j| = 1,\ \ell = p-1 > 1,$$
$$e_i^3 e_j + 3 e_i e_j e_i^2 = 3 e_i^2 e_j e_i + e_j e_i^3 \qquad \text{if } |i-j| = 1,\ \ell = p-1 = 1.$$

Proof In view of Lemma 9.2.3 and coassociativity of Δ, this reduces to checking it on irreducible $\mathcal{H}_n$-modules for $n = 2, 3, 4$ respectively. For this, the character information in Lemmas 6.2.1 and 6.2.2 is sufficient. □

Now consider $K(\lambda)$ for $\lambda \in P_+$. This has a natural structure as $K(\infty)$-comodule: viewing $K(\lambda)$ as a subset of $K(\infty)$, the comodule structure map is the restriction

$$\Delta^\lambda : K(\lambda) \to K(\lambda) \otimes_{\mathbb{Z}} K(\infty) \tag{9.12}$$

of Δ. In other words, each $K(\lambda)$ is a subcomodule of the right regular $K(\infty)$-comodule. This follows from the fact that the restriction from $\mathcal{H}_n$ to $\mathcal{H}_{n-m}$ of an $\mathcal{H}_n$-module, which factors through $\mathcal{H}_n^\lambda$, itself factors through $\mathcal{H}_{n-m}^\lambda$.

The dual maps to $\diamond$, Δ, ι, ε induce on $K(\infty)^*$ the structure of a cocommutative graded Hopf algebra. Moreover, each $K(\lambda)$ is a left $K(\infty)^*$-module in the natural way: $f \in K(\infty)^*$ acts on the left on $K(\lambda)$ as the map $(\mathrm{id} \bar{\otimes} f) \circ \Delta^\lambda$. Similarly, $K(\infty)$ is itself a left $K(\infty)^*$-module, indeed in this case the action is even *faithful*.

Lemma 9.2.5 *The operator $e_i^{(r)}$ acts on $K(\infty)$ (resp. $K(\lambda)$ for any $\lambda \in P_+$) in the same way as the basis element $\delta_{L(i^r)}$ of $K(\infty)^*$.*

Proof Let M be an irreducible module in $\mathrm{Rep}_I \mathcal{H}_n$ or $\mathcal{H}_n^\lambda$-mod. Write

$$\Delta_{n-r,r}[M] = \sum_s [M_s] \otimes [N_s]$$

for irreducible $\mathcal{H}_{n-r}^\lambda$-modules M_s and irreducible $\mathcal{H}_r$-modules N_s. By the definition of the action of $K(\infty)^*$ on $K(\lambda)$, it follows that

$$\delta_{L(i^r)}[M] = \sum_{s \text{ with } N_s \cong L(i^r)} [M_s].$$

Hence, since

$$[\mathrm{res}^r_{1,\dots,1} L(i^r)] = (r!)[L(i)^{\boxtimes r}],$$

we get that $\delta^r_{L(i)}$ acts in the same way as $(r!)\delta_{L(i^r)}$. So in view of (9.6), it just remains to check that $\delta_{L(i)}$ acts in the same way as e_i, which follows by Lemma 9.2.3. □

Lemma 9.2.6 *There is a unique homomorphism $\pi : U_{\mathbb{Z}}^+ \to K(\infty)^*$ of graded Hopf algebras such that $\pi(e_i^{(r)}) = \delta_{L(i^r)}$ for each $i \in I$ and $r \geq 1$.*

Proof Since $K(\infty)$ is a faithful $K(\infty)^*$-module, (9.6) and Lemmas 9.2.4 and 9.2.5 imply that the operators $\delta_{L(i^r)}$ satisfy the same relations as the generators $e_i^{(r)}$ of $U_{\mathbb{Z}}^+$. This implies existence of a unique such algebra homomorphism. The fact that π is a coalgebra map follows because

$$\Delta(\delta_{L(i)}) = \delta_{L(i)} \otimes 1 + 1 \otimes \delta_{L(i)}$$

by the definition of the comultiplication on $K(\infty)^*$. □

9.3 Contravariant form

Now we focus on a fixed $\lambda \in P_+$. For an $\mathcal{H}_n^\lambda$-module M, we let P_M denote its projective cover in the category $\mathcal{H}_n^\lambda$-mod. Since $\mathcal{H}_n^\lambda$ is a finite dimensional algebra, we can identify

$$K(\lambda)^* = \bigoplus_{n \geq 0} K(\mathcal{H}_n^\lambda\text{-proj}) \tag{9.13}$$

so that the basis element δ_M corresponds to the isomorphism class $[P_M]$ for each irreducible $\mathcal{H}_n^\lambda$-module M and each $n \geq 0$. Moreover, under this identification, the canonical pairing

$$(.,.) : K(\lambda)^* \times K(\lambda) \to \mathbb{Z} \tag{9.14}$$

satisfies

$$([P_M], [N]) = \dim \operatorname{Hom}_{\mathcal{H}_n^\lambda}(P_M, N). \tag{9.15}$$

for $\mathcal{H}_n^\lambda$-modules M, N with M irreducible.

There are natural maps

$$K(\mathcal{H}_n^\lambda\text{-proj}) \to K(\mathcal{H}_n^\lambda\text{-mod}),\ [P] \mapsto \sum [P : M][M] \qquad (n \geq 0),$$

where the summation is over all isomorphism classes of irreducible $\mathcal{H}_n^\lambda$-modules M. They induce a homogeneous map

$$\omega : K(\lambda)^* \to K(\lambda). \tag{9.16}$$

In Theorem 9.3.5 we will prove that ω is injective.

In view of Theorem 8.3.2(i), the actions of $(e_i^\lambda)^{(r)}$ and $(f_i^\lambda)^{(r)}$ are defined on $K(\lambda)^*$.

Lemma 9.3.1 *The operators $(e_i^\lambda)^{(r)}, (f_i^\lambda)^{(r)}$ on $K(\lambda)^*$ and $K(\lambda)$ satisfy*

$$((e_i^\lambda)^{(r)} x, y) = (x, (f_i^\lambda)^{(r)} y), \qquad ((f_i^\lambda)^{(r)} x, y) = (x, (e_i^\lambda)^{(r)} y)$$

for each $x \in K(\lambda)^*$ and $y \in K(\lambda)$.

Proof This follows from (9.15) and Theorem 8.3.2(i). □

Corollary 9.3.2 *Suppose*

$$(e_i^\lambda)^{(r)}[M] = \sum_{[N]\in B(\lambda)} a_{M,N}[N], \quad (f_i^\lambda)^{(r)}[M] = \sum_{[N]\in B(\lambda)} b_{M,N}[N].$$

for $[M] \in B(\lambda)$. *Then*

$$(e_i^\lambda)^{(r)}[P_N] = \sum_{[M]\in B(\lambda)} b_{M,N}[P_M], \quad (f_i^\lambda)^{(r)}[P_N] = \sum_{[M]\in B(\lambda)} a_{M,N}[P_M]$$

for $[N] \in B(\lambda)$.

Proof In view of (9.15), we have $([P_M],[N]) = \delta_{[M],[N]}$ for irreducible $\mathcal{H}_n^\lambda$-modules M, N. Now the result follows from Lemma 9.3.1. □

Lemma 9.3.3 *Let* M *be an irreducible* $\mathcal{H}_n^\lambda$*-module, set* $\varepsilon = \varepsilon_i^\lambda(M), \varphi = \varphi_i^\lambda(M)$. *Then for any* $m \geq 0$ *we have*

$$(e_i^\lambda)^{(m)}[P_M] = \sum_{[N] \text{ with } \varepsilon_i^\lambda(N)\geq m} a_N[P_{(\tilde{e}_i^\lambda)^m N}]$$

for coefficients $a_N \in \mathbb{Z}_{\geq 0}$. *Moreover, in case* $m = \varepsilon$ *we have*

$$(e_i^\lambda)^{(\varepsilon)}[P_M] = \binom{\varepsilon+\varphi}{\varepsilon}[P_{(\tilde{e}_i^\lambda)^\varepsilon M}] + \sum_{[N] \text{ with } \varepsilon_i^\lambda(N)>\varepsilon} a_N[P_{(\tilde{e}_i^\lambda)^\varepsilon N}].$$

Proof By Corollary 9.3.2, we have

$$(e_i^\lambda)^{(m)}[P_M] = \sum_{[K]\in B(\lambda)} [(f_i^\lambda)^{(m)}K : M][P_K].$$

So if the term $[P_K]$ appears in the right-hand side with non-zero multiplicity then $\varphi_i^\lambda(K) \geq m$ – otherwise $(f_i^\lambda)^{(m)}K = 0$ by Theorem 8.3.2(v) and the definition (8.17) of φ_i^λ. For such K denote the (non-zero) modules $(\tilde{f}_i^\lambda)^m K$ by N. This gives

$$(e_i^\lambda)^{(m)}[P_M] = \sum_{[N]\in B(\lambda) \text{ with } \varepsilon_i^\lambda(N)\geq m} [(f_i^\lambda)^{(m)}(\tilde{e}_i^\lambda)^m N : M][P_{(\tilde{e}_i^\lambda)^m N}],$$

and the first part of the lemma follows.

Now let $m = \varepsilon$. If $\varepsilon_i^\lambda(N) = \varepsilon$, then we have $\varepsilon_i^\lambda\big((\tilde{e}_i^\lambda)^\varepsilon N\big) = 0$, and so by Lemma 8.5.1(i), the only composition factor of $(f_i^\lambda)^{(m)}(\tilde{e}_i^\lambda)^m N$ whose $\varepsilon_i^\lambda = \varepsilon$ is $(\tilde{f}_i^\lambda)^m(\tilde{e}_i^\lambda)^m N = N$. It follows that $N = M$, and

$$[(f_i^\lambda)^{(\varepsilon)}(\tilde{e}_i^\lambda)^\varepsilon M : M] = \binom{\varepsilon+\varphi}{\varepsilon},$$

thanks to Theorem 8.5.9(i). □

We also need:

Theorem 9.3.4 *Given an irreducible $\mathcal{H}_n^\lambda$-module M, the element $[P_M] \in K(\mathcal{H}_n^\lambda\text{-proj})$ can be written as an integral linear combination of terms of the form $(f_{i_1}^\lambda)^{(r_1)} \dots (f_{i_s}^\lambda)^{(r_s)}[\mathbf{1}_\lambda]$.*

Proof Proceed by induction on n, the conclusion being trivial for $n = 0$. So let $n > 0$ and the result be true for all smaller n. Suppose for a contradiction that we can find an irreducible $\mathcal{H}_n^\lambda$-module M for which the result does not hold. Pick i with $\varepsilon := \varepsilon_i^\lambda(M) > 0$. Since there are only finitely many irreducible $\mathcal{H}_n^\lambda$-modules, we may choose M so that the result holds for all irreducible $\mathcal{H}_n^\lambda$-modules L with $\varepsilon_i^\lambda(L) > \varepsilon$. Write

$$(f_i^\lambda)^{(\varepsilon)}[P_{(\tilde{e}_i^\lambda)^\varepsilon M}] = \sum_{[L] \in \operatorname{Irr} \mathcal{H}_n^\lambda} a_L [P_L]$$

for coefficients $a_L \in \mathbb{Z}$. By Corollary 9.3.2,

$$a_L = [(e_i^\lambda)^{(\varepsilon)} L : (\tilde{e}_i^\lambda)^\varepsilon M].$$

In particular, $a_L = 0$ unless $\varepsilon_i^\lambda(L) \geq \varepsilon$. Moreover, if $a_L \neq 0$ for L with $\varepsilon_i^\lambda(L) = \varepsilon$, then by Theorem 5.5.1(i), we have that

$$(e_i^\lambda)^{(\varepsilon)} L \cong (\tilde{e}_i^\lambda)^\varepsilon L \cong (\tilde{e}_i^\lambda)^\varepsilon M,$$

whence $L \cong M$ and $a_M = 1$. This shows:

$$[P_M] = (f_i^\lambda)^{(\varepsilon)}[P_{(\tilde{e}_i^\lambda)^\varepsilon M}] - \sum_{L \text{ with } \varepsilon_i^\lambda(L) > \varepsilon} a_L [P_L] \tag{9.17}$$

for some $a_L \in \mathbb{Z}$. But the inductive hypothesis and choice of M ensure that all terms on the right-hand side are integral linear combinations of terms $(f_{i_1}^\lambda)^{(r_1)} \dots (f_{i_s}^\lambda)^{(r_s)}[\mathbf{1}_\lambda]$, hence the same is true for $[P_M]$, a contradiction. □

The next two theorems are very important.

Theorem 9.3.5 *The map $\omega : K(\lambda)^* \to K(\lambda)$ from (9.16) is injective.*

Proof We show by induction on n that the map $\omega : K(\mathcal{H}_n^\lambda\text{-proj}) \to K(\mathcal{H}_n^\lambda\text{-mod})$ is injective. This is clear if $n = 0$, so assume $n > 0$ and the result has been proved for all smaller n. Suppose we have a relation

$$\sum a_M \omega([P_M]) = 0$$

where M runs over the isomorphism classes of irreducible $\mathcal{H}_n^\lambda$-modules, and not all coefficients a_M are zero. We may choose $i \in I$ and M such that $a_M \neq 0$, $\varepsilon := \varepsilon_i^\lambda(M) > 0$ and $a_N = 0$ for all N with $\varepsilon_i^\lambda(N) < \varepsilon$.

Apply $(e_i^\lambda)^{(\varepsilon)}$ to the sum. By Lemma 9.3.3, we get

$$\sum_{N \text{ with } \varepsilon_i^\lambda(N)=\varepsilon} \binom{\varepsilon+\varphi_i^\lambda(N)}{\varepsilon} a_N \omega([P_{(\tilde{e}_i^\lambda)^\varepsilon N}]) + X = 0$$

where X is a sum of terms of the form $\omega([P_{(\tilde{e}_i^\lambda)^\varepsilon L}])$ with $\varepsilon_i^\lambda(L) > \varepsilon$. Now the inductive hypothesis shows that $X = 0$ and that $a_N = 0$ for each N with $\varepsilon_i^\lambda(N) = \varepsilon$. In particular, $a_M = 0$, a contradiction. □

In view of Theorem 9.3.5, we may identify $K(\lambda)^*$ with its image under ω, so *from now on*

$$K(\lambda)^* \subseteq K(\lambda)$$

are two lattices in $K(\lambda)_{\mathbb{Q}}$. Extending scalars, the pairing (9.14) induces a bilinear form

$$(.,.) : K(\lambda)_{\mathbb{Q}} \times K(\lambda)_{\mathbb{Q}} \to \mathbb{Q} \tag{9.18}$$

with respect to which the operators e_i^λ and f_i^λ are adjoint.

Theorem 9.3.6 *The form $(.,.) : K(\lambda)_{\mathbb{Q}} \times K(\lambda)_{\mathbb{Q}} \to \mathbb{Q}$ is symmetric and non-degenerate.*

Proof It is non-degenerate, since bases $\{[P_M]\}$ and $\{[N]\}$ are orthonormal to each other, see (9.15). So we just need to check that it is symmetric. Proceed by induction on n to show that

$$(.,.) : K(\mathcal{H}_n^\lambda\text{-mod})_{\mathbb{Q}} \times K(\mathcal{H}_n^\lambda\text{-mod})_{\mathbb{Q}} \to \mathbb{Q}$$

is symmetric. The case $n = 0$ is clear. Let $n > 0$. In view of Theorem 9.3.4, any element of $K(\mathcal{H}_n^\lambda\text{-mod})_{\mathbb{Q}}$ can be written as a linear combination of elements of the form $f_i^\lambda x$ for $x \in K(\mathcal{H}_{n-1}^\lambda\text{-mod})_{\mathbb{Q}}$. So it suffices to show that $(f_i^\lambda x, y) = (y, f_i^\lambda x)$ for any $y \in K(\mathcal{H}_n^\lambda\text{-mod})_{\mathbb{Q}}$. Well, we have

$$(f_i^\lambda x, y) = (x, e_i^\lambda y) = (e_i^\lambda y, x) = (y, f_i^\lambda x)$$

by the induction hypothesis and Lemma 9.3.1. □

9.4 Chevalley relations

Continue working with a fixed $\lambda \in P_+$. We turn now to considering the relations satisfied by the operators e_i^λ, f_i^λ on $K(\lambda)$.

Lemma 9.4.1 *The operators* $e_i^\lambda, f_i^\lambda : K(\lambda) \to K(\lambda)$ *satisfy the Serre relations (7.5).*

Proof We know the e_i satisfy the Serre relations on all of $K(\infty)$ by Lemma 9.2.4, so they certainly satisfy the Serre relations on restriction to $K(\lambda)$. Moreover, e_i^λ and f_i^λ are adjoint operators for the bilinear form $(.,.)$ according to Lemma 9.3.1, and this form is non-degenerate by Theorem 9.3.6. The lemma follows. □

Now we consider relations between the e_i^λ and f_j^λ. For $i \in I$ and an irreducible $\mathcal{H}_n^\lambda$-module M with central character χ_γ for $\gamma \in \Gamma_n$, define

$$h_i^\lambda[M] = \langle h_i, \lambda - \gamma \rangle [M]. \tag{9.19}$$

Recall, according to Lemma 8.5.8 that we have equivalently that

$$h_i^\lambda[M] = (\varphi_i^\lambda(M) - \varepsilon_i^\lambda(M))[M]. \tag{9.20}$$

More generally, define

$$\binom{h_i^\lambda}{r} : K(\lambda) \to K(\lambda), \qquad [M] \mapsto \binom{\varphi_i^\lambda(M) - \varepsilon_i^\lambda(M)}{r}[M]$$

where $\binom{m}{r}$ denotes $m(m-1)\dots(m-r+1)/(r!)$. Extending linearly, each $\binom{h_i^\lambda}{r}$ can be viewed as a diagonal linear operator $K(\lambda) \to K(\lambda)$. The definition (9.19) implies immediately that:

Lemma 9.4.2 *As operators on* $K(\lambda)$,

$$[h_i^\lambda, e_j^\lambda] = \langle h_i, \alpha_j \rangle e_j^\lambda \quad \textit{and} \quad [h_i^\lambda, f_j^\lambda] = -\langle h_i, \alpha_j \rangle f_j^\lambda$$

for all $i, j \in I$.

Lemma 9.4.3 *As operators on* $K(\lambda)$,

$$[e_i^\lambda, f_j^\lambda] = \delta_{i,j} h_i^\lambda \tag{9.21}$$

for all $i, j \in I$.

Proof Let M be an irreducible $\mathcal{H}_n^\lambda$-module. It follows immediately from Theorems 5.5.1(i) and 8.5.9(i) (together with central character considerations in case $i \neq j$) that $[M]$ appears in $e_i^\lambda f_j^\lambda[M] - f_j^\lambda e_i^\lambda[M]$ with multiplicity $\delta_{i,j}(\varphi_i^\lambda(M) - \varepsilon_i^\lambda(M))$. Therefore, it suffices simply to show that $e_i^\lambda f_j^\lambda[M] - f_j^\lambda e_i^\lambda[M]$ is a multiple of $[M]$.

By Lemma 8.2.3, for $m \gg 0$ we have a surjection

$$\mathrm{ind}_{n,1}^{n+1} M \boxtimes L_m(j) \twoheadrightarrow f_j^{\lambda} M.$$

Apply $\mathrm{pr}^{\lambda} \circ e_i$ to get a surjection

$$\mathrm{pr}^{\lambda} e_i \mathrm{ind}_{n,1}^{n+1} M \boxtimes L_m(j) \twoheadrightarrow e_i^{\lambda} f_j^{\lambda} M. \tag{9.22}$$

By the Mackey Theorem, there is an exact sequence

$$0 \to M^{\oplus \delta_{i,j} m} \to e_i \mathrm{ind}_{n,1}^{n+1} M \boxtimes L_m(j) \to \mathrm{ind}_{n-1,1}^{n}(e_i M) \boxtimes L_m(j) \to 0,$$

For sufficiently large m we have

$$\mathrm{pr}^{\lambda} \mathrm{ind}_{n-1,1}^{n}(e_i M) \boxtimes L_m(j) = f_j^{\lambda} e_i^{\lambda} M.$$

So on applying the right exact functor pr^{λ} and using the irreducibility of M, this implies that there is an exact sequence

$$0 \longrightarrow M^{\oplus m_1} \longrightarrow \mathrm{pr}^{\lambda} e_i \mathrm{ind}_{n,1}^{n+1} M \boxtimes L_m(j) \longrightarrow f_j^{\lambda} e_i^{\lambda} M \longrightarrow 0, \tag{9.23}$$

for some m_1. Now let N be any irreducible $\mathcal{H}_n^{\lambda}$-module with $N \not\cong M$. Combining (9.22) and (9.23) shows that

$$[f_j^{\lambda} e_i^{\lambda} M - e_i^{\lambda} f_j^{\lambda} M : N] \geq 0. \tag{9.24}$$

Summing over all i, j and using (8.7) gives $[\mathrm{ind}\,\mathrm{res}\, M - \mathrm{res}\,\mathrm{ind}\, M : N] \geq 0$. But Theorem 7.6.2 shows that equality holds here, hence it must hold in (9.24) for all $i, j \in I$. □

To summarize, we have shown in (9.6), Lemmas 9.4.1, 9.4.2, and 9.4.3 that:

Theorem 9.4.4 *The action of the operators $e_i^{\lambda}, f_i^{\lambda}, h_i^{\lambda}$ on $K(\lambda)$ satisfy the Chevalley relations (7.3), (7.4), and (7.5). Moreover, the actions of $(e_i^{\lambda})^{(r)}$, $(f_i^{\lambda})^{(r)}$, and $\binom{h_i^{\lambda}}{r}$ for all $i \in I, r \geq 1$ make $K(\lambda)_{\mathbb{Q}}$ into a $U_{\mathbb{Q}}$-module so that $K(\lambda)^*, K(\lambda)$ are $U_{\mathbb{Z}}$-submodules.*

Remark 9.4.5 A stronger result than Lemma 9.4.3 holds: the relation (9.21) holds in the module category, not just on the level of the Grothendieck groups. If $i \neq j$ this means that the functors $e_i^{\lambda} f_j^{\lambda}$ and $f_j^{\lambda} e_i^{\lambda}$ are isomorphic (which is easy to see). If $i = j$ this means the following. Let M be an $\mathcal{H}_n^{\lambda}$-module with central character χ_{γ} for $\gamma \in \Gamma_n$. If $\langle h_i, \lambda - \gamma \rangle \geq 0$, then there is a functorial isomorphism

$$e_i^{\lambda} f_i^{\lambda} M \cong f_i^{\lambda} e_i^{\lambda} M \oplus M^{\oplus \langle h_i, \lambda - \gamma \rangle},$$

otherwise there is a functorial isomorphism

$$e_i^\lambda f_i^\lambda M \oplus M^{\oplus - \langle h_i, \lambda - \gamma \rangle} \cong f_i^\lambda e_i^\lambda M.$$

This is proved in [CR, 5.28] (see also [V_2] for the case where M is irreducible).

9.5 Identification of $K(\infty)^*$, $K(\lambda)^*$, and $K(\lambda)$

We recall some basic notions from representation theory of $U_{\mathbb{Q}}$ from [Kc, Chapters 3, 9, 10]. A $U_{\mathbb{Q}}$-module V is called *integrable* if it decomposes as a direct sum of its weight spaces, and all elements e_i, f_i act on V locally nilpotently. Let $\mathcal{O}_{\text{int}}$ be the category of integrable $U_{\mathbb{Q}}$-modules V with the additional property that for every $v \in V$ there exists $N \geq 0$ such that $e_{i_1} \dots e_{i_N} v = 0$ for any $i_1, \dots, i_N \in I$. The category $\mathcal{O}_{\text{int}}$ is semisimple, and its irreducible modules are

$$\{V(\lambda) \mid \lambda \in P_+\},$$

where $V(\lambda)$ is the $U_{\mathbb{Q}}$-module generated by a vector v_λ with defining relations

$$h_i v_\lambda = \langle h_i, \lambda \rangle, \quad e_i v_\lambda = 0, \quad f_i^{1+\langle h_i, \lambda \rangle} v_\lambda = 0. \tag{9.25}$$

We refer to the module $V(\lambda)$ as the irreducible $U_{\mathbb{Q}}$-module of *highest weight* λ and the vector v_λ as its *highest weight vector*. Every $V(\lambda)$ has a unique (up to scalar) non-degenerate symmetric contravariant bilinear form

$$(.,.) : V(\lambda) \times V(\lambda) \to \mathbb{Q},$$

sometimes called the *Shapovalov–Jantzen form*. Thus we have

$$(uv, w) = (v, \kappa(u)w) \qquad (u \in U_{\mathbb{Q}}, \ v, w \in V(\lambda)),$$

where

$$\kappa : U_{\mathbb{Q}} \to U_{\mathbb{Q}}, \quad e_i \mapsto f_i, \ f_i \mapsto e_i, \ h_i \mapsto h_i \qquad (i \in I)$$

is the *Chevalley anti-involution*. The weight spaces of $V(\lambda)$ are orthogonal to each other with respect to this form.

Theorem 9.5.1 *For any $\lambda \in P_+$:*

(i) *$K(\lambda)_{\mathbb{Q}}$ is precisely the irreducible integrable highest weight $U_{\mathbb{Q}}$-module $V(\lambda)$ of highest weight λ, with highest weight vector $[\mathbf{1}_\lambda]$;*
(ii) *the bilinear form $(.,.)$ from (9.18) on the highest weight module $K(\lambda)_{\mathbb{Q}}$ coincides with the usual contravariant form satisfying $([\mathbf{1}_\lambda], [\mathbf{1}_\lambda]) = 1$;*

(iii) *$K(\lambda)^* \subset K(\lambda)$ are integral forms of $K(\lambda)_{\mathbb{Q}}$ containing $[\mathbf{1}_\lambda]$, with $K(\lambda)^*$ being the minimal lattice $U_{\mathbb{Z}}^-[\mathbf{1}_\lambda]$ and $K(\lambda)$ being its dual under the contravariant form;*

(iv) *The classes $[M]$ of the irreducible $\mathcal{H}_n^\lambda$-modules $M \in \mathcal{H}_n^\lambda\text{-mod}[\gamma]$ form a basis of the $(\lambda-\gamma)$-weight space $V(\lambda)_{\lambda-\gamma}$. The same is true for the classes $[P_M]$ of projective indecomposable modules in $\mathcal{H}_n^\lambda\text{-mod}[\gamma]$.*

Proof It makes sense to think of $K(\lambda)_{\mathbb{Q}}$ as a $U_{\mathbb{Q}}$-module according to Theorem 9.4.4. The actions of e_i and f_i are locally nilpotent by Theorems 5.5.1(i) and 8.5.9(i). The action of h_i is diagonal by definition. Thus $K(\lambda)_{\mathbb{Q}}$ is an integrable module. Clearly $[\mathbf{1}_\lambda]$ is a highest weight vector of highest weight λ. Moreover, $K(\lambda)_{\mathbb{Q}} = U_{\mathbb{Q}}^-[\mathbf{1}_\lambda]$ by Theorem 9.3.4. This completes the proof of (i), and (ii) follows immediately from Lemma 9.3.1.

For (iii), we know already that $K(\lambda)^* \subset K(\lambda)$ are dual lattices of $K(\lambda)_{\mathbb{Q}}$, which are invariant under $U_{\mathbb{Z}}$. Moreover, Theorem 9.3.4 again shows $K(\lambda)^* = U_{\mathbb{Z}}^-[\mathbf{1}_\lambda]$. Finally, (iv) follows from (9.19). □

We need the following well-known result:

Lemma 9.5.2 *Let $u \in U_{\mathbb{Z}}^+$ act as zero on every $V(\lambda)$, $\lambda \in P_+$. Then $u = 0$.*

Proof We may assume that u is in the weight component $(U_{\mathbb{Z}}^+)_\mu$ for some μ. Pick λ with $\langle h_i, \lambda\rangle \gg 0$ for every i. By assumption, $uv = 0$ for every $v \in V(\lambda)_{\lambda-\mu}$. It follows that

$$0 = (uv, v_\lambda) = (v, \kappa(u)v_\lambda),$$

where κ is the Chevalley anti-involution. As $(.,.)$ is non-degenerate and $v \in V(\lambda)_{\lambda-\mu}$ is arbitrary, we have $\kappa(u)v_\lambda = 0$. But, since we have chosen λ to be very large, it follows from the definition of $V(\lambda)$ in (9.25) that there is an isomorphism $(U_{\mathbb{Z}}^-)_{-\mu} \to V(\lambda)_{\lambda-\mu}$, given by multiplying v_λ with elements of $(U_{\mathbb{Z}}^-)_{-\mu}$. So $\kappa(u) = 0$, whence $u = 0$. □

Theorem 9.5.3 *The map $\pi : U_{\mathbb{Z}}^+ \to K(\infty)^*$ constructed in Lemma 9.2.6 is an isomorphism.*

Proof Note by Lemma 9.2.5 that the action of the $e_i^{(r)} \in U_{\mathbb{Z}}^+$ on $K(\infty)$ and on each $K(\lambda)$ factors through the map π. So if $x \in \ker \pi$, we have by Theorem 9.5.1 that x acts as zero on all integrable highest weight modules $K(\lambda)$, $\lambda \in P_+$. Hence $x = 0$ in view of Lemma 9.5.2, and π is injective.

To prove surjectivity, take $x \in K(\infty)$. As π is homogeneous and all graded components are finite dimensional, it suffices to show that $(\pi(u), x) = 0$ for all $u \in U_{\mathbb{Z}}^{+}$ implies that $x = 0$. Note

$$(\pi(u), x) = (\pi(1)\pi(u), x) = (\pi(1), \pi(u)x) = (\pi(1), ux),$$

where the second equality follows because the right regular action of $K(\infty)^*$ on itself is precisely the dual action to the left action of $K(\infty)^*$ on $K(\infty)$, and the third equality follows from Lemma 9.2.5. Hence, if $(\pi(u), x) = 0$ for all $u \in U_{\mathbb{Z}}^{+}$, we have that $(\pi(1), ux) = 0$ for all $u \in U_{\mathbb{Z}}^{+}$. Now, $\pi(1)$ is the identity in $K(\infty)^*$, so $(\pi(1), ux)$ is just the coefficient of $[\mathbf{1}]$ in ux, where $[\mathbf{1}] \in K(\infty)$ is the class of the trivial $\mathcal{H}_0$-module $\mathbf{1}$.

Let us choose $\lambda \in P_+$ sufficiently large so that in fact $x \in K(\lambda) \subset K(\infty)$. Then, using identification (9.13), we have $([\mathbf{1}_\lambda], ux) = 0$ for all $u \in U_{\mathbb{Z}}^{+}$, where $(.,.)$ now is a canonical pairing between $K(\lambda)^*$ and $K(\lambda)$. Hence by Lemma 9.3.1, $(v[\mathbf{1}_\lambda], x) = 0$ for all $v \in U_{\mathbb{Z}}^{-}$. But then Theorem 9.3.4 implies that $x = 0$. □

9.6 Blocks

Let $\lambda \in P_+$. We now classify the blocks of the algebras $\mathcal{H}_n^\lambda$, following [G3].

Theorem 9.6.1 *Let M and N be irreducible $\mathcal{H}_n^\lambda$-modules with $M \not\cong N$, and*

$$0 \longrightarrow \mathrm{infl}^\lambda M \longrightarrow X \longrightarrow \mathrm{infl}^\lambda N \longrightarrow 0 \qquad (9.26)$$

be an exact sequence of $\mathcal{H}_n$-modules. Then $\mathrm{pr}^\lambda X = X$.

Proof Let us denote $\mathrm{infl}^\lambda M$ by M' and $\mathrm{infl}^\lambda N$ by N'. We may assume that M' and N' are in the same block for $\mathcal{H}_n$, as otherwise the sequence (9.26) splits, and the result is clear. Recall the ideal $\mathcal{J}_\lambda = \mathcal{H}_n f_\lambda \mathcal{H}_n$ from Section 7.3, and set $\mathcal{J} := \mathcal{J}_\lambda$. Note that

$$\mathrm{pr}^\lambda ? = (\mathcal{H}_n/\mathcal{J}) \otimes_{\mathcal{H}_n} ?.$$

So applying this right exact functor to (9.26) yields the exact sequence

$$\cdots \longrightarrow \mathrm{Tor}_1^{\mathcal{H}_n}(\mathcal{H}_n/\mathcal{J}, N') \xrightarrow{\alpha} M' \longrightarrow \mathrm{pr}^\lambda X \longrightarrow N' \longrightarrow 0.$$

We have to show that $\alpha = 0$. This will follow if we can prove the claim that there is a surjection of left $\mathcal{H}_n$-modules

$$\mathrm{ind}_{n-1,1}^{n} \mathrm{res}_{n-1,1}^{n} N' \twoheadrightarrow \mathrm{Tor}_1^{\mathcal{H}_n}(\mathcal{H}_n/\mathcal{J}, N').$$

Indeed, if the claim is true and $\alpha \neq 0$ then the space

$$\begin{aligned}&\mathrm{Hom}_{\mathcal{H}_{n-1}}(\mathrm{res}^n_{n-1}N', \mathrm{res}^n_{n-1}M')\\ &\supseteq \mathrm{Hom}_{\mathcal{H}_{n-1,1}}(\mathrm{res}^n_{n-1,1}N', \mathrm{res}^n_{n-1,1}M')\\ &\cong \mathrm{Hom}_{\mathcal{H}_n}(\mathrm{ind}^n_{n-1,1}\mathrm{res}^n_{n-1,1}N', M')\end{aligned}$$

is non-zero, which is false in view of Corollary 5.5.2.

Finally, for the claim, consider the resolution

$$\cdots \longrightarrow \mathcal{H}_n \otimes_{\mathcal{H}_{n-1,1}} \mathcal{H}_n \xrightarrow{d_1} \mathcal{H}_n \xrightarrow{d_0} \mathcal{H}_n/\mathcal{J} \longrightarrow 0$$

of the right $\mathcal{H}_n$-module $\mathcal{H}_n/\mathcal{J}$, where d_0 is the natural projection, and $d_1(a \otimes b) = a f_\lambda b$ for $a, b \in \mathcal{H}_n$. Let $\mathcal{K} = \ker d_1$. Then we have two exact sequences of $\mathcal{H}_n$-bimodules

$$0 \longrightarrow \mathcal{J} \longrightarrow \mathcal{H}_n \longrightarrow \mathcal{H}_n/\mathcal{J} \longrightarrow 0,$$

$$0 \longrightarrow \mathcal{K} \longrightarrow \mathcal{H}_n \otimes_{\mathcal{H}_{n-1,1}} \mathcal{H}_n \longrightarrow \mathcal{J} \longrightarrow 0.$$

On tensoring with N' these yield the following exact sequences of left $\mathcal{H}_n$-modules

$$\begin{aligned}0 \longrightarrow \mathrm{Tor}_1^{\mathcal{H}_n}(\mathcal{H}_n/\mathcal{J}, N') \longrightarrow \mathcal{J} \otimes_{\mathcal{H}_n} N'\\ \longrightarrow \mathcal{H}_n \otimes_{\mathcal{H}_n} N' \xrightarrow{\beta} \mathcal{H}_n/\mathcal{J} \otimes_{\mathcal{H}_n} N' \longrightarrow 0,\end{aligned} \tag{9.27}$$

$$\cdots \longrightarrow \mathcal{H}_n \otimes_{\mathcal{H}_{n-1,1}} \mathcal{H}_n \otimes_{\mathcal{H}_n} N' \longrightarrow \mathcal{J} \otimes_{\mathcal{H}_n} N' \longrightarrow 0. \tag{9.28}$$

Now note that β is an isomorphism, as $\mathcal{H}_n \otimes_{\mathcal{H}_n} N' \cong N'$ and

$$\mathcal{H}_n/\mathcal{J} \otimes_{\mathcal{H}_n} N' \cong \mathrm{pr}^\lambda N' \cong N'.$$

Hence from (9.27), $\mathrm{Tor}_1^{\mathcal{H}_n}(\mathcal{H}_n/\mathcal{J}, N') \cong \mathcal{J} \otimes_{\mathcal{H}_n} N'$. Now, the claim follows from (9.28), as

$$\mathcal{H}_n \otimes_{\mathcal{H}_{n-1,1}} \mathcal{H}_n \otimes_{\mathcal{H}_n} N' \cong \mathrm{ind}^n_{n-1,1}\mathrm{res}^n_{n-1,1}N'.$$

□

Recalling the definitions from Sections 4.2 and 8.1 and using Theorem 9.5.1, we immediately deduce the following corollary which determines the blocks of $\mathcal{H}_n^\lambda$:

Corollary 9.6.2 *The blocks of cyclotomic algebras $\mathcal{H}_n^\lambda$ are precisely the subcategories $\mathcal{H}_n^\lambda$-mod$[\gamma]$ for $\gamma \in \Gamma_n$. Moreover, $\mathcal{H}_n^\lambda$-mod$[\gamma]$ is non-trivial if* *and only if the $(\lambda - \gamma)$-weight space of the highest weight module $K(\lambda)_{\mathbb{Q}}$ is non-zero.*

Remark 9.6.3 Recently, Chuang, and Rouquier [CR] have proved that the blocks of the algebras $\mathcal{H}_n^\lambda$ (for various n) corresponding to two weights γ_1 and γ_2 of $V(\lambda)$ in the same W-orbit are derived equivalent. A key notion introduced in [CR] is that of an $\mathfrak{sl}_2$*-categorification*. We refer the reader to the original paper for the precise definition, but, informally speaking, a categorification is a pair of endo-functors E and F on an appropriate category $\mathfrak{A}$ adjoint to each other on both sides, together with natural transformations $X \in \mathrm{End}(E)$ and $T \in \mathrm{End}(E^2)$, and a scalar $a \in F$, satisfying the following additional properties:

- linear operators $e = [E]$ and $f = [F]$ on the Grothendieck group $K(\mathfrak{A})_{\mathbb{Q}}$ give a locally finite $\mathfrak{sl}_2(\mathbb{Q})$-representation;
- the classes of irreducible objects of $\mathfrak{A}$ are weight vectors;
- $T \circ T = \mathbf{1}_{E^2}$ on E^2, $T \circ (\mathbf{1}_E X) \circ T + T = X\mathbf{1}_E$ on E^2, and $(\mathbf{1}_E T) \circ (T\mathbf{1}_E) \circ (\mathbf{1}_E T) = (T\mathbf{1}_E) \circ (\mathbf{1}_E T) \circ (T\mathbf{1}_E)$ on E^3;
- $X - a$ is locally nilpotent.

The main result of [CR] is that an $\mathfrak{sl}_2$-categorification leads to a complex of functors, which induces a self-equivalence of $D^b(\mathfrak{A})$.

This can be applied to the category $\mathfrak{A} = \oplus_{n \geq 0} \mathcal{H}_n^\lambda$-mod, functors $E = e_i$, $F = f_i$, and $a = i$. The natural transformations X and T come from the action of x_n on $e_i M$ and the action of s_n on $e_i^2 M$ for an $\mathcal{H}_n^\lambda$-module M. From what was proved in this chapter it follows easily that this gives us an $\mathfrak{sl}_2$-categorification. The main result of [CR] now gives a self-equivalence of the derived category, which on restriction to blocks gives the desired result.

10

Identification of the crystal

In this chapter we will push one step further the deep connection between degenerate affine and cyclotomic Hecke algebras on the one hand and Kac–Moody algebras on the other. Namely, we will explain how some natural representation theoretic data coming from affine algebras $\mathcal{H}_n$ and cyclotomic algebras $\mathcal{H}_n^\lambda$ can be used to define crystals in the sense of Kashiwara. Moreover, we will identify these crystals with those corresponding to $U_{\mathbb{Q}}^-$ and $V(\lambda)$, respectively. These crystals are explicitly known, which provides us with rich new information on representation theory of affine and cyclotomic algebras. For the case $\lambda = \Lambda_0$, when $\mathcal{H}_n^\lambda$ is the group algebra of the symmetric group, this will be exploited in the next chapter.

10.1 Final properties of $B(\infty)$

Recall the "starred" versions of $\tilde{e}_i$, $\tilde{f}_i$, ε_i, from Section 7.4.

Lemma 10.1.1 *Let $M \in \mathrm{Rep}_I\,\mathcal{H}_m$ be irreducible.*

(i) *For any $i \in I$, either $\varepsilon_i(\tilde{f}_i^* M) = \varepsilon_i(M)$ or $\varepsilon_i(M) + 1$.*
(ii) *For any $i, j \in I$ with $i \neq j$, we have $\varepsilon_i(\tilde{f}_j^* M) = \varepsilon_i(M)$.*

Proof We prove (i), the proof of (ii) being similar. By the Shuffle Lemma, we certainly have that $\varepsilon_i(\tilde{f}_i^* M) \leq \varepsilon_i(M) + 1$. Now let $N = \tilde{f}_i^* M$. Then obviously, $\varepsilon_i(\tilde{e}_i^* N) \leq \varepsilon_i(N)$. Hence, $\varepsilon_i(M) \leq \varepsilon_i(\tilde{f}_i^* M)$, as $\tilde{e}_i^* \tilde{f}_i^* M = M$, see Lemma 5.2.3. □

Lemma 10.1.2 *Let $M \in \mathrm{Rep}_I\,\mathcal{H}_m$ be irreducible and i, j be elements of I (not necessarily distinct). Assume that $\varepsilon_i(\tilde{f}_j^* M) = \varepsilon_i(M)$. Then, writing $\varepsilon := \varepsilon_i(M)$, we have*

$$\tilde{e}_i^\varepsilon \tilde{f}_j^* M \cong \tilde{f}_j^* \tilde{e}_i^\varepsilon M.$$

Proof Set $n = m - \varepsilon$. Let $N = \tilde{e}_i^{\varepsilon} M$, so N is an irreducible $\mathcal{H}_n$-module with $\varepsilon_i(N) = 0$ and $M = \tilde{f}_i^{\varepsilon} N$. For $0 \leq b \leq \varepsilon$, let

$$Q_b = e_i^{\varepsilon - b} \tilde{f}_j^* M.$$

Theorem 5.5.1(i) implies that in the Grothendieck group, Q_b is some number of copies of $\tilde{e}_i^{\varepsilon-b} \tilde{f}_j^* M$ plus terms with strictly smaller ε_i. In particular, $\varepsilon_i(L) \leq b$ for all composition factors L of Q_b, while Q_0 consists only of copies of $\tilde{e}_i^{\varepsilon} \tilde{f}_j^* M$.

We will show by decreasing induction on $b = \varepsilon, \varepsilon - 1, \ldots, 0$ that there is a non-zero $\mathcal{H}_{n+b+1}$-module homomorphism

$$\gamma_b : \mathrm{ind}_{1,n,b}^{n+b+1} L(j) \boxtimes N \boxtimes L(i^b) \to Q_b.$$

In case $b = \varepsilon$, $Q_\varepsilon = \tilde{f}_j^* M$ is a quotient of $\mathrm{ind}_{1,m}^{m+1} L(j) \boxtimes M$ (see (7.10)), and M is a quotient of $\mathrm{ind}_{n,\varepsilon}^{m} N \boxtimes L(i^\varepsilon)$, so the induction starts. Now we suppose by induction that we have proved $\gamma_b \neq 0$ exists for some $b \geq 1$ and construct γ_{b-1}.

Consider $\mathrm{res}_{n+b,1}^{n+b+1} \mathrm{ind}_{1,n,b}^{n+b+1} L(j) \boxtimes N \boxtimes L(i^b)$. By the Mackey Theorem, this has a filtration $0 \subset F_1 \subset F_2 \subset F_3$ with successive quotients

$$\begin{aligned} F_1 &\cong \mathrm{ind}_{1,n,b-1,1}^{n+b,1} \mathrm{res}_{1,n,b-1,1}^{1,n,b} L(j) \boxtimes N \boxtimes L(i^b), \\ F_2/F_1 &\cong \mathrm{ind}_{1,n-1,b,1}^{n+b,1} {}^{w}\mathrm{res}_{1,n-1,1,b}^{1,n,b} L(j) \boxtimes N \boxtimes L(i^b), \\ F_3/F_2 &\cong \mathrm{ind}_{n,b,1}^{n+b,1} N \boxtimes L(i^b) \boxtimes L(j), \end{aligned}$$

where w is the obvious permutation. As $\gamma_b \neq 0$, Frobenius reciprocity implies that there is a copy of the $\mathcal{H}_{1,n,b}$-module $L(j) \boxtimes N \boxtimes L(i^b)$ in the image of γ_b. Now $b > 0$, so the i-eigenspace of x_{n+b+1} acting on $L(j) \boxtimes N \boxtimes L(i^b)$ is non-trivial. We conclude that the map

$$\tilde{\gamma}_b = e_i(\gamma_b) : e_i \mathrm{ind}_{1,n,b}^{n+b+1} L(j) \boxtimes N \boxtimes L(i^b) \to e_i Q_b = Q_{b-1}$$

is non-zero.

If $i \neq j$, then it follows from the description of F_3/F_2 and F_2/F_1 above that $e_i(F_3/F_1) = 0$ (for F_2/F_1 we need to use the fact that $\varepsilon_i(N) = 0$). So in this case, we necessarily have that $\tilde{\gamma}_b(F_1) \neq 0$. Similarly if $i = j$, $e_i(F_2/F_1) = 0$, so if $\tilde{\gamma}_b(F_1) = 0$ we see that $\tilde{\gamma}_b$ factors to a non-zero homomorphism

$$\mathrm{res}_{n+b}^{n+b,1} \mathrm{ind}_{n,b,1}^{n+b,1} N \boxtimes L(i^b) \boxtimes L(i) \to Q_{b-1}.$$

But this implies that Q_{b-1} has a constituent L with $\varepsilon_i(L) = b$, which we know is not the case. Hence we have that $\tilde{\gamma}_b(F_1) \neq 0$ in the case $i = j$ too.

Hence, the restriction of $\tilde{\gamma}_b$ to F_1 gives us a non-zero homomorphism

$$\text{res}^{n+b,1}_{n+b}\,\text{ind}^{b+n,1}_{1,n,b-1,1}\text{res}^{1,n,b}_{1,n,b-1,1}L(j)\boxtimes N\boxtimes L(i^b)\to Q_{b-1}.$$

Now finally as all composition factors of $\text{res}^b_{b-1}L(i^b)$ are isomorphic to $L(i^{b-1})$, this implies the existence of a non-zero homomorphism

$$\gamma_{b-1}:\text{ind}^{b+n}_{1,n,b-1}L(j)\boxtimes N\boxtimes L(i^{b-1})\to Q_{b-1}$$

completing the induction.

Now taking $b=0$ we have a non-zero map $\gamma_0:\text{ind}^{n+1}_{1,n}L(j)\boxtimes N\to Q_0$. But the left-hand side has irreducible head $\tilde{f}^*_jN=\tilde{f}^*_j\tilde{e}^\varepsilon_iM$, while all composition factors of the right-hand side are isomorphic to $\tilde{e}^\varepsilon_i\tilde{f}^*_jM$. This completes the proof. □

We will also need the "starred" versions of Lemmas 10.1.1 and 10.1.2, which are obtained by applying the lemmas to the module M^σ:

Lemma 10.1.3 *Let $M\in\text{Rep}_I\,\mathcal{H}_m$ be irreducible.*

(i) *For any $i\in I$, either $\varepsilon^*_i(\tilde{f}_iM)=\varepsilon^*_i(M)$ or $\varepsilon^*_i(M)+1$.*
(ii) *For any $i,j\in I$ with $i\neq j$, we have $\varepsilon^*_i(\tilde{f}_jM)=\varepsilon^*_i(M)$.*

Lemma 10.1.4 *Let $M\in\text{Rep}_I\,\mathcal{H}_m$ be irreducible and i,j be elements of I (not necessarily distinct). Assume that $\varepsilon^*_i(\tilde{f}_jM)=\varepsilon^*_i(M)$. Then, writing $a:=\varepsilon^*_i(M)$, we have*

$$(\tilde{e}^*_i)^a\tilde{f}_jM\cong\tilde{f}_j(\tilde{e}^*_i)^aM.$$

Corollary 10.1.5 *Let $M\in\text{Rep}_I(\mathcal{H}_m)$ be irreducible and $i\in I$. Let $M_1=\tilde{e}_i^{\varepsilon_i(M)}M$ and $M_2=(\tilde{e}^*_i)^{\varepsilon^*_i(M)}M$. Then $\varepsilon^*_i(M)=\varepsilon^*_i(M_1)$ if and only if $\varepsilon_i(M)=\varepsilon_i(M_2)$.*

Proof Suppose $\varepsilon_i(M)=\varepsilon_i(M_2)$. Lemma 10.1.1 implies that

$$\varepsilon_i((\tilde{f}^*_i)^kM_2)=\varepsilon_i((\tilde{f}^*_i)^{k+1}M_2),\qquad k=0,1,\dots,\varepsilon^*_i(M)-1.$$

Now, using Lemma 10.1.2, we obtain

$$M_1\cong\tilde{e}_i^{\varepsilon_i(M)}M\cong\tilde{e}_i^{\varepsilon_i(M)}(\tilde{f}^*_i)^{\varepsilon^*_i(M)}M_2\cong(\tilde{f}^*_i)^{\varepsilon^*_i(M)}\tilde{e}_i^{\varepsilon_i(M)}M_2. \tag{10.1}$$

Note that $\varepsilon^*_i(M_2)=0$, hence $\varepsilon^*_i(\tilde{e}_i^{\varepsilon_i(M)}M_2)=0$. Now (10.1) shows that $\varepsilon^*_i(M_1)=\varepsilon^*_i(M)$. We have shown that $\varepsilon_i(M)=\varepsilon_i(M_2)$ implies $\varepsilon^*_i(M)=\varepsilon^*_i(M_1)$. The converse is obtained in exactly the same way, but using Lemmas 10.1.3 and 10.1.4 instead of Lemmas 10.1.1 and 10.1.2, respectively. □

Lemma 10.1.6 *Let $M \in \operatorname{Rep}_I \mathcal{H}_m$ be irreducible and $i \in I$ satisfy $\varepsilon_i(\tilde{f}_i^* M) = \varepsilon_i(M) + 1$. Then $\tilde{e}_i \tilde{f}_i^* M = M$.*

Proof Set $\varepsilon := \varepsilon_i(M)$ and $N = \tilde{e}_i^{\varepsilon} M$. By the Shuffle Lemma and Theorem 5.3.1, we have

$$[\operatorname{ind}_{1,m-\varepsilon,\varepsilon}^{m+1} L(i) \boxtimes N \boxtimes L(i^{\varepsilon})] = [\operatorname{ind}_{m-\varepsilon,\varepsilon+1}^{m+1} N \boxtimes L(i^{\varepsilon+1})]$$

(in the Grothendieck group). Hence by Theorem 5.5.1(i),

$$[\operatorname{ind}_{1,m-\varepsilon,\varepsilon}^{m+1} L(i) \boxtimes N \boxtimes L(i^{\varepsilon})]$$

equals $[\tilde{f}_i^{\varepsilon+1} N] = [\tilde{f}_i M]$ plus terms $[L]$ for irreducible L with $\varepsilon_i(L) \leq \varepsilon$. However, $\operatorname{ind}_{1,m-\varepsilon,\varepsilon}^{m+1} L(i) \boxtimes N \boxtimes L(i^{\varepsilon})$ surjects on to $\tilde{f}_i^* M$. So the assumption $\varepsilon_i(\tilde{f}_i^* M) = \varepsilon + 1$ implies $\tilde{f}_i M \cong \tilde{f}_i^* M$, or, applying $\tilde{e}_i$, $M = \tilde{e}_i \tilde{f}_i^* M$. □

Again, we record the "starred" version of the above:

Lemma 10.1.7 *Let $M \in \operatorname{Rep}_I \mathcal{H}_m$ be irreducible and $i \in I$ satisfy $\varepsilon_i^*(\tilde{f}_i M) = \varepsilon_i^*(M) + 1$. Then $\tilde{e}_i^* \tilde{f}_i M = M$.*

10.2 Crystals

Let us now recall some definitions from [Ka]. A *crystal* (of type $\mathfrak{g}$) is a set B endowed with maps

$$\varphi_i, \varepsilon_i : B \to \mathbb{Z} \cup \{-\infty\} \quad (i \in I),$$
$$\tilde{e}_i, \tilde{f}_i : B \to B \cup \{0\} \quad (i \in I),$$
$$\mathrm{wt} : B \to P$$

such that:

(C1) $\varphi_i(b) = \varepsilon_i(b) + \langle h_i, \mathrm{wt}(b) \rangle$ for any $i \in I$;

(C2) if $b \in B$ satisfies $\tilde{e}_i b \neq 0$, then $\varepsilon_i(\tilde{e}_i b) = \varepsilon_i(b) - 1$, $\varphi_i(\tilde{e}_i b) = \varphi_i(b) + 1$, and $\mathrm{wt}(\tilde{e}_i b) = \mathrm{wt}(b) + \alpha_i$;

(C3) if $b \in B$ satisfies $\tilde{f}_i b \neq 0$, then $\varepsilon_i(\tilde{f}_i b) = \varepsilon_i(b) + 1$, $\varphi_i(\tilde{f}_i b) = \varphi_i(b) - 1$, and $\mathrm{wt}(\tilde{e}_i b) = \mathrm{wt}(b) - \alpha_i$;

(C4) for $b_1, b_2 \in B$, $b_2 = \tilde{f}_i b_1$ if and only if $b_1 = \tilde{e}_i b_2$;

(C5) if $\varphi_i(b) = -\infty$, then $\tilde{e}_i b = \tilde{f}_i b = 0$.

Informally, we can think of a crystal as a colored directed graph with the set of vertices B, and an arrow of color $i \in I$ going from vertex b_1 to vertex b_2 if and only if $b_2 = \tilde{f}_i b_1$. Moreover, to every vertex b we associate its weight $\mathrm{wt}(b)$ and a bunch of numbers $\varepsilon_i(b)$, $\varphi_i(b)$ satisfying certain axioms.

For example, for each $i \in I$, we have the crystal B_i defined as a set to be $\{b_i(n) \mid n \in \mathbb{Z}\}$ with

$$\varepsilon_j(b_i(n)) = \begin{cases} -n & \text{if } j = i, \\ -\infty & \text{if } j \neq i; \end{cases} \qquad \varphi_j(b_i(n)) = \begin{cases} n & \text{if } j = i, \\ -\infty & \text{if } j \neq i; \end{cases}$$

$$\tilde{e}_j(b_i(n)) = \begin{cases} b_i(n+1) & \text{if } j = i, \\ 0 & \text{if } j \neq i; \end{cases}$$

$$\tilde{f}_j(b_i(n)) = \begin{cases} b_i(n-1) & \text{if } j = i, \\ 0 & \text{if } j \neq i; \end{cases}$$

and $\mathrm{wt}(b_i(n)) = n\alpha_i$. We abbreviate $b_i(0)$ by b_i. Also for $\lambda \in P$, we have the crystal T_λ equal as a set to $\{t_\lambda\}$, with $\varepsilon_i(t_\lambda) = \varphi_i(t_\lambda) = -\infty$, $\tilde{e}_i t_\lambda = \tilde{f}_i t_\lambda = 0$ and $\mathrm{wt}(t_\lambda) = \lambda$.

A morphism $\psi : B \to B'$ of crystals is a map $\psi : B \cup \{0\} \to B' \cup \{0\}$ such that:

(H1) $\psi(0) = 0$;

(H2) if $\psi(b) \neq 0$ for $b \in B$, then $\mathrm{wt}(\psi(b)) = \mathrm{wt}(b)$, $\varepsilon_i(\psi(b)) = \varepsilon_i(b)$ and $\varphi_i(\psi(b)) = \psi(b)$;

(H3) for $b \in B$ such that $\psi(b) \neq 0$ and $\psi(\tilde{e}_i b) \neq 0$, we have that $\psi(\tilde{e}_i b) = \tilde{e}_i \psi(b)$;

(H4) for $b \in B$ such that $\psi(b) \neq 0$ and $\psi(\tilde{f}_i b) \neq 0$, we have that $\psi(\tilde{f}_i b) = \tilde{f}_i \psi(b)$.

A morphism of crystals is called *strict* if ψ commutes with the $\tilde{e}_i$s and $\tilde{f}_i$s, and an *embedding* if ψ is injective.

Note that the trivial map sending "everything" to $\{0\}$ is a strict morphism. As for embeddings, we can informally think of them as follows: there is an embedding from a crystal B to a crystal B' if and only if we can identify B with a subset of B', so that the data wt, ε, φ are respected, and arrows in B are arrows in B'. Then the embedding is strict if there are no arrows starting in B and going "outside", no arrows starting "outside" and coming to B, and no arrows in $B' \setminus B$ conecting two elements in B.

We also need the notion of a tensor product of two crystals B, B'. As a set, $B \otimes B'$ is just $B \times B'$, but we write $b \otimes b'$ instead of (b, b') for $b \in B$, $b' \in B'$. This is made into a crystal by setting

$$\varepsilon_i(b \otimes b') = \max(\varepsilon_i(b), \varepsilon_i(b') - \langle h_i, \mathrm{wt}(b)\rangle),$$

$$\varphi_i(b \otimes b') = \max(\varphi_i(b) + \langle h_i, \mathrm{wt}(b')\rangle, \varphi_i(b')),$$

$$\tilde{e}_i(b \otimes b') = \begin{cases} \tilde{e}_i b \otimes b' \text{ if } \varphi_i(b) \geq \varepsilon_i(b') \\ b \otimes \tilde{e}_i b' \text{ if } \varphi_i(b) < \varepsilon_i(b') \end{cases},$$

$$\tilde{f}_i(b \otimes b') = \begin{cases} \tilde{f}_i b \otimes b' \text{ if } \varphi_i(b) > \varepsilon_i(b') \\ b \otimes \tilde{f}_i b' \text{ if } \varphi_i(b) \leq \varepsilon_i(b') \end{cases},$$

$$\mathrm{wt}(b \otimes b') = \mathrm{wt}(b) + \mathrm{wt}(b').$$

Here, we understand $b \otimes 0 = 0 = 0 \otimes b$. The definition comes from representation theory of sl_2.

Recall the sets of isomorphism classes of irreducible modules $B(\infty)$ and $B(\lambda)$ from Section 9.1. We now explain how to make them into crystals in the above sense. For $B(\lambda)$, we use the operators $\tilde{e}_i^\lambda$, $\tilde{f}_i^\lambda$ from (5.10), (8.8) and functions ε_i^λ, φ_i^λ from (5.5), (8.17) to define the maps $\tilde{e}_i, \tilde{f}_i, \varepsilon_i, \varphi_i$, respectively. For the corresponding functions on $B(\infty)$ use $\tilde{e}_i$, $\tilde{f}_i$ from (5.10), functions ε_i from (5.5), and φ_i defined below.

For the weight functions on $B(\infty)$ and $B(\lambda)$ set

$$\mathrm{wt}(M) = -\gamma, \tag{10.2}$$

for an irreducible $M \in \operatorname{Rep} \mathcal{H}_n[\gamma]$, and

$$\mathrm{wt}^\lambda(N) = \lambda - \gamma, \tag{10.3}$$

for an irreducible $N \in \mathcal{H}_n^\lambda\text{-mod}[\gamma]$, respectively. Finally, for $[M] \in B(\infty)$, define

$$\varphi_i(M) = \varepsilon_i(M) + \langle h_i, \mathrm{wt}(M)\rangle. \tag{10.4}$$

Note we have defined all data purely in terms of representation theory of $\mathcal{H}_n$ and $\mathcal{H}_n^\lambda$.

Lemma 10.2.1 *The tuples*

$$(B(\infty), \varepsilon_i, \varphi_i, \tilde{e}_i, \tilde{f}_i, \mathrm{wt})$$

and

$$(B(\lambda), \varepsilon_i^\lambda, \varphi_i^\lambda, \tilde{e}_i^\lambda, \tilde{f}_i^\lambda, \mathrm{wt}^\lambda)$$

for $\lambda \in P_+$ are crystals in the sense of Kashiwara.

Proof Property (C1) is Lemma 8.5.8 for the case of $B(\lambda)$ or the definition for the case of $B(\infty)$. Property (C4) is Lemma 5.2.3. The remaining properties are immediate. □

Recall the embedding $\mathrm{infl}^\lambda : B(\lambda) \cup \{0\} \to B(\infty) \cup \{0\}$ from (9.3).

Lemma 10.2.2 *The map*

$$B(\lambda) \hookrightarrow B(\infty) \otimes T_\lambda \quad [N] \mapsto \mathrm{infl}^\lambda[N] \otimes t_\lambda$$

is an embedding of crystals with image

$$\left\{[M] \otimes t_\lambda \in B(\infty) \otimes T_\lambda \,\middle|\, \varepsilon_i^*(M) \le \langle h_i, \lambda\rangle \text{ for each } i \in I\right\}.$$

Proof Since $\tilde{e}_i^\lambda$ and $\tilde{f}_i^\lambda$ are restrictions of $\tilde{e}_i, \tilde{f}_i$ from $B(\infty)$ to $B(\lambda)$, respectively, the first statement is immediate. The second is a restatement of Corollary 7.4.1. □

10.3 Identification of $B(\infty)$ and $B(\lambda)$

In this section we will identify the crystals $B(\infty)$ and $B(\lambda)$ defined purely in terms of modular representation theory with the Kashiwara's crystals of $U_{\mathbb{Q}}^-$ and $V(\lambda)$, respectively.

Lemma 10.3.1 *Let $M \in \mathrm{Rep}_I \mathcal{H}_m$ be irreducible and $i, j \in I$ with $i \neq j$. Set $a = \varepsilon_i^*(M)$.*

(i) $\varepsilon_j(M) = \varepsilon_j((\tilde{e}_i^*)^a M)$.
(ii) *If $\varepsilon_j(M) > 0$, then $\varepsilon_i^*(\tilde{e}_j M) = \varepsilon_i^*(M)$ and $(\tilde{e}_i^*)^a \tilde{e}_j M \cong \tilde{e}_j (\tilde{e}_i^*)^a M$.*

Proof (i) Applying Lemma 10.1.1(ii) (with i and j swapped) repeatedly, we get

$$\varepsilon_j((\tilde{e}_i^*)^a M) = \varepsilon_j(\tilde{f}_i^* (\tilde{e}_i^*)^a M) = \cdots = \varepsilon_j((\tilde{f}_i^*)^a (\tilde{e}_i^*)^a M) = \varepsilon_j(M).$$

(ii) By Lemma 10.1.3(ii), we have $\varepsilon_i^*(\tilde{f}_j N) = \varepsilon_i^*(N)$ for any irreducible N. Applying this to $N = \tilde{e}_j M$ gives the first part of the claim. Further, $a = \varepsilon_i^*(N)$, so by Lemma 10.1.4, we have

$$(\tilde{e}_i^*)^a \tilde{f}_j N = \tilde{f}_j (\tilde{e}_i^*)^a N,$$

whence the second part of the claim. □

Lemma 10.3.2 *Let $M \in \mathrm{Rep}_I \mathcal{H}_m$ be irreducible and $i \in I$. Set $a = \varepsilon_i^*(M)$ and $L = (\tilde{e}_i^*)^a M$.*

(i) $\varepsilon_i(M) = \max\big(\varepsilon_i(L),\ a - \langle h_i, \mathrm{wt}(L)\rangle\big)$.

(ii) *If $\varepsilon_i(M) > 0$,*

$$\varepsilon_i^*(\tilde{e}_i M) = \begin{cases} a & \text{if } \varepsilon_i(L) \geq a - \langle h_i, \mathrm{wt}(L)\rangle, \\ a-1 & \text{otherwise.} \end{cases}$$

(iii) *If $\varepsilon_i(M) > 0$,*

$$(\tilde{e}_i^*)^b \tilde{e}_i M \cong \begin{cases} \tilde{e}_i L & \text{if } \varepsilon_i(L) \geq a - \langle h_i, \mathrm{wt}(L)\rangle, \\ L & \text{otherwise,} \end{cases}$$

where $b = \varepsilon_i^*(\tilde{e}_i M)$.

Proof Let $\varepsilon = \varepsilon_i(M)$, $n = m - \varepsilon$, and $N = (\tilde{e}_i)^\varepsilon M$.

(i) By twisting with σ, it suffices to prove that (for arbitrary M)

$$\varepsilon_i^*(M) = \max\big(\varepsilon_i^*(N),\ \varepsilon - \langle h_i, \mathrm{wt}(N)\rangle\big). \tag{10.5}$$

Define the weights $\lambda(0), \lambda(1), \dots \in P_+$ by taking

$$\langle h_i, \lambda(r)\rangle = \varepsilon_i^*(N) + r \quad \text{and} \quad \langle h_j, \lambda(r)\rangle \gg 0 \quad \text{for } j \neq i.$$

Then $\mathcal{J}_{\lambda(r)} N = 0$ for any $r \geq 0$, see Corollary 7.4.1. Moreover, the same corollary and Lemma 10.1.3 imply that for $k = \varphi_i^{\lambda(r)}(N)$ we have

$$\varepsilon_i^*(\tilde{f}_i^k N) = \langle h_i, \lambda(r)\rangle \quad \text{and} \quad \varepsilon_i^*(\tilde{f}_i^{k+1} N) = \langle h_i, \lambda(r)\rangle + 1.$$

As $\varepsilon_i^{\lambda(r)}(N) = \varepsilon_i(N) = 0$, Lemma 8.5.8 gives

$$\varphi_i^{\lambda(r)}(N) = \langle h_i, \lambda(r) + \mathrm{wt}(N)\rangle = \varepsilon_i^*(N) + r + \langle h_i, \mathrm{wt}(N)\rangle.$$

Moreover, by Lemma 10.1.3(i) we have $\varepsilon_i^*(\tilde{f}_i^s N) \geq \varepsilon_i^*(N)$ for any s. All of these applied consecutively to $\lambda(0), \lambda(1), \dots$ imply that

$$\varepsilon_i^*(\tilde{f}_i^s N) = \begin{cases} \varepsilon_i^*(N) & \text{if } s \leq \varepsilon_i^*(N) + \langle h_i, \mathrm{wt}(N)\rangle, \\ s - \langle h_i, \mathrm{wt}(N)\rangle & \text{if } s \geq \varepsilon_i^*(N) + \langle h_i, \mathrm{wt}(N)\rangle \end{cases}$$

for all $s \geq 0$. For $s = \varepsilon$ this gives (10.5).

(ii) As $\varepsilon_i(\tilde{e}_i M) = \varepsilon_i(M) - 1$, it follows from (10.5) and (10.5) applied to $\tilde{e}_i M$ in place of M that

$$\varepsilon_i^*(\tilde{e}_i M) = \begin{cases} \varepsilon_i^*(M) & \text{if } \varepsilon \leq \langle h_i, \mathrm{wt}(N)\rangle + \varepsilon_i^*(N), \\ \varepsilon_i^*(M) - 1 & \text{otherwise.} \end{cases}$$

Note that $\mathrm{wt}(N) = \mathrm{wt}(M) + \varepsilon\alpha_i$, so

$$\langle h_i, \mathrm{wt}(N)\rangle = \langle h_i, \mathrm{wt}(M)\rangle + 2\varepsilon. \tag{10.6}$$

Therefore

$$\varepsilon_i^*(\tilde{e}_i M) = \begin{cases} a & \text{if } \langle h_i, \mathrm{wt}(M)\rangle + \varepsilon + \varepsilon_i^*(N) \geq 0, \\ a-1 & \text{otherwise.} \end{cases}$$

Next, $\mathrm{wt}(L) = \mathrm{wt}(M) + a\alpha_i$, so

$$\langle h_i, \mathrm{wt}(L)\rangle = \langle h_i, \mathrm{wt}(M)\rangle + 2a. \tag{10.7}$$

Hence $\varepsilon_i(L) \geq a - \langle h_i, \mathrm{wt}(L)\rangle$ is equivalent to $\langle h_i, \mathrm{wt}(M)\rangle + \varepsilon_i(L) + a \geq 0$. So (ii) follows if we show that

$$\langle h_i, \mathrm{wt}(M)\rangle + \varepsilon_i^*(N) + \varepsilon < 0$$

is equivalent to

$$\langle h_i, \mathrm{wt}(M)\rangle + \varepsilon_i(L) + a < 0.$$

Now, by (i) and (10.7), we have

$$\begin{aligned} &\langle h_i, \mathrm{wt}(M)\rangle + \varepsilon_i^*(N) + \varepsilon \\ &= \max(\langle h_i, \mathrm{wt}(M)\rangle + \varepsilon_i^*(N) + \varepsilon_i(L),\ \varepsilon_i^*(N) - a), \end{aligned}$$

and, by (10.5) and (10.6), we have

$$\begin{aligned} &\langle h_i, \mathrm{wt}(M)\rangle + \varepsilon_i(L) + a \\ &= \max(\langle h_i, \mathrm{wt}(M)\rangle + \varepsilon_i(L) + \varepsilon_i^*(N),\ \varepsilon_i(L) - \varepsilon). \end{aligned}$$

Moreover, obviously $\varepsilon_i^*(N) - a \leq 0$ and $\varepsilon_i(L) - \varepsilon \leq 0$, and it remains to observe that $\varepsilon_i^*(N) - a = 0$ if and only if $\varepsilon_i(L) - \varepsilon = 0$, thanks to Corollary 10.1.5.

(iii) If $\varepsilon_i(L) \geq a - \langle h_i, \mathrm{wt}(L)\rangle$ then (ii) implies $b = a$, and so by Lemma 10.1.4 we have

$$(\tilde{e}_i^*)^b \tilde{e}_i M = \tilde{e}_i (\tilde{e}_i^*)^b = \tilde{e}_i L.$$

However, if $\varepsilon_i(L) < a - \langle h_i, \mathrm{wt}(L)\rangle$, then by (ii) implies $b = a - 1$, and so by Lemma 10.1.7 we have

$$(\tilde{e}_i^*)^b \tilde{e}_i M = (\tilde{e}_i^*)^b \tilde{e}_i^* M = (\tilde{e}_i^*)^a M = L.$$

□

Now for each $i \in I$, define a map

$$\Psi_i : B(\infty) \to B(\infty) \otimes B_i \tag{10.8}$$

mapping each $[M] \in B(\infty)$ to $[(\tilde{e}_i^*)^a M] \otimes \tilde{f}_i^a b_i$, where $a = \varepsilon_i^*(M)$.

Lemma 10.3.3 *The following properties hold:*

(i) *for every* $[M] \in B(\infty)$, $\mathrm{wt}(M)$ *is a negative sum of simple roots;*
(ii) $[\mathbf{1}]$ *is the unique element of* $B(\infty)$ *with weight* 0;
(iii) $\varepsilon_i(\mathbf{1}) = 0$ *for every* $i \in I$;
(iv) $\varepsilon_i(M) \in \mathbb{Z}$ *for every* $[M] \in B(\infty)$ *and every* $i \in I$;
(v) *for every i, the map* $\Psi_i : B(\infty) \to B(\infty) \otimes B_i$ *defined above is a strict embedding of crystals;*
(vi) $\Psi_i(B(\infty)) \subseteq B(\infty) \times \{\tilde{f}_i^n b_i \mid n \geq 0\}$;
(vii) *for any* $[M] \in B(\infty)$ *other than* $[\mathbf{1}]$, *there exists* $i \in I$ *such that* $\Psi_i([M]) = [N] \otimes \tilde{f}_i^n b_i$ *for some* $[N] \in B(\infty)$ *and* $n > 0$.

Proof Properties (i)–(iv) are immediate from our construction of $B(\infty)$. The information required to verify (v) is exactly contained in Lemmas 10.3.1 and 10.3.2. Finally, (vi) is immediate from the definition of Ψ_i, and (vii) holds because every such M has $\varepsilon_i^*(M) > 0$ for at least one $i \in I$. □

The properties in Lemma 10.3.3 exactly characterize the crystal $B(\infty)$ by [KS, Proposition 3.2.3]. Hence, we have proved:

Theorem 10.3.4 *The crystal* $B(\infty)$ *is isomorphic to Kashiwara's crystal* $B(\infty)$ *associated to the crystal base of* $U_{\mathbb{Q}}^-$.

In view of [Ka, Theorem 8.2], we can also identify our maps Ψ_i with those of [Ka]. Taking into account [Ka, Proposition 8.1] we can then identify our functions ε_i^* on $B(\infty)$ with those in [Ka]. It follows from this, Lemma 10.2.2 and [Ka, Proposition 8.2] that:

Theorem 10.3.5 *For each* $\lambda \in P_+$, *the crystal* $B(\lambda)$ *is isomorphic to Kashiwara's crystal* $B(\lambda)$ *associated to the integrable highest weight* $U_{\mathbb{Q}}$*-module of highest weight* λ.

Remark 10.3.6 Recent result of Berenstein and Kazhdan [BeKa] might make much of the work of this section, as well as Section 10.1, unnecessary, providing [BeKa] is generalized from locally finite modules over finite dimensional semisimple Lie algebras to integrable modules over Kac–Moody algebras (which actually seems to be more or less automatic).

Indeed, let V be an integrable $\mathfrak{g}$-module, and $\mathbf{B}$ be a basis of V, which consists of weight vectors. For any $i \in I$ define $V_i^{<r} := \{v \in V \mid e_i^r v = 0\}$ and $V_{-i}^{<r} := \{v \in V \mid f_i^r v = 0\}$. Also for every $b \in \mathbf{B}$ and $b \in I$ define $\varepsilon_i(b) := \max\{r \mid e_i^r b \neq 0\}$ and $\varphi_i(b) := \max\{r \mid f_i^r b \neq 0\}$. Following [BeKa], we say that the basis $\mathbf{B}$ is *perfect* if for every $b \in \mathbf{B}$ and $i \in I$ such that $e_i b \neq 0$ there exists a unique element $b' \in \mathbf{B}$ such that $\varepsilon_i(b') = \varepsilon_i(b) - 1$ and

$$e_i b \in \mathbb{C}^{\times} \cdot b' + V_i^{<\varepsilon_i(b)-1},$$

and also for every $b \in \mathbf{B}$ and $i \in I$ such that $f_i b \neq 0$ there exists a unique element $b'' \in \mathbf{B}$ such that $\varphi_i(b') = \varphi_i(b) - 1$ and

$$f_i b \in \mathbb{C}^{\times} \cdot b'' + V_{-i}^{<\varphi_i(b)-1}$$

To a perfect basis $\mathbf{B}$ we can associate a crystal $\tilde{\mathbf{B}}$ in the obvious way (edge of color i connecting b and b' as above, etc.) It follows from the main result of [BeKa] that no matter which perfect basis we take, the crystal $\tilde{\mathbf{B}}$ is the same and it is isomorphic to Kashiwara's crystal associated to the crystal base of V.

Now, by Theorems 5.5.1(i) and 8.5.9(i), the classes of the irreducible $\mathcal{H}_n^{\lambda}$-modules form a perfect basis of $K(\lambda)$. And the result above (providing it holds for Kac–Moody algebras), together with Theorem 9.5.1, imply Theorem 10.3.5, which in turn implies Theorem 10.3.4.

11
Symmetric groups II

In this chapter we specialize to the case $\lambda = \Lambda_0$. In this case $\mathcal{H}_n^\lambda \cong FS_n$, so we will be getting results on symmetric groups. In particular, we obtain a classification of the irreducible FS_n-modules, describe how the irreducibles split into blocks ("Nakayama's Conjecture"), and prove some of the branching rules of [K$_1$, K$_2$, K$_5$, K$_4$, BK$_1$]. As $\lambda = \Lambda_0$ is fixed throughout the chapter, we will not use the superscripts and just write e_i for $e_i^{\Lambda_0}$, $\tilde{f}_i$ for $\tilde{f}_i^{\Lambda_0}$, etc. These should not be confused with the corresponding notions for the affine Hecke algebra $\mathcal{H}_n$.

11.1 Description of the crystal graph

In the case $\lambda = \Lambda_0$, Misra and Miwa [MiMi] described the crystal $B(\Lambda_0)$ in terms of Young diagrams, which we now explain.

We will use the terminology concerning partitions introduced in Chapter 1. In particular, recall the definition of the residue content $\mathrm{cont}(\alpha)$ of a partition α from (1.2).

Fix a partition α of n. We define the important notions of *normal* and *good* nodes. Label all i-addable nodes of the diagram α by $+$ and all i-removable nodes by $-$. Then the *i-signature* of α is the sequence of pluses and minuses obtained by going along the rim of the Young diagram from bottom left to top right and reading off all the signs. The *reduced i-signature* of α is obtained from the i-signature by successively erasing all neighboring pairs of the form $-+$.

Note the reduced i-signature always looks like a sequence of $+$s followed by $-$s. Nodes corresponding to a "$-$" in the reduced i-signature are called *i-normal*, nodes corresponding to a "$+$" are called *i-conormal*. The leftmost i-normal node (corresponding to the leftmost $-$ in the reduced i-signature) is called *i-good*, and, similarly, the rightmost i-conormal node is called *i-cogood*. A node is called *normal* (resp. *conormal*, *good*, *cogood*) if it is i-normal (resp. i-conormal, i-good, i-cogood) for some i.

Note that the notions just introduced are pretty boring for the case $p = 0$: every removable node is normal and good and every addable node is conormal and cogood.

Example 11.1.1 Let $p = 5$, $\alpha = (14, 11, 10, 10, 9, 4, 1)$. The residues are as follows:

0	1	2	3	4	0	1	2	3	4	0	1	2	3
4	0	1	2	3	4	0	1	2	3	4			
3	4	0	1	2	3	4	0	1	2				
2	3	4	0	1	2	3	4	0	1				
1	2	3	4	0	1	2	3	4					
0	1	2	3										
4													

The residue content of α is $\gamma = (\gamma_i)_{i \in I}$ where

$$\gamma_0 = 11, \ \ \gamma_1 = 12, \ \ \gamma_2 = 12, \ \ \gamma_3 = 12, \ \ \gamma_4 = 12.$$

The 4-addable and 4-removable nodes are labeled in the diagram:

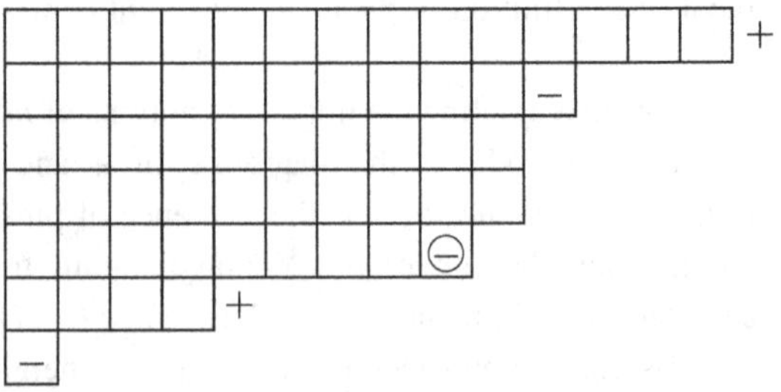

Hence, the 4-signature of α is

$$-, +, -, -, +$$

and the reduced 4-signature is $-$. The corresponding node is circled in the above diagram. So, there is one 4-normal node, which is also 4-good; there are no 4-conormal or 4-cogood nodes.

In general, we define

$$\begin{aligned} \varepsilon_i(\alpha) &= \sharp\{i\text{-normal nodes in } \alpha\} \\ &= \sharp\{-\text{'s in the reduced } i\text{-signature of } \alpha\}, \end{aligned} \tag{11.1}$$

$$\begin{aligned} \varphi_i(\alpha) &= \sharp\{i\text{-conormal nodes in } \alpha\} \\ &= \sharp\{+\text{'s in the reduced } i\text{-signature of } \alpha\}. \end{aligned} \tag{11.2}$$

Also set

$$\tilde{e}_i(\alpha) = \begin{cases} \alpha_A & \text{if } \varepsilon_i(\alpha) > 0 \text{ and } A \text{ is the (unique) } i\text{-good node,} \\ 0 & \text{if } \varepsilon_i(\alpha) = 0, \end{cases} \tag{11.3}$$

$$\tilde{f}_i(\alpha) = \begin{cases} \alpha^B & \text{if } \varphi_i(\alpha) > 0 \text{ and } B \text{ is the (unique)} \\ & \quad i\text{-cogood node,} \\ 0 & \text{if } \varphi_i(\alpha) = 0. \end{cases} \tag{11.4}$$

The definitions imply that $\tilde{e}_i(\alpha), \tilde{f}_i(\alpha)$ are p-regular (or zero) in case α is itself p-regular. Finally define

$$\mathrm{wt}(\alpha) = \Lambda_0 - \sum_{i \in I} \gamma_i \alpha_i \tag{11.5}$$

where $\mathrm{cont}(\alpha) = (\gamma_i)_{i \in I}$. We have now defined a datum

$$(\mathcal{P}_p, \varepsilon_i, \varphi_i, \tilde{e}_i, \tilde{f}_i, \mathrm{wt})$$

which makes the set $\mathcal{P}_p$ of all p-regular partitions into a crystal in the sense of Section 10.2 (O.K., the axiom (C1) is perhaps not so obvious. If you cannot prove it directly, no problem – it will follow from Theorem 11.1.3 anyway).

Example 11.1.2 The first nine levels of the crystal graph $\mathcal{P}_2$ are as follows:

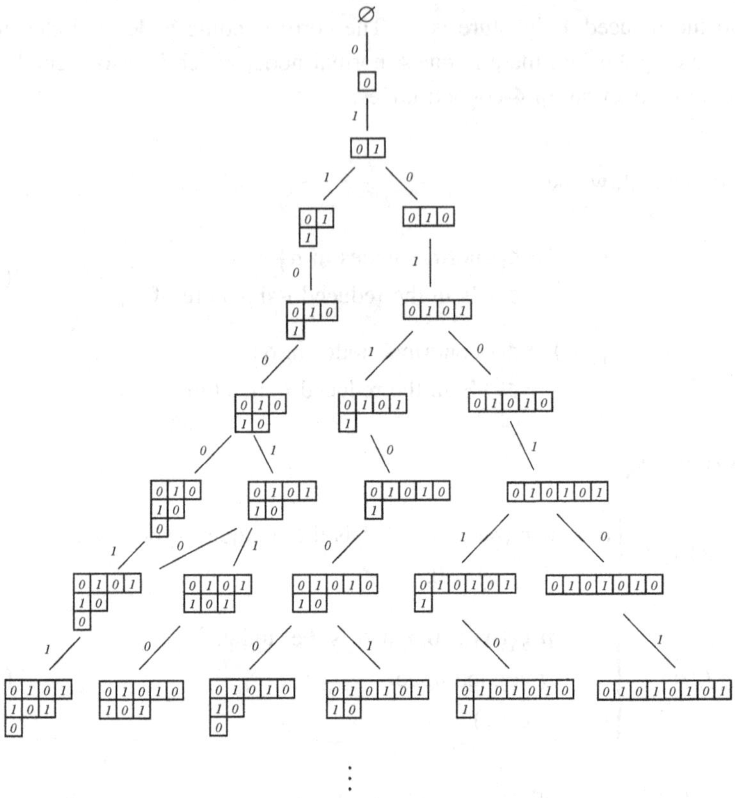

We now state the result of Misra and Miwa [MiMi]:

Theorem 11.1.3 *The set* $\mathcal{P}_p$ *equipped with* $\varepsilon_i, \varphi_i, \tilde{e}_i, \tilde{f}_i, \mathrm{wt}$ *as above is isomorphic (in the unique way) to the crystal* $B(\Lambda_0)$ *associated to the integrable highest weight* $U_{\mathbb{Q}}$*-module of fundamental highest weight* Λ_0.

We now introduce some more combinatorial notions relevant to modular representation theory of S_n, see [JK, 2.7]. Until the end of the section we assume that $p > 0$. The *rim* of a Young diagram α is its south-east border. A *rim p-hook* of α is a connected part of the rim with p nodes and such that its removal leaves a Young diagram of a partition. If α has no rim p-hooks, it is called a *p-core*. In general, the p-core $\tilde{\alpha}$ of α is obtained by successively removing rim p-hooks, until it is reduced to a core (we check that $\tilde{\alpha}$ is well defined). The *p-weight* of α, denoted $w(\alpha)$, is then the total number of rim p-hooks that need to be removed from α before we reach the p-core.

It is easy to see using the *abacus* notation (see [JK, 2.7] for details) that, for $\beta, \alpha \in \mathcal{P}(n)$,

$$\operatorname{cont}(\beta) = \operatorname{cont}(\alpha) \text{ if and only if } \tilde{\beta} = \tilde{\alpha}. \tag{11.6}$$

We now give a Lie theoretic interpretation of the combinatorial notions introduced in the previous paragraph, following the ideas of [LLT, Section 5.3]. Let W be the (affine) Weyl group corresponding to $\mathfrak{g}$. Recall from [Kc, Section 12] that the set of weights of $V(\Lambda_0)$ is

$$P(\Lambda_0) = \{w\Lambda_0 - k\delta \mid w \in W,\ k \in \mathbb{Z}_{\geq 0}\}.$$

Moreover, let $\operatorname{Par}_m(n)$ denote the number of partitions of n as a sum of positive integers of m different colors. Then the multiplicity of the weight $w\Lambda_0 - k\delta$ in $V(\Lambda_0)$ equals $\operatorname{Par}_{p-1}(k)$, see [Kc, Section 12.13]. As W leaves δ invariant, the W-orbits on $P(\Lambda_0)$ are precisely $X_0, X_1, X_2, \dots$, where

$$X_k := \{w\Lambda_0 - k\delta \mid w \in W\}. \tag{11.7}$$

Elements of X_0 are called *extremal weights*.

Lemma 11.1.4 *The weight $\nu = \Lambda_0 - \sum_{i\in I}\gamma_i\alpha_i \in P(\Lambda_0)$ belongs to X_k if and only if $k = \gamma_0 + \sum_{i\in I}(\gamma_i\gamma_{i+1} - \gamma_i^2)$.*

Proof We have $\nu \in X_k$ if and only if $\nu = w\Lambda_0 - k\delta$. Let $(\cdot \mid \cdot)$ be the normalized invariant form from [Kc, Section 6.2]. Note that

$$k = -(w\Lambda_0 - k\delta \mid w\Lambda_0 - k\delta)/2.$$

So it remains to notice that

$$-(\Lambda_0 - \sum_{i\in I}\gamma_i\alpha_i \mid \Lambda_0 - \sum_{i\in I}\gamma_i\alpha_i)/2 = \gamma_0 + \sum_{i\in I}(\gamma_i\gamma_{i+1} - \gamma_i^2).$$

□

Proposition 11.1.5 *Let α be a p-regular partition. Then $w(\alpha) = k$ if and only if $\operatorname{wt}(\alpha) \in X_k$. In particular, α is a p-core if and only if $\operatorname{wt}(\alpha)$ is an extremal weight.*

Proof Let $w(\alpha) = k$ and $\operatorname{wt}(\alpha) \in X_l$. As the removal of a rim p-hook from a p-regular partition α leads to the subtraction of δ from $\operatorname{wt}(\alpha)$, it follows that $l \geq k$. In particular, $l = 0$ implies that α is a p-core. So there is a p-core β with $\operatorname{wt}(\beta) = \operatorname{wt}(\alpha) - l\delta$. Attaching l rim p-hooks horizontally to the first row of β we get a partition γ with $\tilde{\gamma} = \beta$ and $\operatorname{wt}(\gamma) = \operatorname{wt}(\alpha)$, whence $\tilde{\gamma} = \tilde{\alpha}$, and so $k = l$. □

Now, Theorem 11.1.3 and Kac's formula [Kc, (12.13.5)] for the character of the highest weight $U_{\mathbb{Q}}$-module of highest weight Λ_0 imply a classical combinatorial result: for $\alpha \in \mathcal{P}_p(n)$,

$$\sharp\{\beta \in \mathcal{P}_p(n) \mid \mathrm{cont}(\beta) = \mathrm{cont}(\alpha)\} = \mathrm{Par}_{p-1}(w(\alpha)). \qquad (11.8)$$

11.2 Main results on S_n

By Theorem 10.3.5, the isomorphism classes of irreducible FS_n-modules are parametrized by the nodes of the crystal graph $B(\Lambda_0)$. By Theorems 11.1.3 and 10.3.5, we can identify $B(\Lambda_0)$ with $\mathcal{P}_p$. In other words, we can use the set $\mathcal{P}_p(n)$ of p-regular partitions of n to label the irreducible modules over FS_n for each $n \geq 0$.

Let us write D^α for the irreducible FS_n-module corresponding to $\alpha \in \mathcal{P}_p(n)$. To be precise

$$D^\alpha := L(i_1, \ldots, i_n)$$

if $\alpha = \tilde{f}_{i_n} \ldots \tilde{f}_{i_1} \varnothing$, see (5.14). Here the operator $\tilde{f}_i$ is as defined in (11.4), corresponding under the identification $\mathcal{P}_p(n) = B(\Lambda_0)$ to the crystal operator denoted $\tilde{f}_i^{\Lambda_0}$ in earlier sections, and $\varnothing$ denotes the empty partition, corresponding to $[\mathbf{1}_\lambda] \in B(\Lambda_0)$.

Theorem 11.2.1 *The modules $\{D^\alpha \mid \alpha \in \mathcal{P}_p(n)\}$ form a complete set of pairwise non-isomorphic irreducible FS_n-modules. Moreover, for $\alpha, \beta \in \mathcal{P}_p(n)$ we have:*

(i) *D^α is self-dual;*
(ii) *modules D^α and D^β belong to the same block of FS_n if and only if $\mathrm{cont}(\alpha) = \mathrm{cont}(\beta)$;*
(iii) *D^α is a projective module if and only if α is p-core.*

Proof We have already discussed the first statement of the theorem, being a consequence of our main results combined with Theorem 11.1.3. Now, (i) follows from Corollary 5.3.2, and (ii) is a special case of Corollary 9.6.2. For (iii), note that if D^α is projective then it is the only irreducible in its block, hence by (11.8), $\mathrm{Par}_{p-1}(w(\alpha)) = 1$. So either $w(\alpha) = 0$, or $p = 2$ and $w(\alpha) = 1$. Now if $w(\alpha) = 0$ then α is p-core so the contravariant form on the (1-dimensional) $\mathrm{wt}(\alpha)$-weight space of $K(\Lambda_0)_{\mathbb{Z}}$ is 1 (since $\mathrm{wt}(\alpha)$ is conjugate to Λ_0 under the action of the affine Weyl group). Hence, D^α is projective by

Theorem 9.5.1(ii). To rule out the remaining possibility $p = 2$, $w(\alpha) = 1$, we check in that case that the contravariant form on the wt(α)-weight space of $K(\Lambda_0)_{\mathbb{Z}}$ is 2. □

Remark 11.2.2 Our characterization of the irreducible FS_n-module D^α corresponding to a p-regular partition α is explicit but rather unpractical. Indeed, in order to say what D^α is, we need to do the following. First, find a path of length n in the crystal graph $B(\Lambda_0)$ from α to the empty partition $\varnothing$ – in practice this amounts to "destroying" α by successively removing good nodes, say, of residues $i_n, \dots, i_1$, where i_n is the residue of the first removed node, i_{n-1} is the residue of the second removed node, etc. This is cumbersome but explicit combinatorics. Moreover, some remarkable special sequences $i_n, \dots, i_1$ can be read off the formal character of D^α – this will be explained in Remark 11.2.14. The second step however is more unpleasant: this sequence of residues is then used to define D^α as $\tilde{f}_{i_n} \dots \tilde{f}_{i_1} D^\varnothing$ where $D^\varnothing$ is the trivial FS_0-module **1** (recall that FS_0 is interpreted as F). This means that the operation $\tilde{f}_i$ should be performed n times, that is n times we should induce, project to a block and take the head of the resulting module.

A much more direct approach to the construction of D^α is the classical approach of James [J]. For each partition α of n there is an explicit construction of a module S^α, the corresponding *Specht module*. Its head D^α is proved to be irreducible when α is p-regular, and the D^α form a complete set of irreducible FS_n-modules. Thus we only have to deal with taking the head once. The natural question arises if our D^α are the same as James' D^α. The answer to this question is "yes", but unfortunately the only proof we know is somewhat unsatisfactory. Namely, we compare the "socle branching rule" for James' D^α, established in [K_1] by completely different methods, with Theorem 11.2.7(i),(ii) below, and observe that the two branching rules are exactly the same. This is enough to identify the two labelings of irreducible modules. It would be interesting to find a more direct proof.

Now, consider in more detail the natural surjection

$$\pi : \mathcal{H}_n \twoheadrightarrow \mathcal{H}_n / \mathcal{J}_{\Lambda_0} = \mathcal{H}_n^{\Lambda_0} \cong FS_n$$

(see Remark 7.5.7). Under this surjection the Coxeter generator s_m of $\mathcal{H}_n$ maps on to the Coxeter generators s_m of FS_n, for any $1 \leq m < n$. We now describe the images of the polynomial generators $x_k \in \mathcal{H}_n$. Recall the JM-elements L_k from (2.1).

Lemma 11.2.3 *For any $1 \le k \le n$, we have $\pi(x_k) = L_k$.*

Proof Induction on k. For $k = 1$ we know that $\pi(x_1) = 0$ because $\mathcal{J}_{\Lambda_0} = \mathcal{H}_n x_1 \mathcal{H}_n$. However, L_1 is also zero. For the induction step, just observe that $L_k = s_k L_{k-1} s_k + s_k$ (clear) and $x_k = s_k x_{k-1} s_k + s_k$ (see (3.6)). □

We can now use JM-elements to "avoid" affine Hecke algebras in definitions concerning symmetric groups. We noted in Section 2.1 that the JM-elements commute (it also follows from the fact that they are images of commuting elements under π). Lemma 7.3.1 implies that JM-elements act on FS_n-modules with "integral" eigenvalues, that is eigenvalues from $I = \mathbb{Z} \cdot 1 \subset F$. If $\underline{i} = (i_1, \dots, i_n) \in I^n$ and $M \in FS_n$-mod we define its *$\underline{i}$-weight space*

$$M_{\underline{i}} := \{v \in M \mid (L_k - i_k)^N v = 0 \text{ for } N \gg 0 \text{ and } k = 1, \dots, n\}.$$

Lemma 11.2.4 *Any $M \in FS_n$-*mod *decomposes as $M = \oplus_{\underline{i} \in I^n} M_{\underline{i}}$.*

Now define the *formal character* of an FS_n-module M by

$$\operatorname{ch} M := \sum_{\underline{i} \in I^n} (\dim M_{\underline{i}}) e^{\underline{i}}, \tag{11.9}$$

an element of the free $\mathbb{Z}$-module on the formal basis $\{e^{\underline{i}} \mid \underline{i} \in I^n\}$. From Theorem 5.3.1 we get:

Lemma 11.2.5 *Let $D_1, \dots, D_k$ be non-isomorphic simple FS_n-modules. then* $\operatorname{ch} D_1, \dots, \operatorname{ch} D_k$ *are $\mathbb{Z}$-linearly independent.*

It follows from Theorem 3.3.1 that the symmetric polynomials in the JM-elements form a central subalgebra in FS_n (Murphy [Mu_2, 1.9] proves that this central subalgebra actually equals the center of FS_n, but we will never need this fact). The characters of this central subalgebra are captured by the following data. Let $\gamma = \sum_{i \in I} \gamma_i \alpha_i \in \Gamma_n$ (see Section 8.1). For an FS_n-module M set

$$M[\gamma] := \sum_{\underline{i} \in I^n \text{ with } \operatorname{cont}(\underline{i}) = \gamma} M_{\underline{i}} \tag{11.10}$$

(see (8.1)).

Theorem 11.2.6 *For $\gamma = \sum_{i \in I} \gamma_i \alpha_i \in \Gamma_n$ the following are equivalent:*

(i) *there exists a finite dimensional FS_n-module M with $M[\gamma] \ne 0$;*
(ii) *$\Lambda_0 - \gamma$ is a weight of $V(\Lambda_0)$;*
(iii) *there exists a p-regular partition λ of n whose residue content is $(\gamma_i)_{i \in I}$.*

Moreover, two FS_n-modules M and N are in the same block if and only if $M = M[\delta]$ and $N = N[\delta]$ for some $\delta \in \Gamma_n$.

Proof Follows from Theorem 11.2.1(ii) and Corollary 9.6.2. □

Now, the functor

$$e_i : FS_n\text{-mod} \to FS_{n-1}\text{-mod}$$

is defined on a module M as the generalized i-eigenspace of the last JM-element L_n acting on M. This is considered as an FS_{n-1}-module via restriction to FS_{n-1}, see (5.1) and (5.6). The more general divided power functor

$$e_i^{(r)} : FS_n\text{-mod} \to FS_{n-r}\text{-mod}$$

on a module M is defined as follows (see Section 8.3). Let S'_r be the subgroup of S_n from (8.12). Now $e_i^{(r)}M$ is the space of S'_r-invariants in the simultaneous generalized i-eigenspace of the last r JM-elements $L_{n-r+1}, \dots, L_n$. Again, this is considered as an FS_{n-r}-module via restriction. As explained in Section 8.3, taking S'_r-coinvariants or S'_r-anti(co)invariants instead of invariants yields isomorphic functors.

As in Section 8.3, in the definitions above, instead of taking generalized eigenspaces, we could just project to appropriate blocks of FS_{n-r}. More precisely, assume first that M belongs to a fixed block of FS_n, i.e. $M = M[\gamma]$ for $\gamma \in \Gamma_n$. Then the simultaneous generalized i-eigenspace of the last r JM-elements on M equals $\mathrm{res}^{S_n}_{S_{n-r}} M[\gamma - r\alpha_i]$ (restriction followed by the projection to the block of S_{n-r} corresponding to $\gamma - r\alpha_i$). Finally, we extend $e_i^{(r)}$ to an arbitrary M by additivity. This approach is used to define the dual functors

$$f_i^{(r)} : FS_n\text{-mod} \to FS_{n+r}\text{-mod}.$$

On a module $M \in FS_n\text{-mod}[\gamma]$, we have

$$f_i^{(r)}M = \left(\mathrm{ind}^{S_{n+r}}_{S_n \times S_r} M \boxtimes \mathbf{1}_{S_r}\right)[\gamma + r\alpha_i],$$

see (8.15). Then the functor is extended to an arbitrary module by additivity. Taking the sign module $\mathbf{sgn}_{S_r}$ instead of the trivial module $\mathbf{1}_{S_r}$ leads to an isomorphic functor. Set $f_i := f_i^{(1)}$.

The next two theorems summarize earlier results concerning restriction and induction in the special case of the symmetric groups.

Theorem 11.2.7 *Let $\alpha \in \mathcal{P}_p(n)$. We have*

$$\mathrm{res}_{FS_{n-1}} D^\alpha \cong e_0 D^\alpha \oplus e_1 D^\alpha \oplus \cdots \oplus e_{p-1} D^\alpha,$$

and, for each $i \in I$, $e_i D^\alpha \neq 0$ if and only if α has an i-good node A, in which case $e_i D^\alpha$ is a self-dual indecomposable module with irreducible socle and head isomorphic to D^{α_A}. Moreover, if $i \in I$ and α has an i-good node A, then:

(i) *the multiplicity of D^{α_A} in $e_i D^\alpha$ is $\varepsilon_i(\alpha)$, $\varepsilon_i(\alpha_A) = \varepsilon_i(\alpha) - 1$, and $\varepsilon_i(\beta) < \varepsilon_i(\alpha) - 1$ for all other composition factors D^β of $e_i D^\alpha$;*

(ii) *The algebra $\mathrm{End}_{FS_{n-1}}(e_i D^\alpha)$ is isomorphic to the algebra of truncated polynomials $F[x]/(x^{\varepsilon_i(\alpha)})$. In particular,*

$$\dim \mathrm{End}_{FS_{n-1}}(e_i D^\alpha) = \varepsilon_i(\alpha);$$

(iii) *$\mathrm{Hom}_{FS_{n-1}}(e_i D^\alpha, e_i D^\beta) = 0$ for all $\beta \in \mathcal{P}_p(n)$ with $\beta \neq \alpha$;*

(iv) *$e_i D^\alpha$ is irreducible if and only if $\varepsilon_i(\alpha) = 1$. In particular, the restriction $\mathrm{res}_{FS_{n-1}} D^\alpha$ is completely reducible if and only if $\varepsilon_i(\alpha) \leq 1$ for every $i \in I$, and $\mathrm{res}_{FS_{n-1}} D^\alpha$ is irreducible if and only if $\sum_{i \in I} \varepsilon_i(\alpha) = 1$.*

Proof The first statement follows from (8.7) and Theorem 8.2.5(i), combined as usual with Theorem 11.1.3. For the remaining properties, (i),(ii), and (iii) follow from Theorem 5.5.1 and Corollary 5.5.2. Finally, (iv) follows from (i) as $e_i D^\alpha$ is a module with irreducible socle and head both isomorphic to D^{α_A}. □

Theorem 11.2.8 *Let $\alpha \in \mathcal{P}_p(n)$. We have*

$$\mathrm{ind}^{FS_{n+1}} D^\alpha \cong f_0 D^\alpha \oplus f_1 D^\alpha \oplus \cdots \oplus f_{p-1} D^\alpha,$$

and, for each $i \in I$, $f_i D^\alpha \neq 0$ if and only if α has an i-cogood node B, in which case $f_i D^\alpha$ is a self-dual indecomposable module with irreducible socle and head isomorphic to D^{α^B}. Moreover, if $i \in I$ and α has an i-cogood node B, then:

(i) *the multiplicity of D^{α^B} in $f_i D^\alpha$ is $\varphi_i(\alpha)$, $\varphi_i(\alpha^B) = \varphi_i(\alpha) - 1$, and $\varphi_i(\beta) < \varphi_i(\alpha) - 1$ for all other composition factors D^β of $f_i D^\alpha$;*

(ii) *the algebra $\mathrm{End}_{FS_{n+1}}(f_i D^\alpha)$ is isomorphic to the algebra of truncated polynomials $F[x]/(x^{\varphi_i(\alpha)})$. In particular,*

$$\dim \mathrm{End}_{FS_{n+1}}(f_i D^\alpha) = \varphi_i(\alpha);$$

(iii) *$\mathrm{Hom}_{FS_{n+1}}(f_i D^\alpha, f_i D^\beta) = 0$ for all $\beta \in \mathcal{P}_p(n)$ with $\beta \neq \alpha$;*

(iv) *$f_i D^\alpha$ is irreducible if and only if $\varphi_i(\alpha) = 1$. In particular, $\mathrm{ind}^{FS_{n+1}} D^\alpha$ is completely reducible if and only if $\varphi_i(\alpha) \leq 1$ for every $i \in I$.*

Proof The argument is the same as in Theorem 11.2.7, but using (8.7), Theorem 8.2.5(ii), Corollary 8.5.2, and Theorem 8.5.9. □

Remark 11.2.9 We state without proof some further branching results mentioned in the introduction, which do not seem to follow from the methods developed here. Let $\alpha \in \mathcal{P}_p(n)$. The following is proved in [K$_4$] (resp. [BK$_1$]). Suppose that A is an i-removable (resp. i-addable) node of α such that α_A (resp. α^A) is p-regular. Then, $[e_i D^\alpha : D^{\alpha_A}]$ (resp. $[f_i D^\alpha : D^{\alpha^A}]$ is the number of i-normal (resp. i-conormal) nodes to the right (resp. left) of A, counting A itself, or 0 if A is not i-normal (resp. i-conormal).

We also have somewhat less strong results on $e_i^{(r)}$ and $f_i^{(r)}$.

Theorem 11.2.10 *Let $\alpha \in \mathcal{P}_p(n)$, $i \in I$ and $r \geq 1$. We have*

$$e_i^r D^\alpha \cong (e_i^{(r)} D^\alpha)^{\oplus r!},$$

and $e_i^{(r)} D^\alpha \neq 0$ if and only if α has at least r i-normal nodes, in which case $e_i^{(r)} D^\alpha$ is a self-dual indecomposable module with irreducible socle and head isomorphic to D^β, where β is obtained from α by removing r bottom i-normal nodes. Moreover:

(i) *the multiplicity of D^β in $e_i^{(r)} D^\alpha$ is $\binom{\varepsilon_i(\alpha)}{r}$, $\varepsilon_i(\beta) = \varepsilon_i(\alpha) - r$, and $\varepsilon_i(\gamma) < \varepsilon_i(\alpha) - r$ for all other composition factors D^γ of $e_i^{(r)} D^\alpha$;*
(ii) *The endomorphism algebra $\mathrm{End}_{FS_{n-r}}(e_i^{(r)} D^\alpha)$ is isomorphic to the algebra $\mathcal{Z}_r/(\mathcal{Z}_r \cap \mathcal{P}_{\varepsilon_i(\alpha)} \mathcal{Z}^+_{\varepsilon_i(\alpha)})$ of (8.24). In particular,*

$$\dim \mathrm{End}_{FS_{n-r}}(e_i^{(r)} D^\alpha) = \binom{\varepsilon_i(\alpha)}{r};$$

(iii) $\mathrm{Hom}_{FS_{n-r}}(e_i^{(r)} D^\alpha, e_i^{(r)} D^\gamma) = 0$ *for all $\gamma \in \mathcal{P}_p(n)$ with $\gamma \neq \alpha$;*
(iv) *$e_i^{(r)} D^\alpha$ is irreducible if and only if $r = \varepsilon_i(\alpha)$.*

Proof The first statement follows from Theorem 8.3.2. For the remaining properties, (i) and (iii) follow from Proposition 8.5.10(i) and (ii) respectively; (ii) comes from Remark 8.6.3 and Proposition 8.6.2, and (iv) follows from (i). □

Theorem 11.2.11 *Let $\alpha \in \mathcal{P}_p(n)$, $i \in I$ and $r \geq 1$. We have*

$$f_i^r D^\alpha \cong (f_i^{(r)} D^\alpha)^{\oplus r!},$$

and $f_i^{(r)} D^\alpha \neq 0$ if and only if α has at least r i-conormal nodes, in which case $f_i^{(r)} D^\alpha$ is a self-dual indecomposable module with irreducible socle and head *isomorphic to* D^β*, where* β *is obtained from* α *by adding* r *top i-conormal nodes. Moreover:*

(i) *the multiplicity of* D^β *in* $f_i^{(r)}D^\alpha$ *is* $\binom{\varphi_i(\alpha)}{r}$, $\varphi_i(\beta) = \varphi_i(\alpha) - r$, *and* $\varphi_i(\gamma) < \varphi_i(\alpha) - r$ *for all other composition factors* D^γ *of* $f_i^{(r)}D^\alpha$*;*
(ii) $\dim \mathrm{End}_{FS_{n+r}}(f_i^{(r)}D^\alpha) = \binom{\varphi_i(\alpha)}{r}$;
(iii) $\mathrm{Hom}_{FS_{n+r}}(f_i^{(r)}D^\alpha, f_i^{(r)}D^\gamma) = 0$ *for all* $\gamma \in \mathcal{P}_p(n)$ *with* $\gamma \neq \alpha$*;*
(iv) $f_i^{(r)}D^\alpha$ *is irreducible if and only if* $r = \varphi_i(\alpha)$.

Proof Similar to the proof of Theorem 11.2.10 using Theorem 8.3.2, Proposition 8.5.10(i′),(ii′), and (8.25). □

Let M be an FS_n-module. Define

$$\begin{aligned}\varepsilon_i(M) &= \max\{r \geq 0 \mid e_i^r M \neq 0\},\\ \varphi_i(M) &= \max\{r \geq 0 \mid f_i^r M \neq 0\}.\end{aligned} \tag{11.11}$$

From Theorems 11.2.10 and 11.2.11 we have:

Lemma 11.2.12 *Let* $\alpha \in \mathcal{P}_p(n)$*. Then:*

(i) $\varepsilon_i(D^\alpha) = \max\{r \geq 0 \mid \tilde{e}_i^r M \neq 0\} = \varepsilon_i(\alpha)$.
(ii) $\varphi_i(D^\alpha) = \max\{r \geq 0 \mid \tilde{f}_i^r M \neq 0\} = \varphi_i(\alpha)$.

Note that $\varepsilon_i(M)$ can be computed just from knowledge of the character of M: it is the maximal r such that $e^{(\dots,i^r)}$ appears with non-zero coefficient in $\mathrm{ch}\, M$. Less obviously, $\varphi_i(M)$ can also be read off from the character of M. By additivity of f_i, we may assume that $M = M[\gamma]$ for $\gamma \in \Gamma_n$. Then, from (9.20) we get:

$$\varphi_i(M) = \varepsilon_i(M) + \delta_{i,0} - 2\gamma_i + \gamma_{i-1} + \gamma_{i+1}, \tag{11.12}$$

We record the effect of $e_i^{(r)}$ on formal characters:

Lemma 11.2.13 *Let* $M \in FS_n$-mod *and* $\mathrm{ch}\, M = \sum_{\underline{i} \in I^n} a_{\underline{i}} e^{\underline{i}}$*. Then* $\mathrm{ch}\,(e_i^{(r)}M) = \sum_{\underline{i} \in I^{n-r}} a_{(i_1,\dots,i_{n-r},i^r)} e^{\underline{i}}$.

Proof For $r = 1$ this follows from the definition. For $r > 1$, use the fact that $e_i^r = (e_i^{(r)})^{\oplus r!}$. □

Remark 11.2.14 We describe an inductive algorithm to determine the label of an irreducible FS_n-module D purely from knowledge of its character $\mathrm{ch}\, D$. Pick $i \in I$ such that $\varepsilon := \varepsilon_i(D)$ is non-zero. Let $E = e_i^{(\varepsilon)}D$. In view of

Theorem 11.2.10(iv), E is an irreducible $FS_{n-\varepsilon}$-module, isomorphic to $\tilde{e}_i^{\varepsilon} D$. By Lemma 5.2.3, $D \cong \tilde{f}_i^{(\varepsilon)} E$. Moreover, the formal character of E is explicitly known by Lemma 11.2.13. By induction, the label β of E can be computed purely from knowledge of its character. Then, $D \cong \tilde{f}_i^{\varepsilon} D^{\beta} \cong D^{\alpha}$, where α is obtained from β by adding the rightmost ε of the i-conormal nodes.

We would of course like to be able to reverse this process: given a p-regular partition α of n, we would like to be able to compute the character of the irreducible FS_n-module D^{α}. We can compute a quite effective *lower bound* for this character inductively using Remark 11.2.9. But only over $\mathbb{C}$ is this lower bound always correct: indeed if $p = 0$ then from combinatorics we always have $\varepsilon_i \leq 1$ and so

$$\operatorname{ch} D^{\alpha} = \sum_{(i_1,\dots,i_n)} e^{(i_1,\dots,i_n)} \tag{11.13}$$

summing over all paths $\varnothing \xrightarrow{i_1} \alpha^{(1)} \xrightarrow{i_2} \cdots \xrightarrow{i_n} \alpha$ in the characteristic zero crystal graph (that is Young's partition lattice) from $\varnothing$ to α. Of course, we already know this result from Chapter 2.

Remark 11.2.15 There is an interesting class of modules in characteristic p where the lower bound just mentioned gives a correct answer for the formal character. These modules are called *completely splittable* and can be characterized as FS_n-modules whose restriction to any standard subgroup in S_n is completely reducible. For more information on this see [K_3, M, R].

Remark 11.2.16 Reducing the entries i_k in (11.13) modulo p gives the formal characters of the Specht module in characteristic p – this follows from the branching rule for Specht modules [J, 9.3].

Next we explain a useful inductive method for finding some composition factors of FS_n-modules using their formal characters and induction. It is based on the following:

Lemma 11.2.17 *Let M be an FS_n-module and set $\varepsilon = \varepsilon_i(M)$. If $[e_i^{(\varepsilon)} M : D^{\alpha}] = m > 0$ then $\tilde{f}_i^{\varepsilon} D^{\alpha} \neq 0$ and $[M : \tilde{f}_i^{\varepsilon} D^{\alpha}] = m$.*

Proof Follows from Theorem 11.2.10(iv) and Lemmas 11.2.5, 5.2.3. □

Using the known characters of Specht modules (see Remark 11.2.16) and Lemma 11.2.17 provides new non-trivial information on decomposition numbers, which is difficult to obtain by other methods.

Example 11.2.18 Let $p = 3$. By [J, Tables], the composition factors of the Specht module $S^{(6,4,2,1)}$ are $D^{(12,1)}$, $D^{(9,4)}$, $D^{(9,2^2)}$, $D^{(7,4,2)}$, $D^{(6,5,2)}$, $D^{(6,4,3)}$, and $D^{(6,4,2,1)}$, all appearing with multiplicity 1. As $\varepsilon_1(S^{(6,4,2^2)}) = 1$ and $e_1 S^{(6,4,2^2)} = S^{(6,4,2,1)}$ by Remark 11.2.16, application of Lemma 11.2.17 implies that the following composition factors appear in $S^{(6,4,2^2)}$ with multiplicity 1: $D^{(12,1^2)}$, $D^{(9,4,1)}$, $D^{(9,3,2)}$, $D^{(8,4,2)}$, $D^{(6^2,2)}$, $D^{(6,4^2)}$, and $D^{(6,4,2^2)}$.

Given $\underline{i} = (i_1, \dots, i_n) \in I^n$ we can gather consecutive equal terms to write it in the form

$$\underline{i} = (j_1^{m_1} \dots j_r^{m_r}) \tag{11.14}$$

where $j_s \neq j_{s+1}$ for all $1 \leq s < r$. For example $(2, 2, 2, 1, 1) = (2^3 1^2)$. Now, for an FS_n-module M, the weight (11.14) is called *extremal* if

$$m_s = \varepsilon_{j_s}(e_{j_{s+1}}^{m_{s+1}} \dots e_{j_r}^{m_r} M)$$

for all $s = r, r-1, \dots, 1$ (do not confuse with extremal weights for $\mathfrak{g}$ as in Section 11.1). Informally speaking this means that among all weights $\underline{i}$ of M we first choose those with the longest j_r-string in the end, then among these we choose the ones with the longest j_{r-1}-string preceding the j_r-string in the end, etc. By definition $M_{\underline{i}} \neq 0$ if $\underline{i}$ is extremal for M.

Example 11.2.19 The formal character of the Specht module $S^{(5,2)}$ in characteristic 3 is

$$e^{(0210201)} + 2e^{(0120201)} + 2e^{(02120^2 1)} + 4e^{(012^2 0^2 1)}$$
$$+ e^{(0212010)} + 2e^{(012^2 010)} + e^{(0120210)} + e^{(0120120)}.$$

The extremal weights are $(012^2 0^2 1)$, $(012^2 010)$, (0120210), (0120120).

Our main result about extremal weights is:

Theorem 11.2.20 *Let $\underline{i} = (i_1, \dots, i_n) = (j_1^{m_1} \dots j_r^{m_r})$ be an extremal weight for an irreducible FS_n-module D^α written in the form (11.14). Then $D^\alpha = \tilde{f}_{i_n} \dots \tilde{f}_{i_1} D^\varnothing$, and $\dim D^\alpha_{\underline{i}} = m_1! \dots m_r!$. In particular, the weight $\underline{i}$ is not extremal for any irreducible $D^\beta \not\cong D^\alpha$.*

Proof We apply induction on r. If $r = 1$, then by considering possible weights appearing in the Specht module S^α, of which D^α is a quotient, we conclude that $n = 1$ and $D = D^{(1)}$. So for $r = 1$ the result is obvious. Let $r > 1$. By definition of an extremal weight, $m_r = \varepsilon_{j_r}(D^\alpha)$. So, in view of Theorem 11.2.10, we have

$$e_{j_r}^{m_r} D^\alpha = (\tilde{e}_{j_r}^{m_r} D^\alpha)^{\oplus m_r!}.$$

Moreover, $(j_1^{m_1} \dots j_{r-1}^{m_{r-1}})$ is clearly an extremal weight for the irreducible module $\tilde{e}_{j_r}^{m_r} D^\alpha$. So the inductive step follows: □

Corollary 11.2.21 *If M is an FS_n-module and $\underline{i} = (i_1, \dots, i_n) = (j_1^{m_1} \dots j_r^{m_r})$ is an extremal weight for M written in the form (11.14), then the multiplicity of $D^\alpha := \tilde{f}_{i_n} \dots \tilde{f}_{i_1} D^\varnothing$ as a composition factor of M is* $\dim M_{\underline{i}}/(m_1! \dots m_r!)$.

Example 11.2.22 In view of Corollary 11.2.21, the extremal weight $(012^2 0^2 1)$ in Example 11.2.19 yields the composition factor $D^{(5,2)}$ of $S^{(5,2)}$, while the extremal weight (0120120) yields the composition factor $D^{(7)}$. It turns out that these are exactly the composition factors of $S^{(5,2)}$, see For example [J, Tables].

For more non-trivial examples let us consider a couple of Specht modules for $n = 11$ in characteristic 3. For $S^{(6,3,1^2)}$, Corollary 11.2.21 yields composition factors $D^{(6,3,1^2)}$, $D^{(7,3,1)}$, and $D^{(8,2,1)}$ but 'misses' $D^{(11)}$, and for $S^{(4,3,2^2)}$ we get hold of $D^{(4,3,2^2)}$, $D^{(5,3,2,1)}$, $D^{(8,2,1)}$, and $D^{(8,3)}$, but 'miss' $2D^{(11)}$ and $D^{(5,4,1^2)}$, cf. [J, Tables].

We record here some other useful general facts about formal characters:

Lemma 11.2.23 *For any weight $\underline{i}$ represented in the form (11.14) and any FS_n-module M we have that* $\dim M_{\underline{i}}$ *is divisible by $m_1! \dots m_r!$.*

Proof We can lift M to an $\mathcal{H}_n$-module. By Theorem 4.3.2, each composition factors of $\mathrm{res}^n_{m_1,\dots,m_r} M$ isomorphic to $L(j_1^{m_1}) \boxtimes \dots \boxtimes L(j_r^{m_r})$ contributes the multiplicity of $m_1! \dots m_r!$ to the $\underline{i}$-weight space of M, and no other composition factors contribute to this weight. □

Lemma 11.2.24 *Let M be an FS_n-module. Assume $i, j, i_1, \dots, i_{n-2} \in I$ and $i \neq j$:*

(i) *Assume that $|i - j| > 1$. Then for any $1 \le r \le n - 2$ we have*

$$\dim M_{(i_1,\dots,i_r,i,j,i_{r+1},\dots,i_{n-2})} = \dim M_{(i_1,\dots,i_r,j,i,i_{r+1},\dots,i_{n-2})}.$$

(ii) *Assume that* $|i-j|=1$ *and* $p>2$. *Then for any* $1\le r\le n-3$ *we have*

$$\begin{aligned}&2\dim M_{(i_1,\dots,i_r,i,j,i,i_{r+1},\dots,i_{n-3})}\\&=\dim M_{(i_1,\dots,i_r,i,i,j,i_{r+1},\dots,i_{n-3})}\\&\quad+\dim M_{(i_1,\dots,i_r,j,i,i,i_{r+1},\dots,i_{n-3})}.\end{aligned}$$

(iii) *Assume that* $|i-j|=1$ *and* $p=2$. *Then for any* $1\le r\le n-4$ *we have*

$$\begin{aligned}&\dim M_{(i_1,\dots,i_r,i,i,i,j,i_{r+1},\dots,i_{n-4})}\\&+3\dim M_{(i_1,\dots,i_r,i,j,i,i,i_{r+1},\dots,i_{n-4})}\\&=\dim M_{(i_1,\dots,i_r,j,i,i,i,i_{r+1},\dots,i_{n-4})}\\&\quad+3\dim M_{(i_1,\dots,i_r,i,i,j,i,i_{r+1},\dots,i_{n-4})}.\end{aligned}$$

Proof Follows from the Serre relations satisfied by the operators e_i, see Lemma 9.2.4. □

Combining Lemma 11.2.24 with Corollary 11.2.21 and Theorem 11.2.9 we obtain further non-trivial results on branching, which are difficult to get by other methods. To illustrate:

Example 11.2.25 We explain how to see that $D^{(7,3,2^2)}$ appears as a composition factor of $e_2D^{(6,4,3,2)}$. We have $\varepsilon_2(D^{(6,4,3,2)})=2$. Note that $\tilde{e}_2^2D^{(6,4,3,2)}=D^{(5,3^2,2)}$. So by Theorem 11.2.10(iv), the weights ending on $2,2$ appearing in the character of $D^{(6,4,3,2)}$ are all obtained just by adding $2,2$ to the end of the weights appearing in the character of $D^{(5,3^2,2)}$. Next, $\varepsilon_0(D^{(5,3^2,2)})=1$. So there is an extremal weight in the character of $D^{(5,3^2,2)}$ which ends at 0, say $(i_1,\dots,i_{12},0)$. Then the weight $(i_1,\dots,i_{12},0,2,2)$ appears in the character of $D^{(6,4,3,2)}$. By Lemma 11.2.24(ii), the weight $(i_1,\dots,i_{12},2,0,2)$ also appears in the character. This weight contributes $(i_1,\dots,i_{12},2,0)$ to the character of $e_2D^{(6,4,3,2)}$. Note that the character of $e_2D^{(6,4,3,2)}$ could not involve weights ending on $0,0$ because then the character of $D^{(6,4,3,2)}$ would involve a weight ending on $0,0,2$ and then, by Lemma 11.2.24(ii), the weight ending on $0,2,0$, which contradicts the fact that $\varepsilon_0(D^{(6,4,3,2)})=0$. Moreover, the character of $e_2D^{(6,4,3,2)}$ could not involve weights ending on $2,2,0$ because then the character of $D^{(6,4,3,2)}$ would involve a weight ending on $2,2,0,2$ and then, by Lemma 11.2.24(ii), the weight ending on $2,0,2,2$, which contradicts the fact that $\varepsilon_2(\tilde{e}_0\tilde{e}_2^2D^{(6,4,3,2)})=0$. The two facts just observed and the choice of $i_1,\dots,i_{12}$ imply that $(i_1,\dots,i_{12},2,0)$ is an extremal weight for the module $e_2 D^{(6,4,3,2)}$. So, by Corollary 11.2.21, this module has $D^{(7,3,2^2)} = \tilde{f}_0 \tilde{f}_2 \tilde{f}_{i_{12}} \cdots \tilde{f}_{i_1} D^{\varnothing}$ as a composition factor.

Finally we discuss some properties of blocks, assuming now that $p \neq 0$. The affine Weyl group W acts on the $\mathfrak{g}$-module

$$V(\Lambda_0) = \oplus_{n \geq 0} K(FS_n\text{-mod})_{\mathbb{Q}},$$

the generator s_i of W acting by the familiar formula

$$s_i = \exp(-e_i)\exp(f_i)\exp(-e_i) \qquad (i \in I)). \tag{11.15}$$

The resulting action preserves the Shapovalov form, and leaves the lattices $\oplus_{n\geq 0} K(FS_n\text{-mod})$ and $\oplus_{n \geq 0} K(FS_n\text{-proj})$ invariant. Moreover, W permutes the weight spaces of $V(\Lambda_0)$ in the same way as its defining action on the weight lattice P, the orbits being X_k, see (11.7) and the discussion at the end of Section 11.1. So using Proposition 11.1.5 we see:

Theorem 11.2.26 *Let B and B' be blocks of symmetric groups with the same p-weight. Then B and B' are isometric in the sense that there is an isomorphism between their Grothendieck groups that is an isometry with respect to the Cartan form.*

Remark 11.2.27 The existence of such isometries was first noticed by Enguehard [E]. Implicit in Enguehard's paper is the following conjecture, made formally by Rickard: *blocks B and B' of symmetric groups with the same p-weight should be derived equivalent.* Moreover, it is known by work of Marcus [Ma] and Chuang–Kessar [CK] that the Abelian Defect Group Conjecture of Broué for symmetric groups follows from the Rickard's conjecture above. The conjecture has been proved by Rickard for blocks of p-weight ≤ 5 and in full generality recently by Chuang and Rouquier [CR], cf. Remark 9.6.3. The complex of functors leading to the derived equivalence can be guessed by looking at the formula (11.15).

There is one situation when there is actually a *Morita* equivalence between blocks of the same p-weight. This is a theorem of Scopes [Sc], though we are stating the result in a more Lie theoretic way following [LM, Section 8]:

Theorem 11.2.28 *Let $\Lambda, \Lambda + \alpha_i, \ldots, \Lambda + r\alpha_i$ be an α_i-string of weights of $V(\Lambda_0)$ (so $\Lambda - \alpha_i$ and $\Lambda + (r+1)\alpha_i$ are not weights of $V(\Lambda_0)$). Then the functors $f_i^{(r)}$ and $e_i^{(r)}$ define mutually inverse Morita equivalences between the blocks parametrized by Λ and by $\Lambda + r\alpha_i$.*

Proof Since $e_i^{(r)}$ and $f_i^{(r)}$ are both left and right adjoint to one another, it suffices to check that $e_i^{(r)}$ and $f_i^{(r)}$ induce mutually inverse bijections between the isomorphism classes of irreducible modules belonging to the respective blocks. This follows from Theorems 11.2.10(iv) and 11.2.11(iv). □

Remark 11.2.29 We can use Lie Theory to explicitly compute the determinant of the Cartan matrix of a block. The details of the proof appear in [BK_6] (see also [BO_2] for a different approach). Note that, in view of Theorem 11.2.26, the determinant of the Cartan matrix only depends on the p-weight of the block. Moreover, by Theorem 9.5.1, we can work instead in terms of the Shapovalov form on $V(\Lambda_0)$. Using the explicit construction of the latter module over $\mathbb{Z}$ given in [CKK], we show: *if B is a block of p-weight w of FS_n, then the determinant of the Cartan matrix of B is p^N, where*

$$N = \sum_{\lambda=(1^{r_1}2^{r_2}\dots)\vdash w} \frac{r_1+r_2+\cdots}{p-1}\binom{p-2+r_1}{r_1}\binom{p-2+r_2}{r_2}\cdots.$$

PART II

Projective representations

Throughout Part II of the book F will stand for an algebraically closed field of characteristic $p \neq 2$.

The main goal of Part II is to develop a theory of *projective* or *spin* representations of symmetric and alternating groups which is parallel to the "classical" theory developed in Part I. Although in some places we do not go quite as far as in Part I, the similarity of the two theories is compelling. Informally speaking we just have to replace the Kac–Moody algebra of type $A_{p-1}^{(1)}$ with the twisted Kac–Moody algebra of type $A_{p-1}^{(2)}$.

Having this in mind, we will always indicate which chapter in Part I a given chapter in Part II is parallel to (starting with Chapter 14). The reader is then advised to first browse the corresponding chapter of Part I and especially read a motivation and informal explanations in the beginning of that chapter, since those will not be duplicated in Part II.

Out of two synonymous terms *projective representation* and *spin representation* we will stick with *spin representation* (or *spin module*), while the term *projective representation* or *projective supermodule* will be reserved for the usual homological algebra notion (direct summand of a free module).

12

Generalities on superalgebra

The language of superalgebra will be used throughout Part II of this book – this is convenient when dealing with spin representations of symmetric and alternating groups. In this chapter we review some known results concerning superalgebras and their modules. We recommend [Le, Chapter I], [Man, Chapter 3, Sections 1 and 2], and especially [Jos] as basic references. Alternatively, for reader's convenience we sketch the proofs of the results which are going to be used later.

When dealing with superalgebras, it is natural to assume that the characteristic of the ground field is different from 2. This restriction will not be a problem later when we study spin representations of S_n and A_n, because in characteristic 2 such representations are linear and hence have been treated in the first part of this book.

12.1 Superalgebras and supermodules

By a *(vector) superspace* we mean a $\mathbb{Z}_2$-graded vector space $V = V_{\bar{0}} \oplus V_{\bar{1}}$ over F. If $\dim V_{\bar{0}} = m$ and $\dim V_{\bar{1}} = n$ we write $\operatorname{sdim} V = (m, n)$ and $\dim V = m+n$. Elements of $V_{\bar{0}}$ are called *even* and elements of $V_{\bar{1}}$ are called *odd*. A vector is called *homogeneous* if it is either even or odd. Given a homogeneous vector $0 \neq v \in V$, we denote its *degree* by $\bar{v} \in \mathbb{Z}_2$. A subspace W of V is called a *subsuperspace* if it is homogeneous, that is $W = (W \cap V_{\bar{0}}) + (W \cap V_{\bar{1}})$. Define the linear map

$$\delta_V : V \to V, \quad v \mapsto (-1)^{\bar{v}} v$$

for homogeneous vectors v. Note it is typical in superalgebra to write expressions which only make sense for homogeneous elements, and the expected meaning for arbitrary elements is obtained by extending linearly from the homogeneous case. It is easy to see that a subspace of V is a subsuperspace if and only if it is stable under δ_V.

Given superspaces V and W, we view the direct sum $V \oplus W$ and the tensor product $V \otimes W$ as superspaces with $(V \oplus W)_i = V_i \oplus W_i$, and $(V \otimes W)_{\bar{0}} = V_{\bar{0}} \otimes W_{\bar{0}} \oplus V_{\bar{1}} \otimes W_{\bar{1}}$, $(V \otimes W)_{\bar{1}} = V_{\bar{0}} \otimes W_{\bar{1}} \oplus V_{\bar{1}} \otimes W_{\bar{0}}$. Also, we make the vector space $\mathrm{Hom}_F(V, W)$ of all linear maps from V to W into a superspace by declaring that $\mathrm{Hom}_F(V, W)_i$ consists of the *homogeneous maps of degree i* for each $i \in \mathbb{Z}_2$, that is, the linear maps $\theta : V \to W$ with $\theta(V_j) \subseteq W_{i+j}$ for $j \in \mathbb{Z}_2$. Elements of $\mathrm{Hom}_F(V, W)_{\bar{0}}$ will be referred to as *even* linear maps. The dual superspace V^* is $\mathrm{Hom}_F(V, F)$, where we view F as a superspace concentrated in degree $\bar{0}$.

A *superalgebra* is a vector superspace $\mathcal{A}$ with the additional structure of an associative unital F-algebra such that $\mathcal{A}_i\mathcal{A}_j \subseteq \mathcal{A}_{i+j}$ for $i, j \in \mathbb{Z}_2$. By forgetting the grading we may consider any superalgebra $\mathcal{A}$ as a usual algebra–this algebra will be denoted $|\mathcal{A}|$. By a *superideal* of $\mathcal{A}$ we mean a homogeneous ideal. Left and right superideals are defined similarly. The superalgebra $\mathcal{A}$ is called *simple* if if it has no non-trivial superideals. A *superalgebra homomorphism* $\theta : \mathcal{A} \to \mathcal{B}$ is an even linear map that is an algebra homomorphism in the usual sense; its kernel is a superideal. An *antiautomorphism* $\tau : \mathcal{A} \to \mathcal{A}$ of a superalgebra $\mathcal{A}$ is an even linear map which satisfies $\tau(ab) = \tau(b)\tau(a)$ (note there is no sign).

Given two superalgebras $\mathcal{A}$ and $\mathcal{B}$, we view the tensor product of superspaces $\mathcal{A} \otimes \mathcal{B}$ as a superalgebra with multiplication defined by

$$(a \otimes b)(a' \otimes b') = (-1)^{\bar{b}\bar{a}'}(aa') \otimes (bb') \qquad (a, a' \in \mathcal{A},\ b, b' \in \mathcal{B}). \quad (12.1)$$

We note that $\mathcal{A} \otimes \mathcal{B} \cong \mathcal{B} \otimes \mathcal{A}$, an isomorphism being given by the *supertwist map*

$$T_{\mathcal{A},\mathcal{B}} : \mathcal{A} \otimes \mathcal{B} \to \mathcal{B} \otimes \mathcal{A}, \quad a \otimes b \mapsto (-1)^{\bar{a}\bar{b}} b \otimes a \qquad (a \in \mathcal{A},\ b \in \mathcal{B}).$$

It is important that the tensor product of two superalgebras $\mathcal{A}$ and $\mathcal{B}$ is not the same as the tensor product of $\mathcal{A}$ and $\mathcal{B}$ as usual algebras (with natural $\mathbb{Z}_2$-grading) – the product rule is different!

Example 12.1.1 If V is a superspace with $\mathrm{sdim}\, V = (m, n)$ then

$$\mathcal{M}(V) := \mathrm{End}_F(V)$$

is a superalgebra with $\mathrm{sdim}\, \mathcal{M}(V) = (m^2 + n^2, 2mn)$. Moreover, if W is another finite dimensional superspace, we have an isomorphism of superalgebras

$$\mathcal{M}(V) \otimes \mathcal{M}(W) \cong \mathcal{M}(V \otimes W). \quad (12.2)$$

Under this isomorphism $f \otimes g$ corresponds to the endomorphism of $V \otimes W$, mapping $v \otimes w$ to $(-1)^{\bar{g}\bar{v}} f(v) \otimes g(w)$. In what follows we identify the algebras $\mathcal{M}(V) \otimes \mathcal{M}(W)$ and $\mathcal{M}(V \otimes W)$ using this isomorphism. In particular, it makes sense to speak of the element $f \otimes g \in \mathrm{End}_F(V \otimes W)$.

Moreover, the algebra $\mathcal{M}(V)$ is defined uniquely up to an isomorphism by the superdimension (m, n) of V. So we can speak of the superalgebra $\mathcal{M}_{m,n}$, which can also be identified with the obvious superalgebra of matrices. Now, (12.2) becomes

$$\mathcal{M}_{m,n} \otimes \mathcal{M}_{k,l} \cong \mathcal{M}_{mk+nl,ml+nk}. \tag{12.3}$$

Finally, we note that $\mathcal{M}_{m,n}$ is a simple superalgebra, as $|\mathcal{M}_{m,n}| = \mathcal{M}_{m+n}$, the algebra of $(m+n) \times (m+n)$ matrices, which is simple.

Example 12.1.2 Let V be a finite dimensional superspace and J be a degree $\bar{1}$ involution in $\mathrm{End}_F(V)$. Such one exists if and only if $\dim V_{\bar{0}} = \dim V_{\bar{1}}$. Consider the superalgebra

$$\mathcal{Q}(V, J) := \{f \in \mathrm{End}_F(V) \mid fJ = (-1)^{\bar{f}} Jf\}.$$

Note that all degree $\bar{1}$ involutions in $\mathrm{End}_F(V)$ are conjugate to each other by an (invertible) element in $\mathrm{End}_F(V)_{\bar{0}}$. Hence another choice of J will yield an isomorphic superalgebra. So we can speak of the superalgebra $\mathcal{Q}(V)$, defined up to an isomorphism. Let $\mathrm{sdim}\, V = (n, n)$. Pick a basis $\{v_1, \dots, v_n\}$ of $V_{\bar{0}}$, and set $v_i' = J(v_i)$ for $1 \le i \le n$. Then $\{v_1', \dots, v_n'\}$ is a basis of $V_{\bar{1}}$. With respect to the basis $\{v_1, \dots, v_n; v_1', \dots, v_n'\}$, the elements of $\mathcal{Q}(V, J)$ have matrices of the form

$$\begin{pmatrix} A & B \\ -B & A \end{pmatrix}, \tag{12.4}$$

where A and B are arbitrary $n \times n$ matrices, with $B = 0$ for even endomorphisms and $A = 0$ for odd ones. In particular, $\mathrm{sdim}\, \mathcal{Q}(V) = (n^2, n^2)$. The superalgebra $\mathcal{Q}(V, J)$ can be identified with the superalgebra $\mathcal{Q}_n$ of all matrices of the form (12.4).

Moreover, it is easy to see that the isomorphism (12.2) restricts to an isomorphism:

$$\mathcal{M}(V) \otimes \mathcal{Q}(W, J) \cong \mathcal{Q}(V \otimes W, \mathrm{id}_V \otimes J) \tag{12.5}$$

or, in terms of matrix superalgebras,

$$\mathcal{M}_{m,n} \otimes \mathcal{Q}_k \cong \mathcal{Q}_{(m+n)k}. \tag{12.6}$$

Let ε be a primitive 4th root of 1 in F. It is easy to check explicitly that the map

$$\begin{pmatrix} a & b \\ -b & a \end{pmatrix} \otimes \begin{pmatrix} a' & b' \\ -b' & a' \end{pmatrix} \mapsto \begin{pmatrix} aa' - \varepsilon bb' & a'b - \varepsilon ab' \\ -a'b - \varepsilon ab' & aa' + \varepsilon bb' \end{pmatrix}$$

from $\mathcal{Q}_1 \otimes \mathcal{Q}_1$ to $\mathcal{M}_{1,1}$ is a superalgebra isomorphism. So (12.3) and (12.6) imply by induction that

$$\mathcal{Q}_m \otimes \mathcal{Q}_n \cong \mathcal{M}_{mn,mn}. \tag{12.7}$$

Finally, $\mathcal{Q}_n$ is easily checked to be a simple superalgebra, even though it is not simple as a usual algebra. In fact, $|\mathcal{Q}_n| = \mathcal{M}_n \times \mathcal{M}_n$.

Example 12.1.3 Define the *Clifford superalgebra* $\mathcal{C}_n$ to be the superalgebra given by odd generators $c_1, \dots, c_n$, subject to the relations

$$c_i^2 = 1 \qquad (1 \le i \le n), \tag{12.8}$$

$$c_i c_j = -c_j c_i \qquad (1 \le i \ne j \le n). \tag{12.9}$$

It is easy to see that

$$\mathcal{C}_{n+m} \cong \mathcal{C}_n \otimes \mathcal{C}_m.$$

The isomorphism is defined as follows: the generators $c_1, \dots, c_n$ are mapped to $c_1 \otimes 1, \dots, c_n \otimes 1$, and $c_{n+1}, \dots, c_{n+m}$ are mapped to $1 \otimes c_1, \dots, 1 \otimes c_m$, respectively. It follows that $\mathcal{C}_n \cong \mathcal{C}_1^{\otimes n}$. In particular, sdim $\mathcal{C}_n = (2^{n-1}, 2^{n-1})$. Note that $\mathcal{C}_1 \cong \mathcal{Q}_1$. So from (12.3), (12.6), and (12.7) we have $\mathcal{C}_{2k} \cong \mathcal{M}_{2^{k-1},2^{k-1}}$ and $\mathcal{C}_{2k-1} \cong \mathcal{Q}_{2^{k-1}}$ for $k = 1, 2, \dots$. In particular, $\mathcal{C}_n$ is a simple superalgebra.

Example 12.1.4 Define the *Grassman superalgebra* $\mathcal{L}_n$ to be the superalgebra given by odd generators $d_1, \dots, d_n$, subject to the relations

$$d_i^2 = 0 \qquad (1 \le i \le n), \tag{12.10}$$

$$d_i d_j = -d_j d_i \qquad (1 \le i \ne j \le n). \tag{12.11}$$

As for Clifford algebra, it is easy to see that there is a natural isomorphism $\mathcal{L}_n \cong \mathcal{L}_1^{\otimes n}$.

More generally, let $\underline{j} = (j_1, \dots, j_n) \in F^n$ be fixed, and $\mathcal{A}(\underline{j})$ be the superalgebra given by odd generators $a_1, \dots, a_n$, subject to the relations

$$a_i^2 = j_i \qquad (1 \le i \le n), \tag{12.12}$$

$$a_i a_j = -a_j a_i \qquad (1 \le i \ne j \le n). \tag{12.13}$$

We call $\mathcal{A}(\underline{j})$ the *Clifford–Grassman superalgebra*. Using the fact that F is algebraically closed, it is easy to see that

$$\mathcal{A}(\underline{j}) \cong \mathcal{L}_1^{\otimes \gamma_0} \otimes \mathcal{C}_1^{\otimes n-\gamma_0}, \tag{12.14}$$

where $\gamma_0 = \sharp\{i \mid 1 \le i \le n,\ j_i = 0\}$.

Let $\mathcal{A}$ be a superalgebra. A *(left) $\mathcal{A}$-supermodule* is a vector superspace V which is a left $\mathcal{A}$-module in the usual sense, such that $\mathcal{A}_i V_j \subseteq V_{i+j}$ for $i, j \in \mathbb{Z}_2$. Right supermodules are defined similarly. A *homomorphism* $f: V \to W$ of (left) $\mathcal{A}$-supermodules V and W means a (not necessarily homogeneous) linear map such that

$$f(av) = (-1)^{\bar{f}\bar{a}} af(v) \qquad (a \in \mathcal{A},\ v \in V).$$

The category of finite dimensional $\mathcal{A}$-supermodules is denoted $\mathcal{A}$-smod. A *homomorphism* $f: V \to W$ of right $\mathcal{A}$-supermodules V and W means a (not necessarily homogeneous) linear map such that

$$f(va) = f(v)a \qquad (a \in \mathcal{A},\ v \in V).$$

Note there is no sign here as a "does not go past f".

If $\mathcal{B}$ is a subsuperalgebra of $\mathcal{A}$, and W is a $\mathcal{B}$-supermodule we write $\mathrm{ind}_{\mathcal{B}}^{\mathcal{A}} W$ for the induced supermodule $\mathcal{A} \otimes_{\mathcal{B}} W$. We may consider $\mathrm{ind}_{\mathcal{B}}^{\mathcal{A}}$ as a functor from the category of $\mathcal{B}$-supermodules to the category of $\mathcal{A}$-supermodules. This functor is left adjoint to the restriction functor $\mathrm{res}_{\mathcal{B}}^{\mathcal{A}}$ going in the other direction. If $\mathcal{A}$ is free as a right $\mathcal{B}$-supermodule the induction functor is exact.

Any $\mathcal{A}$-supermodule V can be considered as a usual $|\mathcal{A}|$-module denoted $|V|$. Thus, for $V \in \mathcal{A}$-smod we have $|V| \in |\mathcal{A}|$-mod. The following result establishes an isomorphism of vector spaces between $\mathrm{Hom}_{\mathcal{A}}(V, W)$ and $\mathrm{Hom}_{|\mathcal{A}|}(|V|, |W|)$.

Lemma 12.1.5 *Let $V, W \in \mathcal{A}$-smod, and $f: V \to W$ be a linear map. Define a linear map*

$$f^- : V \to W, \quad v \mapsto (-1)^{\bar{f}\bar{v}} f(v).$$

Then $f \in \mathrm{Hom}_{\mathcal{A}}(V, W)$ if and only if $f^- \in \mathrm{Hom}_{|\mathcal{A}|}(|V|, |W|)$.

Let τ be an antiautomorphism of the superalgebra $\mathcal{A}$. If V is a finite dimensional $\mathcal{A}$-supermodule, then we can use τ to make the dual space V^* into an $\mathcal{A}$-supermodule by defining

$$(af)(v) = f(\tau(a)v) \qquad (a \in \mathcal{A}, f \in V^*, v \in V). \tag{12.15}$$

We will denote the resulting module by V^τ. There is a natural even isomorphism

$$\mathrm{Hom}_{\mathcal{A}}(V, W) \to \mathrm{Hom}_{\mathcal{A}}(W^\tau, V^\tau) \tag{12.16}$$

for any $V, W \in \mathcal{A}$-smod. The isomorphism sends $\theta \in \mathrm{Hom}_{\mathcal{A}}(V, W)$ to the dual map $\theta^* \in \mathrm{Hom}_{\mathcal{A}}(W^\tau, V^\tau)$ defined by $(\theta^* f)(v) = f(\theta v)$ for all $f \in W^\tau$.

We have the *(left) parity change functor*

$$\Pi : \mathcal{A}\text{-smod} \to \mathcal{A}\text{-smod}. \tag{12.17}$$

For an object V, ΠV is the same underlying vector space but with the opposite $\mathbb{Z}_2$-grading. The new action of $a \in \mathcal{A}$ on $v \in \Pi V$ is defined in terms of the old action by $a \cdot v := (-1)^{\bar{a}} av$. On a morphism f, Πf is the same underlying linear map as f. Note that the identity map on V defines an *odd* isomorphism from V to ΠV. It is more subtle however whether V and ΠV are *evenly* isomorphic, see for example Lemma 12.2.8 below. We will write

$$V \cong W$$

if the $\mathcal{A}$-supermodules V and W are isomorphic, and

$$V \simeq W$$

if V and W are *evenly* isomorphic.

Given two superalgebras $\mathcal{A}, \mathcal{B}$, an $(\mathcal{A}, \mathcal{B})$-bisupermodule is a left $\mathcal{A}$-supermodule V, which is also a right $\mathcal{B}$-supermodule (with respect to the same grading of V) such that $(av)b = a(vb)$ for all $a \in \mathcal{A}, b \in \mathcal{B}, v \in V$. A *homomorphism* $f : V \to W$ of $(\mathcal{A}, \mathcal{B})$-bisupermodules is a map which is both a homomorphism of left $\mathcal{A}$-supermodules and a homomorphism of right supermodules. If V is an $(\mathcal{A}, \mathcal{B})$-bisupermodule, ΠV denotes the $\mathcal{A}$-supermodule, defined as in the previous paragraph, with the right $\mathcal{B}$-action on ΠV being the same as the original action on V.

For a superalgebra $\mathcal{A}$, the subcategory $\mathcal{A}\text{-smod}_{\mathrm{ev}}$ of $\mathcal{A}$-smod, consisting of the same objects but only even morphisms, is an abelian category in the usual sense. This allows us to make use of all the basic notions of homological algebra by restricting our attention to even morphisms. For example, by a short exact sequence in $\mathcal{A}$-smod, we mean a sequence

$$0 \longrightarrow V_1 \longrightarrow V_2 \longrightarrow V_3 \longrightarrow 0, \tag{12.18}$$

with all the maps being *even*. All functors between categories of superobjects that we will ever consider send even morphisms to even morphisms. So they will give rise to the corresponding functors between the underlying even subcategories.

We define the *Grothendieck group* $K(\mathcal{A}\text{-smod})$ to be the quotient of the free $\mathbb{Z}$-module with generators given by all finite dimensional $\mathcal{A}$-supermodules by the $\mathbb{Z}$-submodule generated by:

(1) $V_1 - V_2 + V_3$ for every short exact sequence of the form (12.18);
(2) $V - \Pi V$ for every $\mathcal{A}$-supermodule V.

We will write $[V]$ for the image of the $\mathcal{A}$-module V in $K(A\text{-smod})$. Similarly, we define the Grothendieck group $K(\mathcal{A}\text{-proj})$, where $\mathcal{A}$-proj denotes the full subcategory of $\mathcal{A}$-smod consisting of projective $\mathcal{A}$-modules.

As usual, the Grothendieck groups $K(\mathcal{A}\text{-smod})$ and $K(\mathcal{A}\text{-proj})$ are free $\mathbb{Z}$-modules with canonical bases corresponding to the isomorphism classes of irreducible and projective indecomposable supermodules, respectively (see below). The embedding $\mathcal{A}\text{-proj} \subset \mathcal{A}\text{-smod}$ induces the natural *Cartan map*

$$\omega : K(\mathcal{A}\text{-proj}) \to K(\mathcal{A}\text{-smod}).$$

12.2 Schur's Lemma and Wedderburn's Theorem

By a *subsupermodule* of an $\mathcal{A}$-supermodule we mean a subsuperspace which is $\mathcal{A}$-stable. An $\mathcal{A}$-supermodule is *irreducible* (or *simple*) if it is non-zero and has no non-zero proper $\mathcal{A}$-subsupermodules. An $\mathcal{A}$-supermodule M is called *completely reducible* if any subsupermodule of M is a direct summand of M.

It might happen that a supermodule V is irreducible, but the module $|V|$ is reducible – in this case $|V|$ will have a non-trivial proper $\mathcal{A}$-invariant subspace which is *not homogeneous*. We need to understand this situation better. First of all, we say that an irreducible $\mathcal{A}$-supermodule V is of *type* M if the $|\mathcal{A}|$-module $|V|$ is irreducible, and otherwise we say that V is of *type* Q.

Lemma 12.2.1 *Let V be a finite dimensional irreducible $\mathcal{A}$-supermodule of type* Q. *Then there exist bases $\{v_1, \dots, v_n\}$ of $V_{\bar{0}}$ and $\{v_1', \dots, v_n'\}$ of $V_{\bar{1}}$ such that*

$$|V| = \text{span}\{v_1 + v_1', \dots, v_n + v_n'\} \oplus \text{span}\{v_1 - v_1', \dots, v_n - v_n'\},$$

a direct sum of two non-isomorphic irreducible $|\mathcal{A}|$-submodules. Moreover, the linear map $J_V : V \to V$ defined by $v_i \mapsto -v_i'$, $v_i' \mapsto v_i$ is an endomorphism of V as an $\mathcal{A}$-supermodule.

Proof Denote $\delta := \delta_V$. Let W be an irreducible $|\mathcal{A}|$-submodule of $|V|$. Since V is an irreducible supermodule, W is not δ-stable. Moreover, $\delta(W)$

is also an irreducible $|\mathcal{A}|$-submodule of $|V|$. We have $W \cap \delta(W) = \{0\}$ and $W + \delta(W)$ is δ-stable. Hence $V = W \oplus \delta(W)$. Let $\{w_1, \ldots, w_n\}$ be a basis for W. Then $\{\delta(w_1), \ldots, \delta(w_n)\}$ is a basis for $\delta(W)$. Take $v_i = w_i + \delta(w_i)$ and $v'_i = w_i - \delta(w_i)$ for $1 \le i \le n$. To verify that J_V is an endomorphism of V, note that $J_V^- = (-\mathrm{id}_W) \oplus \mathrm{id}_{\delta(W)}$ is an endomorphism of $|V|$ and use Lemma 12.1.5.

Finally, assume that $W \cong \delta(W)$ as $|\mathcal{A}|$-modules. Then in view of Lemma 12.1.5, $\dim \mathrm{End}_{\mathcal{A}}(V) = \dim \mathrm{End}_{|\mathcal{A}|}(|V|) = 4$. It follows that $\dim \mathrm{End}_{\mathcal{A}}(V)_{\bar{0}} = \dim \mathrm{End}_{\mathcal{A}}(V)_{\bar{1}} = 2$. Now we can construct a non-zero *homogeneous* endomorphism of V with non-trivial kernel as a linear combination of two linearly independent elements of $\mathrm{End}_{\mathcal{A}}(V)_{\bar{0}}$. The existence of such endomorphism contradicts the irreducibility of V. □

Remark 12.2.2 Let ε be a primitive 4th root of 1 in F, V be a type $\mathtt{Q}$ irreducible $\mathcal{A}$-supermodule, and J_V be the map constructed in Lemma 12.2.1. Then εJ_V is a degree $\bar{1}$ involution in $\mathrm{End}_{\mathcal{A}}(V)$.

Now we have the following analogue of Schur's lemma:

Lemma 12.2.3 (Schur's lemma) *Let V be a finite dimensional irreducible $\mathcal{A}$-supermodule. Then*

$$\mathrm{End}_{\mathcal{A}}(V) = \begin{cases} \mathrm{span}\{\mathrm{id}_V\} & \text{if } V \text{ is of type } \mathtt{M}, \\ \mathrm{span}\{\mathrm{id}_V, J_V\} & \text{if } V \text{ is of type } \mathtt{Q}, \end{cases}$$

where J_V is as in Lemma 12.2.1. Moreover, if W is another irreducible $\mathcal{A}$-supermodule with $V \not\cong W$, then $\mathrm{Hom}_{\mathcal{A}}(V, W) = 0$.

Proof Apply Lemmas 12.1.5, 12.2.1, and usual Schur's Lemma. □

Example 12.2.4 Let V be a superspace. Then V is naturally an irreducible type $\mathtt{M}$ supermodule over $\mathcal{M}(V)$. Moreover, if $\mathrm{sdim}\, V = (n, n)$, then V is naturally an irreducible type $\mathtt{Q}$ supermodule over $\mathcal{Q}(V)$. This explains our terminology of types.

So, in view of Example 12.1.3, $\mathcal{C}_n$ has the irreducible supermodule U_n of dimension $2^{n/2}$ and type $\mathtt{M}$ if n is even, and of dimension $2^{(n+1)/2}$ and type $\mathtt{Q}$ if n is odd. This supermodule is called the *Clifford supermodule*. As $\mathcal{C}_n$ is a simple superalgebra, U_n is the unique irreducible $\mathcal{C}_n$-supermodule up to isomorphism.

Lemma 12.2.5 *Let $V \in \mathcal{A}$-smod. Then V is completely reducible if and only if $|V| \in |\mathcal{A}|$-mod is completely reducible.*

Proof If V is completely reducible then $|V|$ is completely reducible by Lemma 12.2.1. Conversely, let $|V|$ be completely reducible, and $W \subseteq V$ be a subsupermodule. We have to show that there exists a subsupermodule $X \subseteq V$ with $V = W \oplus X$. By the complete reducibility of $|V|$, we have an $|\mathcal{A}|$-submodule $Y \subseteq |V|$ with $V = W \oplus Y$. However, Y might not be homogeneous. Let π be the projection to W along Y, and consider the linear map

$$\pi' := (\pi + \delta_V \pi \delta_V)/2 : V \to V.$$

Set $X := \ker \pi'$. Note that $\pi|_W = \mathrm{id}_W$, and $\mathrm{im}\, \pi' = \mathrm{im}\, \pi = W$. So $V = W \oplus X$. Moreover, it is easy to see that X is δ_V-invariant, so homogeneous, and that X is $\mathcal{A}$-invariant. □

A finite dimensional superalgebra $\mathcal{A}$ is called *semisimple* if the left regular $\mathcal{A}$-supermodule ${}_{\mathcal{A}}\mathcal{A}$ is semisimple.

Corollary 12.2.6 *Let $\mathcal{A}$ be a finite dimensional superalgebra. Then $\mathcal{A}$ is semisimple if and only if $|\mathcal{A}|$ is semisimple.*

Lemma 12.2.7 *Let $\mathcal{A}$ be a finite dimensional superalgebra. Then the Jacobson radical $J(|\mathcal{A}|)$ can be characterized as the unique smallest superideal $\mathcal{K}$ of $\mathcal{A}$ such that $\mathcal{A}/\mathcal{K}$ is a semisimple superalgebra.*

Proof Set $\mathcal{J} := J(|\mathcal{A}|)$. Observe that $\mathcal{J}$ is a superideal since $\mathcal{J}$ is invariant under the algebra automorphism $\delta_{\mathcal{A}}$ of $|\mathcal{A}|$. Let $\mathcal{K}$ be any superideal of $\mathcal{A}$ that is minimal with respect to the property that $\mathcal{A}/\mathcal{K}$ is a semisimple superalgebra. By Corollary 12.2.6, $|\mathcal{A}|/|\mathcal{K}|$ is semisimple, so $\mathcal{J} \subseteq \mathcal{K}$. However, $\mathcal{A}/\mathcal{J}$ is a superalgebra that is semisimple as an algebra. So, by Corollary 12.2.6, it is a semisimple superalgebra, so $\mathcal{J} = \mathcal{K}$ by minimality of $\mathcal{K}$. □

The superideal $\mathcal{K} = J(|\mathcal{A}|)$ from Lemma 12.2.7 is called the *Jacobson radical* of the superalgebra $\mathcal{A}$ and denoted $J(\mathcal{A})$.

Lemma 12.2.8 *Let $V \in \mathcal{A}$-smod be an irreducible supermodule.*

(i) *V is evenly isomorphic to ΠV if and only if V is of type* Q.
(ii) *Assume that V is of type* M, *and let $W = V^{\oplus m} \oplus (\Pi V)^{\oplus n}$. Then* $\mathrm{End}_{\mathcal{A}}(W) \cong \mathcal{M}_{m,n}$.
(iii) *Assume that V is of type* Q, *and let $W = V^{\oplus n}$. Then* $\mathrm{End}_{\mathcal{A}}(W) \cong \mathcal{Q}_n$.

Proof (i) V is evenly isomorphic to ΠV if and only if V is oddly isomorphic to itself, which by virtue of Lemma 12.2.3, is equivalent to V being of type Q.

(ii) follows from the analogous result for the usual modules, as $|V| \cong |\Pi V|$ is irreducible $|\mathcal{A}|$-module, and Schur's Lemma 12.1.5.

(iii) It is clear from Schur's Lemma that $\text{End}_{\mathcal{A}}(V) \cong \mathcal{Q}_1$. So, if we write the superspace W in the form $W \cong V \otimes E$, where E is a superspace of dimension $(n, 0)$, then we get an embedding of the algebra $\mathcal{Q}_1 \otimes \mathcal{M}_{n,0}$ into $\text{End}_{\mathcal{A}}(W)$. Now the result follows by dimensions and (12.6). □

Theorem 12.2.9 (Wedderburn's Theorem) *Any finite dimensional simple superalgebra is isomorphic to some $\mathcal{M}_{m,n}$ or $\mathcal{Q}_n$. Moreover, the following conditions on a finite dimensional superalgebra $\mathcal{A}$ are equivalent:*

(i) *$\mathcal{A}$ is semisimple;*

(ii) *every $\mathcal{A}$-supermodule is completely reducible;*

(iii) *$\mathcal{A}$ is a direct product of finitely many simple F-superalgebras;*

(iv) *$J(\mathcal{A}) = 0$.*

Proof If $\mathcal{A}$ is a finite dimensional simple superalgebra, take an irreducible $\mathcal{A}$-supermodule V with sdim $V = (m, n)$. The action of $\mathcal{A}$ on V must be faithful by simplicity of $\mathcal{A}$. So, by Schur's Lemma 12.2.3, we get an embedding of $\mathcal{A}$ into $\mathcal{M}(V)$, if V is of type M, and into $\mathcal{Q}(V)$ if V is of type Q. Now, the usual Wedderburn Theorem applied to the algebra $|\mathcal{A}|$ and module $|V|$ shows that the dimension of $\mathcal{A}$ equals $\dim \mathcal{M}(V)$ in the first case and $\dim \mathcal{Q}(V)$ in the second case.

For the second part of the theorem, the usual proof for algebras more or less works. Indeed, (i)⇔(ii) is clear, for every $\mathcal{A}$-supermodule is a quotient of a free $\mathcal{A}$-supermodule.

(ii)⇒(iii) Decompose ${}_{\mathcal{A}}\mathcal{A} = n_1 L_1 \oplus \cdots \oplus n_r L_r$ where $L_1, \dots, L_r$ are pairwise non-isomorphic irreducible supermodules. For every $a \in \mathcal{A}$, we define $f_a \in \text{End}_{\mathcal{A}}({}_{\mathcal{A}}\mathcal{A})$ by $f_a(b) = (-1)^{\bar{a}\bar{b}} ba$ for $b \in A$. Then $a \mapsto f_a$ is an isomorphism of superalgebras between $\mathcal{A}$ and $\text{End}_{\mathcal{A}}({}_{\mathcal{A}}\mathcal{A})$. Now (iii) follows from Lemma 12.2.3 and Lemma 12.2.8.

(iii)⇒(iv) is clear for example by Lemma 12.2.7.

(iv)⇒(i) follows from Lemma 12.2.7 and Corollary 12.2.6. □

As $\delta = \delta_{\mathcal{A}} : |\mathcal{A}| \to |\mathcal{A}|$ is an algebra automorphism, we can twist any $|\mathcal{A}|$-module V with δ to get a new $\mathcal{A}$-module V^{δ}, which is the same vector space V with the new action $a \cdot v = \delta(a)v$.

Corollary 12.2.10 *Let $\mathcal{A}$ be a finite dimensional superalgebra, and $\{V_1, \dots, V_n\}$ be a complete set of pairwise non-isomorphic irreducible $\mathcal{A}$-supermodules such that $V_1, \dots, V_m$ are of type* M *and $V_{m+1}, \dots, V_n$ are of type* Q. *For $i = m+1, \dots, n$, write $|V_i| = V_i^+ \oplus V_i^-$ as a direct sum of irreducible $|\mathcal{A}|$-modules. Then*

$$\{|V_1|, \dots, |V_m|, V_{m+1}^{\pm}, \dots, V_n^{\pm}\}$$

is a complete set of pairwise non-isomorphic irreducible $|\mathcal{A}|$-modules. Moreover, $|V_i|^{\delta} \cong |V_i|$ and $(V_i^{\pm})^{\delta} \cong V_i^{\mp}$.

Proof In view of Lemma 12.2.7 and Corollary 12.2.6 the result reduces to the semisimple case, and then to the simple case, when it follows from Wedderburn's Theorem. □

If $s \in \mathcal{A}$ is a homogeneous invertible element, conjugation by s defines an automorphism φ_s of the algebra $\mathcal{A}_{\bar{0}}$. Given an $\mathcal{A}_{\bar{0}}$-module V we write V^s for the twisted $\mathcal{A}_{\bar{0}}$-module V^{φ_s}. The following result is similar to Clifford Theory for index 2 subgroups.

Proposition 12.2.11 *Let $\mathcal{A}$ be a finite dimensional superalgebra, and $\{V_1, \dots, V_n\}$ be a complete set of pairwise non-isomorphic irreducible $\mathcal{A}$-supermodules such that $V_1, \dots, V_m$ are of type* M *and $V_{m+1}, \dots, V_n$ are of type* Q. *Let $\{|V_1|, \dots, |V_m|, V_{m+1}^{\pm}, \dots, V_n^{\pm}\}$ be a complete set of pairwise non-isomorphic irreducible $|\mathcal{A}|$-modules constructed as in Corollary 12.2.10. Assume that $\mathcal{A}_{\bar{1}}$ contains an invertible element s. Then, on restriction to $\mathcal{A}_{\bar{0}}$, the modules $|V_i|$, $1 \le i \le m$, split as a direct sum $W_i^+ \oplus W_i^-$ of two non-isomorphic irreducible modules such that $W_i^- \cong (W_i^+)^s$, and the modules $V_i^{\pm}$, $m < i \le n$, are irreducible with $V_i^+ \cong V_i^- =: W_i$. Moreover,*

$$\{W_1^{\pm}, \dots, W_m^{\pm}, W_{m+1}, \dots, W_n\}$$

is a complete set of pairwise non-isomorphic irreducible $\mathcal{A}_{\bar{0}}$-modules.

Proof From Wedderburn's Theorem, the restriction of an irreducible $\mathcal{A}$-supermodule to $\mathcal{A}_{\bar{0}}$ always has at most two composition factors. So if we can decompose an irreducible $\mathcal{A}$-supermodule as a direct sum of two non-trivial $\mathcal{A}_{\bar{0}}$-modules, those modules must be irreducible.

Let $1 \le i \le m$. By assumption, $s(V_i)_{\bar{0}} = (V_i)_{\bar{1}}$. Moreover, $(V_i)_{\bar{0}}$ and $(V_i)_{\bar{1}}$ are $\mathcal{A}_{\bar{0}}$-invariant, and $s(V_i)_{\bar{0}} \cong ((V_i)_{\bar{0}})^s$. So it remains to notice that $(V_i)_{\bar{0}}$ and $(V_i)_{\bar{1}}$ are non-isomorphic by Wedderburn and irreducible by the previous paragraph.

Let $m < i \le n$. As $V_i^+ \cong (V_i^-)^\delta$ and δ is trivial on $\mathcal{A}_{\bar{0}}$, the modules V_i^+ and V_i^- are isomorphic on restriction to $\mathcal{A}_{\bar{0}}$. Now use the first paragraph to see that they are irreducible $\mathcal{A}_{\bar{0}}$-modules.

That $W_i^{(\pm)}$ form a complete set of irreducible $\mathcal{A}_{\bar{0}}$-modules up to isomorphism again follows from Wedderburn. □

Let $\mathcal{A}$ be a finite dimensional superalgebra. Indecomposable summands of the left regular module ${}_{\mathcal{A}}\mathcal{A}$ are called *principal indecomposable* $\mathcal{A}$-supermodules. Theory of principal indecomposable supermodules is analogous to the usual one. In particular, we can show (we leave this as an exercise) that principal indecomposable supermodules are projective in the category $\mathcal{A}$-smod, have irreducible heads, and are determined up to an isomorphism by their heads. We refer to the principal indecomposable supermodule with head L as the *projective cover* of L, and denote it by P_L.

Proposition 12.2.12 *Let $\mathcal{A}$ be a finite dimensional superalgebra, and $\{V_1, \dots, V_n\}$ be a complete set of pairwise non-isomorphic irreducible $\mathcal{A}$-supermodules such that $V_1, \dots, V_m$ are of type* M *and $V_{m+1}, \dots, V_n$ are of type* Q. *Set* $\operatorname{sdim} V_i = (d_i, d_i')$. *Then*

$$ {}_{\mathcal{A}}\mathcal{A} \simeq \bigoplus_{i=1}^{m} (d_i P_{V_i} \oplus d_i' \Pi P_{V_i}) \oplus \bigoplus_{i=m+1}^{n} d_i P_{V_i}. $$

Proof Reduce to the semisimple case and use Wedderburn. □

Given left supermodules V and W over superalgebras $\mathcal{A}$ and $\mathcal{B}$ respectively, the *(outer) tensor product* $V \boxtimes W$ is the superspace $V \otimes W$ considered as an $\mathcal{A} \otimes \mathcal{B}$-supermodule via

$$ (a \otimes b)(v \otimes w) = (-1)^{\bar{b}\bar{v}} av \otimes bw \quad (a \in \mathcal{A}, b \in \mathcal{B}, v \in V, w \in W). \tag{12.19} $$

If $f: V \to V'$ (resp. $g: W \to W'$) is a homomorphism of $\mathcal{A}$- (resp. $\mathcal{B}$-) supermodules, then $f \otimes g : V \otimes W \to V' \otimes W'$ is a homomorphism of $\mathcal{A} \otimes \mathcal{B}$-supermodules.

If $\tau_{\mathcal{A}}$ is an antiautomorphism of $\mathcal{A}$ and $\tau_{\mathcal{B}}$ is an antiautomorphism of $\mathcal{B}$, then $\tau : \mathcal{A} \otimes \mathcal{B} : a \otimes b \mapsto (-1)^{\bar{a}\bar{b}} \tau(a) \otimes \tau(b)$ is an antiautomorphism of $\mathcal{A} \otimes \mathcal{B}$. Under these circumstances there is a natural isomorphism isomorphism

$$ (V \boxtimes W)^\tau \xrightarrow{\sim} V^{\tau_{\mathcal{A}}} \otimes W^{\tau_{\mathcal{B}}}, \ f \otimes g \mapsto (-1)^{\bar{f}\bar{g}} f \otimes g. \tag{12.20} $$

Lemma 12.2.13 *Let V be an irreducible $\mathcal{A}$-supermodule and N be an irreducible $\mathcal{B}$-supermodule.*

(i) If both V and W are of type M, *then $V \boxtimes W$ is an irreducible $\mathcal{A} \otimes \mathcal{B}$-supermodule of type* M.

(ii) If one of V or W is of type M *and the other is of type* Q, *then $V \boxtimes W$ is an irreducible $\mathcal{A} \otimes \mathcal{B}$-supermodule of type* Q.

(iii) If both V and W are of type Q, *then $V \boxtimes W \simeq X \oplus \Pi X$ for a type* M *irreducible $\mathcal{A} \otimes \mathcal{B}$-supermodule X.*

Moreover, all irreducible $\mathcal{A} \otimes \mathcal{B}$-supermodules arise as constituents of $V \boxtimes W$ for some choice of irreducibles V, W.

Proof Wedderburn's Theorem reduces the lemma to the situation where $\mathcal{A}$ and $\mathcal{B}$ are simple, in which case the result follows from (12.3), (12.6), and (12.7). □

If $V \in \mathcal{A}$-smod and $W \in \mathcal{B}$-smod are irreducible, denote by $V \circledast W$ an irreducible component of $V \boxtimes W$. Thus,

$$V \boxtimes W \simeq \begin{cases} V \circledast W \oplus \Pi(V \circledast W), & \text{if } V \text{ and } W \text{ are both of type Q}, \\ V \circledast W, & \text{otherwise.} \end{cases}$$

We stress that $V \circledast W$ is in general only well-defined up to isomorphism.

Example 12.2.14 Recall the Clifford–Grassman superalgebra $\mathcal{A}(\underline{j})$ from Example 12.1.4. We already know that the Clifford algebra $\mathcal{C}_1$ has only one irreducible supermodule up to isomorphism, namely U_1, and it is of type Q. It is also easy to see that the only irreducible supermodule over $\mathcal{L}_1$ is the trivial supermodule F (in degree $\bar{0}$ or $\bar{1}$, it does not matter up isomorphism) on which the generator d_1 acts as 0. It follows from the isomorphism (12.14) and Lemma 12.2.13 that $\mathcal{A}(\underline{j})$ has only one irreducible supermodule

$$U(\underline{j}) \cong F^{\circledast \gamma_0} \circledast U_1^{\circledast n-\gamma_0}.$$

In particular, $U_n \cong U_1^{\circledast n}$. Note that $\dim U(\underline{j}) = 2^{\lfloor \frac{n-\gamma_0+1}{2} \rfloor}$. Also $U(\underline{j})$ is of type M if and only if $n - \gamma_0$ is even.

The following simple observation explains why the operation $\circledast$ is convenient:

Lemma 12.2.15 *Let $\mathcal{A}$ and $\mathcal{B}$ be finite dimensional superalgebras. Then there is an isomorphism*

$$\begin{aligned} K(\mathcal{A}\text{-smod}) \otimes_{\mathbb{Z}} K(\mathcal{B}\text{-smod}) &\to K(\mathcal{A} \otimes \mathcal{B}\text{-smod}), \\ [L] \otimes [L'] &\mapsto [L \circledast L'] \end{aligned} \tag{12.21}$$

Proof Follows using Lemma 12.2.13. □

Lemma 12.2.16 *P_L admits an odd involution if and only if L does, that is if and only if L is of type* Q.

Proof If P_L admits an odd involution J, then J factors through the radical of P_L to give an odd involution of L. Conversely, assume that L is of type Q, that is L admits an odd involution J. Then we can define the action of the Clifford algebra $\mathcal{C}_1$ on V by requiring that the generator c_1 acts as J. This makes V into an irreducible $\mathcal{A} \otimes \mathcal{C}_1$-supermodule of type M. In fact, it is clear that this supermodule is isomorphic to $V \circledast U_1$.

Now, we claim that $\operatorname{res}^{\mathcal{A}\otimes\mathcal{C}_1}_{\mathcal{A}} P_{V\circledast U_1} \cong P_V$. This completes the proof since then the odd involution on P_V comes from the action of $1 \otimes c_1 \in \mathcal{A} \otimes \mathcal{C}_1$ on $P_{V\circledast U_1}$. For the claim, note first that $\operatorname{res}^{\mathcal{A}\otimes\mathcal{C}_1}_{\mathcal{A}} P_{V\circledast U_1}$ is projective as an $\mathcal{A}$-supermodule, so it remains to prove that its head is isomorphic to V. This follows from the following calculation for any irreducible $\mathcal{A}$-supermodule W:

$$\begin{aligned} &\operatorname{Hom}_{\mathcal{A}}(\operatorname{res}^{\mathcal{A}\otimes\mathcal{C}_1}_{\mathcal{A}} P_{V\circledast U_1}, W) \\ \simeq &\operatorname{Hom}_{\mathcal{A}\otimes\mathcal{C}_1}(P_{V\circledast U_1}, \operatorname{Hom}_{\mathcal{A}}(\mathcal{A} \otimes \mathcal{C}_1, W)) \\ \simeq &\operatorname{Hom}_{\mathcal{A}\otimes\mathcal{C}_1}(P_{V\circledast U_1}, \operatorname{Hom}_{\mathcal{A}}(\mathcal{A}, W) \boxtimes \operatorname{Hom}_F(\mathcal{C}_1, F)) \\ \simeq &\operatorname{Hom}_{\mathcal{A}\otimes\mathcal{C}_1}(P_{V\circledast U_1}, W \boxtimes U_1) \\ \simeq &\operatorname{Hom}_{\mathcal{A}\otimes\mathcal{C}_1}(V \circledast U_1, W \boxtimes U_1) \\ \simeq &\operatorname{Hom}_{\mathcal{A}}(V, W). \end{aligned}$$

□

13
Sergeev superalgebras

By general theory, studying spin representations of the symmetric group S_n is equivalent to studying representations of its twisted group algebra $\mathcal{T}_n$. It is convenient to consider $\mathcal{T}_n$ as a superalgebra with respect to the natural grading, where even (resp. odd) elements come from even (resp. odd) permutations. Even if we are only interested in the usual $\mathcal{T}_n$-modules, Corollary 12.2.10 shows that, at least as far as irreducibles are concerned, we do not loose anything by working in the category of supermodules, providing we keep track of types of irreducible supermodules. Moreover, we even gain an additional insight into the usual irreducible modules, in view of Proposition 12.2.11. This additional information is exactly what we need in order to deal with spin representations of the alternating groups. It is interesting that the superalgebra approach is not useful for the linear representations of S_n, while spin representation theory of S_n has intrinsic features of a "supertheory".

An important idea due to Sergeev is that instead of the superalgebra $\mathcal{T}_n$ it is more convenient to consider $\mathcal{Y}_n := \mathcal{T}_n \otimes \mathcal{C}_n$, where $\mathcal{C}_n$ is the Clifford superalgebra, and $\otimes$ is the tensor product of superalgebras. On the one hand, nothing much is going to happen to representation theory when we tensor our superalgebra with a simple superalgebra (classically we get a Morita equivalence and in the "superworld" we get either a Morita equivalence or something almost as good as a Morita equivalence). On the other hand, it is well known that the Clifford algebra plays a special role in the theory of spin representations of symmetric groups, so why not bring it in voluntarily? Finally, the new superalgebra $\mathcal{Y}_n$ turns out to have at least two important advantages over $\mathcal{T}_n$: it has a nice q-analogue (which we will not pursue here) and it has a natural "affinization", so we have a chance to work out a theory parallel to the one developed in the first part of this book for symmetric groups.

13.1 Twisted group algebras

It is well known that $H^2(S_n, F^\times) \cong \mathbb{Z}/2Z$ for $n \geq 5$, see e.g. [Su, Chapter 3 (2.21)] (recall that throughout Part II we assume $p > 2$). So there are two twisted group algebras of the symmetric group S_n up to isomorphism. One of them is the usual group algebra FS_n. Studying FS_n-modules is of course equivalent to studying representations of S_n over F. The *non-trivial* twisted group algebra $\mathcal{T}_n$ of S_n is the associative F-algebra with basis $\{t_g \mid g \in S_n\}$ and multiplication $t_g t_h = \alpha(g, h) t_{gh}$ for any $g, h \in S_n$, where α is a non-trivial 2-cocycle in $Z^2(S_n, F^\times)$. Studying $\mathcal{T}_n$-modules is equivalent to studying *spin* representations of S_n.

It is shown in [Su, p. 303] that the twisted group algebra $\mathcal{T}_n$ is generated by the elements $t_1, \dots, t_{n-1}$ subject only to the relations

$$t_i^2 = 1 \qquad (1 \leq i < n), \tag{13.1}$$

$$t_i t_{i+1} t_i = t_{i+1} t_i t_{i+1} \qquad (1 \leq i \leq n-2), \tag{13.2}$$

$$t_i t_j = -t_j t_i \qquad (1 \leq i, j < n, \ |i-j| > 1). \tag{13.3}$$

(Actually, Suzuki has the relations $T_i^2 = -1$, $T_i T_{i+1} T_i = -T_{i+1} T_i T_{i+1}$, $T_i T_j = -T_j T_i$. To get the relations above, take $t_1 = \sqrt{-1} T_1, t_2 = -\sqrt{-1} T_2, t_3 = \sqrt{-1} T_3, t_4 = -\sqrt{-1} T_4$, etc.)

Inside the algebra $\mathcal{T}_n$ we have a subalgebra

$$\mathcal{U}_n := \mathrm{span}\{t_g \mid g \in A_n\}.$$

If $n > 7$, this is the only non-trivial twisted group algebra of the alternating group A_n, see [Su, p. 304], and the exceptional cases (for $n = 6, 7$) are of course easy to understand, see for example [At, MAt].

We consider $\mathcal{T}_n$ as a superalgebra with respect to the following grading

$$(\mathcal{T}_n)_{\bar{0}} = \mathcal{U}_n, \quad (\mathcal{T}_n)_{\bar{1}} = \mathrm{span}\{t_g \mid g \in S_n \setminus A_n\}.$$

It follows from the defining relations that there exists a superalgebra anti-automorphism

$$\tau : \mathcal{T}_n \to \mathcal{T}_n, \ t_i \mapsto -t_i \qquad (1 \leq i < n). \tag{13.4}$$

In fact, $\tau(t_g) = (-1)^{\bar{g}} t_{g^{-1}}$. The anti-involution τ can be used to define duality on $\mathcal{T}_n$-modules and supermodules as in (12.15). For $1 \leq i < j \leq n$, define "transpositions"

$$[i, j] = -[j, i] = (-1)^{j-i-1} t_{j-1} \dots t_{i+1} t_i t_{i+1} \dots t_{j-1}. \tag{13.5}$$

The defining relations of $\mathcal{T}_n$ imply

$$[i, j]^2 = 1,$$
$$[i, j][k, l] = -[k, l][i, j] \quad \text{if } \{i, j\} \cap \{k, l\} = \varnothing,$$
$$[i, j][j, k][i, j] = [j, k][i, j][j, k] = [k, i] \quad \text{for distinct } i, j, k.$$

Finally, for distinct $1 \le i_1, \dots, i_r \le n$, define '$r$-cycles'

$$[i_1, i_2, \dots, i_r] = (-1)^{r-1}[i_2, \dots, i_r, i_1] = [i_{r-1}, i_r][i_{r-2}, i_r] \dots [i_1, i_r].$$

For $1 \le k \le n$, the analogue of the Jucys–Murphy element (2.1) is

$$M_k := \sum_{i=1}^{k-1} [i, k], \tag{13.6}$$

in particular, $M_1 = 0$. Note that

$$M_k^2 = (k-1) + \sum_{\substack{i,j=1 \\ i \ne j}}^{k-1} [i, j, k]. \tag{13.7}$$

and

$$t_i M_k = \begin{cases} -M_k t_i & \text{if } i \ne k-1, k, \\ -M_{k-1} t_i + 1 & \text{if } i = k-1, \\ -M_{k+1} t_i + 1 & \text{if } i = k. \end{cases} \tag{13.8}$$

It follows that $M_k M_l = -M_l M_k$ if $k \ne l$. Now using these facts, it is easy to show:

Lemma 13.1.1

(i) *for* $1 \le k, l \le n$, M_k^2 *and* M_l^2 *commute;*
(ii) t_i *commutes with* M_k^2 *for* $k \ne i, i+1$*;*
(iii) t_i *commutes with* $M_i^2 + M_{i+1}^2$ *and* $M_i^2 M_{i+1}^2$.

This implies:

Lemma 13.1.2 *The symmetric polynomials in* $M_1^2, M_2^2, \dots, M_n^2$ *belong to the center of* $\mathcal{T}_n$.

Remark 13.1.3 It can be shown (see [BK$_5$]) that the space of all *even* central elements of $\mathcal{T}_n$ *equals* the set of symmetric polynomials in the $M_1^2, \dots, M_n^2$. But $\mathcal{T}_n$ could also have odd central elements. We will not need these facts.

13.2 Sergeev superalgebras

We define the *Sergeev superalgebra* $\mathcal{Y}_n$ (cf. [S₁, N]) to be the tensor product of superalgebras

$$\mathcal{Y}_n := \mathcal{T}_n \otimes \mathcal{C}_n,$$

where $\mathcal{T}_n$ is the twisted group superalgebra defined in Section 13.1 and $\mathcal{C}_n$ is the Clifford superalgebra defined in Section 12.1. Let us write t_i for $t_i \otimes 1 \in \mathcal{Y}_n$ and c_j for $1 \otimes c_j \in \mathcal{Y}_n$. The following result is clear:

Lemma 13.2.1 *The superalgebra $\mathcal{Y}_n$ is generated by (odd generators) $t_1, \dots, t_{n-1}$ and $c_1, \dots, c_n$, subject only to the relations (13.1)–(13.3), (12.8), (12.9), together with $t_i c_j = -c_j t_i$ for all admissible i and j.*

The antiautomorphism τ of $\mathcal{T}_n$ defined in (13.4) can be extended to an antiautomorphism of $\mathcal{Y}_n$, which we also denote by τ:

$$\tau : \mathcal{Y}_n \to \mathcal{Y}_n, \quad t_i \mapsto -t_i, \ c_j \mapsto c_j \quad (1 \le i < n, \ 1 \le j \le n). \tag{13.9}$$

As usual, τ defines a duality on $\mathcal{Y}_n$-supermodules.

Recall the Clifford supermodule U_n from Example 12.2.4, which is the unique irreducible $\mathcal{C}_n$-supermodule up to isomorphism, is of type M if n is even, type Q if n is odd, and $\dim U_n = 2^{\lfloor (n+1)/2 \rfloor}$. Consider the exact functors

$$\mathfrak{F}_n : \mathcal{T}_n\text{-smod} \to \mathcal{Y}_n\text{-smod}, \qquad \mathfrak{F}_n := ? \boxtimes U_n,$$

$$\mathfrak{G}_n : \mathcal{Y}_n\text{-smod} \to \mathcal{T}_n\text{-smod}, \qquad \mathfrak{G}_n := \mathrm{Hom}_{\mathcal{C}_n}(U_n, ?).$$

So, given a $\mathcal{T}_n$-supermodule W, $\mathfrak{F}_n(W)$ is just the outer tensor product $W \boxtimes U_n$ of (12.19), and, for a $\mathcal{Y}_n$-supermodule V, $\mathfrak{G}_n(V)$ is the superspace $\mathrm{Hom}_{\mathcal{C}_n}(U_n, V)$ considered as a $\mathcal{T}_n$-supermodule with respect to the action $(t\theta)(u) = t\theta(u)$ for $t \in \mathcal{T}_n$, $u \in U_n$, $\theta \in \mathrm{Hom}_{\mathcal{C}_n}(U_n, V)$.

Also let

$$\mathrm{res}^{\mathcal{Y}_n}_{\mathcal{Y}_{n-1}} : \mathcal{Y}_n\text{-smod} \to \mathcal{Y}_{n-1}\text{-smod}, \quad \mathrm{ind}^{\mathcal{Y}_n}_{\mathcal{Y}_{n-1}} : \mathcal{Y}_{n-1}\text{-smod} \to \mathcal{Y}_n\text{-smod},$$

$$\mathrm{res}^{\mathcal{T}_n}_{\mathcal{T}_{n-1}} : \mathcal{T}_n\text{-smod} \to \mathcal{T}_{n-1}\text{-smod}, \quad \mathrm{ind}^{\mathcal{T}_n}_{\mathcal{T}_{n-1}} : \mathcal{T}_{n-1}\text{-smod} \to \mathcal{T}_n\text{-smod}$$

denote the induction and restriction functors, where $\mathcal{T}_{n-1} \subset \mathcal{T}_n$ and $\mathcal{Y}_{n-1} \subset \mathcal{Y}_n$ are the natural subalgebras generated by all but the last generators. These functors are exact, which is clear for restriction, and for induction we need to use the fact that the right supermodules $(\mathcal{Y}_n)_{\mathcal{Y}_{n-1}}$ and $(\mathcal{T}_n)_{\mathcal{T}_{n-1}}$ are free.

The following proposition shows that $\mathfrak{F}_n$ and $\mathfrak{G}_n$ establish Morita equivalence between $\mathcal{T}_n$ and $\mathcal{Y}_n$ when n is even and "almost" Morita equivalence when n is odd. It also relates restriction/induction functors for $\mathcal{T}$s and $\mathcal{Y}$s. Recall that Π denotes the parity change functor (12.17).

Proposition 13.2.2 *The functors $\mathfrak{F}_n$ and $\mathfrak{G}_n$ are exact, commute with τ-duality, and are left and right adjoint to one another. Moreover:*

(i) *Suppose that n is even. Then $\mathfrak{F}_n$ and $\mathfrak{G}_n$ are inverse equivalences of categories, so induce a type-preserving bijection between the isomorphism classes of irreducible $\mathcal{T}_n$-supermodules and irreducible $\mathcal{Y}_n$-supermodules. Also,*

$$\mathfrak{F}_{n-1} \circ \operatorname{res}^{\mathcal{T}_n}_{\mathcal{T}_{n-1}} \simeq \operatorname{res}^{\mathcal{Y}_n}_{\mathcal{Y}_{n-1}} \circ \mathfrak{F}_n, \tag{13.10}$$

$$\mathfrak{G}_{n-1} \circ \operatorname{res}^{\mathcal{Y}_n}_{\mathcal{Y}_{n-1}} \simeq \operatorname{res}^{\mathcal{T}_n}_{\mathcal{T}_{n-1}} \circ \mathfrak{G}_n \oplus \Pi \circ \operatorname{res}^{\mathcal{T}_n}_{\mathcal{T}_{n-1}} \circ \mathfrak{G}_n, \tag{13.11}$$

$$\mathfrak{F}_{n+1} \circ \operatorname{ind}^{\mathcal{T}_{n+1}}_{\mathcal{T}_n} \simeq \operatorname{ind}^{\mathcal{Y}_{n+1}}_{\mathcal{Y}_n} \circ \mathfrak{F}_n, \tag{13.12}$$

$$\mathfrak{G}_{n+1} \circ \operatorname{ind}^{\mathcal{Y}_{n+1}}_{\mathcal{Y}_n} \simeq \operatorname{ind}^{\mathcal{T}_{n+1}}_{\mathcal{T}_n} \circ \mathfrak{G}_n \oplus \Pi \circ \operatorname{ind}^{\mathcal{T}_{n+1}}_{\mathcal{T}_n} \circ \mathfrak{G}_n. \tag{13.13}$$

(ii) *Suppose that n is odd. Then*

$$\mathfrak{F}_n \circ \mathfrak{G}_n \simeq \mathrm{Id} \oplus \Pi \quad \textit{and} \quad \mathfrak{G}_n \circ \mathfrak{F}_n \simeq \mathrm{Id} \oplus \Pi.$$

Furthermore, the functor $\mathfrak{F}_n$ induces a bijection between isomorphism classes of irreducible $\mathcal{T}_n$-modules of type `M` *and irreducible $\mathcal{Y}_n$-modules of type* `Q`*, while the functor $\mathfrak{G}_n$ induces a bijection between isomorphism classes of irreducible $\mathcal{Y}_n$-modules of type* `M` *and irreducible $\mathcal{T}_n$-modules of type* `Q`*. Finally,*

$$\operatorname{res}^{\mathcal{Y}_n}_{\mathcal{Y}_{n-1}} \circ \mathfrak{F}_n \simeq \mathfrak{F}_{n-1} \circ \operatorname{res}^{\mathcal{T}_n}_{\mathcal{T}_{n-1}} \oplus \Pi \circ \mathfrak{F}_{n-1} \circ \operatorname{res}^{\mathcal{T}_n}_{\mathcal{T}_{n-1}}, \tag{13.14}$$

$$\operatorname{res}^{\mathcal{T}_n}_{\mathcal{T}_{n-1}} \circ \mathfrak{G}_n \simeq \mathfrak{G}_{n-1} \circ \operatorname{res}^{\mathcal{Y}_n}_{\mathcal{Y}_{n-1}}, \tag{13.15}$$

$$\operatorname{ind}^{\mathcal{Y}_{n+1}}_{\mathcal{Y}_n} \circ \mathfrak{F}_n \simeq \mathfrak{F}_{n+1} \circ \operatorname{ind}^{\mathcal{T}_{n+1}}_{\mathcal{T}_n} \oplus \Pi \circ \mathfrak{F}_{n+1} \circ \operatorname{ind}^{\mathcal{T}_{n+1}}_{\mathcal{T}_n}, \tag{13.16}$$

$$\operatorname{ind}^{\mathcal{T}_{n+1}}_{\mathcal{T}_n} \circ \mathfrak{G}_n \simeq \mathfrak{G}_{n+1} \circ \operatorname{ind}^{\mathcal{Y}_{n+1}}_{\mathcal{Y}_n}. \tag{13.17}$$

Proof It is clear that $\mathfrak{F}_n$ is exact, and for $\mathfrak{G}_n$ this follows from the fact that $\mathcal{C}_n$ is simple.

Let n be even. For any $\mathcal{Y}_n$-supermodule V define the map

$$\alpha_V : \mathfrak{F}_n(\mathfrak{G}_n(V)) = \operatorname{Hom}_{\mathcal{C}_n}(U_n, V) \boxtimes U_n \to V, \quad \theta \otimes u \mapsto \theta(u),$$

and for any $\mathcal{T}_n$-supermodule W define the map

$$\beta_W : W \to \mathfrak{G}_n \circ \mathfrak{F}_n(W) = \operatorname{Hom}_{\mathcal{C}_n}(U_n, W \boxtimes U_n), \quad w \mapsto \theta_w,$$

where $\theta_w(u) = w \otimes u$ for $u \in U_n$. It is easy to see that α_V and β_W are natural isomorphisms, so they define isomorphisms of functors $\mathfrak{F}_n \circ \mathfrak{G}_n \simeq \mathrm{Id}$ and $\mathrm{Id} \simeq \mathfrak{G}_n \circ \mathfrak{F}_n$. Then $\mathfrak{F}_n$ and $\mathfrak{G}_n$ are clearly adjoint to each other.

Now, assume n is odd. Let J be an odd involution in $\mathrm{End}_{\mathcal{C}_n}(U_n)$, see Remark 12.2.2. For any $\mathcal{Y}_n$-supermodule V define the map

$$\alpha_V : \mathfrak{F}_n(\mathfrak{G}_n(V)) = \mathrm{Hom}_{\mathcal{C}_n}(U_n, V) \boxtimes U_n \to V \oplus \Pi(V),$$
$$\theta \otimes u \mapsto (\theta(u), (-1)^{\bar{\theta}}\theta(Ju)),$$

and for any $\mathcal{T}_n$-supermodule W define the map

$$\beta_W : W \oplus \Pi(W) \to \mathfrak{G}_n(\mathfrak{F}_n(W)) = \mathrm{Hom}_{\mathcal{C}_n}(U_n, W \boxtimes U_n),$$
$$(w, w') \mapsto \theta_{w,w'},$$

where

$$\theta_{w,w'}(u) = w \otimes u + (-1)^{\bar{w}'} w' \otimes Ju$$

for each $u \in U_n$. As $\mathcal{C}_n$ is simple, any $\mathcal{C}_n$-supermodule is just a direct sum of U_n's. Using this, we can easily see that α_V is surjective, β_W is injective, and

$$\mathrm{sdim}\,\mathrm{Hom}_{\mathcal{C}_n}(U_n, V) \boxtimes U_n = \mathrm{sdim}\,(V \oplus \Pi(V)),$$
$$\mathrm{sdim}\,(W \oplus \Pi(W)) = \mathrm{sdim}\,\mathrm{Hom}_{\mathcal{C}_n}(U_n, W \boxtimes U_n).$$

Finally, observe that α_V and β_W are supermodule homomorphisms, natural with respect to V and W, respectively. Thus,

$$\mathfrak{F}_n \circ \mathfrak{G}_n \simeq \mathrm{Id} \oplus \Pi \quad \text{and} \quad \mathfrak{G}_n \circ \mathfrak{F}_n \simeq \mathrm{Id} \oplus \Pi. \tag{13.18}$$

Let W be an irreducible $\mathcal{T}_n$-supermodule. If W is type `M`, then $\mathfrak{F}_n W$ is irreducible of type `Q`, thanks to Lemma 12.2.13(ii). Moreover, by Lemma 12.2.13 (iii), if W is type `Q`, then $\mathfrak{F}_n W \simeq V \oplus \Pi V$ for a type `M` irreducible $\mathcal{Y}_n$-supermodule V. It now follows from (13.18) that $\mathfrak{F}_n$ induces a bijection between isoclasses of irreducible $\mathcal{T}_n$-supermodules of type `M` and irreducible $\mathcal{Y}_n$-supermodules of type `Q`, while $\mathfrak{G}_n$ induces a bijection between isoclasses of irreducible $\mathcal{Y}_n$-supermodules of type `M` and irreducible $\mathcal{T}_n$-supermodules of type `Q`.

Now, let

$$\iota : \mathrm{Id} \longrightarrow \mathrm{Id} \oplus \Pi, \qquad \pi : \mathrm{Id} \oplus \Pi \longrightarrow \mathrm{Id}$$

be the obvious natural transformations. Then it is easy to see $\beta \circ \iota$ and $\pi \circ \alpha$ give the unit and the counit of the adjointness needed to prove that $\mathfrak{F}_n$ is left adjoint to $\mathfrak{G}_n$ (cf. [ML, IV.1, Theorem 2(v)]). A more tedious check shows that $\tau := \sqrt{2}\alpha^{-1} \circ \iota$ and $\sigma := \sqrt{2}\pi \circ \beta^{-1}$ give the unit and the counit of the adjointness needed to prove that $\mathfrak{F}_n$ is right adjoint to $\mathfrak{G}_n$. Indeed, let us write $\mathfrak{G} = \mathfrak{G}_n$, $\mathcal{C} = \mathcal{C}_n$, etc., and prove, for example, that the composition of natural transformations

$$\mathfrak{G} \xrightarrow{\mathfrak{G}\circ\tau} \mathfrak{G}\mathfrak{F}\mathfrak{G} \xrightarrow{\sigma\circ\mathfrak{G}} \mathfrak{G}$$

is the identical natural transformation. For any $\mathcal{Y}$-supermodule V, let $\varphi \in \mathfrak{G}(V) = \mathrm{Hom}_{\mathcal{C}}(U, V)$. Then the first arrow takes φ to the function

$$\psi : u \mapsto \sqrt{2}\alpha_V^{-1}(\varphi(u), 0) \in \mathfrak{G}\mathfrak{F}\mathfrak{G}(V) = \mathrm{Hom}_{\mathcal{C}}(U, \mathrm{Hom}_{\mathcal{C}}(U, V) \boxtimes U).$$

It suffices to prove that

$$\psi = \frac{1}{\sqrt{2}}\beta_{\mathfrak{G}(V)}(\varphi, 0) - (-1)^{\bar{\varphi}}\frac{1}{\sqrt{2}}\beta_{\mathfrak{G}(V)}(0, \varphi \circ J), \tag{13.19}$$

since the application of σ to the right-hand side gives φ, as desired. To prove (13.19), evaluate the right-hand side at u to get

$$\frac{1}{\sqrt{2}}\theta_{\varphi,0}(u) - (-1)^{\bar{\varphi}}\frac{1}{\sqrt{2}}\theta_{0,\varphi\circ J}(u) = \frac{1}{\sqrt{2}}\varphi \otimes u + \frac{1}{\sqrt{2}}\varphi \circ J \otimes Ju,$$

which is indeed equal to $\sqrt{2}\alpha_V^{-1}(\varphi, 0)$, since

$$\alpha_V\left(\frac{1}{\sqrt{2}}\varphi \otimes u + \frac{1}{\sqrt{2}}\varphi \circ J \otimes Ju\right) = \sqrt{2}(\varphi(u), 0).$$

Next, for general n we prove that $\mathfrak{F}_n$ commutes with duality. It is enough to show that $\mathfrak{F}_n$ commutes with duality, that is $\tau \circ \mathfrak{F}_n \circ \tau \cong \mathfrak{F}_n$, since then, using (12.16) and the fact that $\mathfrak{G}_n$ is left adjoint to $\mathfrak{F}_n$, the composite functor $\tau \circ \mathfrak{G}_n \circ \tau$ is right adjoint to $\mathfrak{F}_n$, but we already know that $\mathfrak{G}_n$ is right adjoint to $\mathfrak{F}_n$, so uniqueness of adjoints gives that $\tau \circ \mathfrak{G}_n \circ \tau \cong \mathfrak{G}_n$, that is $\mathfrak{G}_n$ commutes with duality too. To prove that $\mathfrak{F}_n$ commutes with duality, note that the antiautomorphism τ of $\mathcal{Y}_n$ induces the antiautomorphism τ of the subalgebra $\mathcal{C}_n$ with $\tau(c_i) = c_i$ for each $i = 1, \dots, n$. As U_n is the only irreducible $\mathcal{C}_n$-supermodule up to isomorphism, there exists a homogeneous isomorphism $\varphi : U_n \to U_n^\tau$. Then, using (12.20) and $\mathrm{id}_{W^\tau} \otimes \varphi$, we get natural isomorphisms

$$(W \boxtimes U_n)^\tau \xrightarrow{\sim} W^\tau \boxtimes U_n^\tau \xrightarrow{\sim} W^\tau \boxtimes U_n.$$

It just remains to check the isomorphisms (13.10)–(13.17). Well, (13.10) follows from definitions noting that $\mathrm{res}^{\mathcal{C}_n}_{\mathcal{C}_{n-1}} U_n \simeq U_{n-1}$ if n is even. Then (13.11) follows from (13.10) on composing on the left with $\mathfrak{G}_{n-1}$ and on the right with $\mathfrak{G}_n$. Next, (13.15) follows from the definition and an application of Frobenius reciprocity, using the observation that $U_n \simeq \mathrm{ind}^{\mathcal{C}_n}_{\mathcal{C}_{n-1}} U_{n-1}$ if n is odd.

To prove (13.15), note that $U_n \simeq \mathrm{ind}_{\mathcal{C}_{n-1}}^{\mathcal{C}_n} U_{n-1}$ when n is odd. So for a $\mathcal{Y}_n$-supermodule V, using Frobenius reciprocity, we get

$$\begin{aligned}\mathfrak{G}_{n-1} \circ \mathrm{res}^{\mathcal{Y}_n}_{\mathcal{Y}_{n-1}}(V) &\simeq \mathrm{Hom}_{\mathcal{C}_{n-1}}(U_{n-1}, \mathrm{res}^{\mathcal{Y}_n}_{\mathcal{Y}_{n-1}} V) \\ &\simeq \mathrm{Hom}_{\mathcal{C}_n}(\mathrm{ind}^{\mathcal{C}_n}_{\mathcal{C}_{n-1}} U_{n-1}, \mathrm{res}^{\mathcal{Y}_n}_{\mathcal{T}_{n-1}\otimes\mathcal{C}_n} V) \\ &\simeq \mathrm{res}^{\mathcal{T}_n}_{\mathcal{T}_{n-1}} \mathrm{Hom}_{\mathcal{C}_n}(U_n, V) \simeq \mathrm{res}^{\mathcal{T}_n}_{\mathcal{T}_{n-1}} \circ \mathfrak{G}_n(V).\end{aligned}$$

Now (13.14) follows from (13.15) by composing with $\mathfrak{F}_{n-1}$ and $\mathfrak{F}_n$. Finally, (13.12), (13.13), (13.16), and (13.17) follow from (13.15), (13.14), (13.11), and (13.10), respectively, by uniqueness of adjoints. □

Now we describe another reincarnation of $\mathcal{Y}_n$ (cf. [S$_2$]). The symmetric group S_n acts on the generators $c_1, \dots, c_n$ of the Clifford algebra $\mathcal{C}_n$ by place permutations: $c_i \cdot w = c_{w^{-1}(i)}$. This action can be extended to the action of S_n on $\mathcal{C}_n$ by superalgebra automorphisms. Recall that $\mathcal{G}_n$ denotes the group algebra FS_n, which will be considered as a superalgebra concentrated in degree $\bar{0}$. Now denote by $\mathcal{G}_n \ltimes \mathcal{C}_n$ the superalgebra, which as a superspace is just $\mathcal{G}_n \otimes \mathcal{C}_n$, where the Clifford algebra $\mathcal{C}_n$ is graded as usual, and the multiplication is given by

$$(w \otimes c)(w' \otimes c') = (ww' \otimes (c \cdot w')c') \qquad (w, w' \in S_n,\ c, c' \in \mathcal{C}_n).$$

Lemma 13.2.3 *There is an isomorphism of superalgebras*

$$\varphi : \mathcal{Y}_n \xrightarrow{\sim} \mathcal{G}_n \ltimes \mathcal{C}_n, \quad 1 \otimes c_j \mapsto 1 \otimes c_j,\ t_i \otimes 1 \mapsto \frac{1}{\sqrt{-2}} s_i \otimes (c_i - c_{i+1})$$

for $j = 1, \dots, n$, $i = 1, \dots, n-1$.

Proof Obviously, the elements $1 \otimes c_j \in \mathcal{G}_n \ltimes \mathcal{C}_n$ satisfy the defining relations of $\mathcal{C}_n$. A simple calculation shows that the $\varphi(t_i \otimes 1)$ satisfy the relations (13.1)–(13.3). Finally, we have

$$\varphi(1 \otimes c_j)\varphi(t_i \otimes 1) = -\varphi(t_i \otimes 1)\varphi(1 \otimes c_j)$$

for all admissible i and j. By Lemma 13.2.1, φ as above exists. Now, φ is surjective, as $c_i - c_{i+1}$ is invertible in $\mathcal{C}_n$, and it remains to compare dimensions. □

We will from now on *identify* $\mathcal{Y}_n$ with $\mathcal{G}_n \ltimes \mathcal{C}_n$ according to the lemma. Let us write s_i for $s_i \otimes 1 \in \mathcal{G}_n \ltimes \mathcal{C}_n = \mathcal{Y}_n$ and c_j for $1 \otimes c_j \in \mathcal{G}_n \ltimes \mathcal{C}_n = \mathcal{Y}_n$ for $1 \le i < n, 1 \le i \le n$. Then the following is easy to check:

Lemma 13.2.4 *The superalgebra $\mathcal{Y}_n$ is generated by (even generators) $s_1, \dots, s_{n-1}$ and (odd generators) $c_1, \dots, c_n$, subject only to the relations (3.2), (3.3), (12.8), (12.9), together with*

$$s_i c_i = c_{i+1} s_i, \quad s_i c_j = c_j s_i \tag{13.20}$$

for all admissible i and j with $j \neq i, i+1$.

The antiautomorphism τ of $\mathcal{Y}_n$ defined in (13.9) can now be defined via:

$$\tau : \mathcal{Y}_n \to \mathcal{Y}_n, \quad s_i \mapsto s_i, \ c_j \mapsto c_j \quad (1 \le i < n, \ 1 \le j \le n). \tag{13.21}$$

The JM-elements $M_k \in \mathcal{T}_n$ from (13.6) can be considered as elements of $\mathcal{Y}_n$ because $\mathcal{T}_n$ is a subalgebra of $\mathcal{Y}_n$. We will need to know the image of M_k under φ. Define the *Jucys–Murphy elements* of $\mathcal{Y}_n$ as follows:

$$L_k := \sum_{1 \le j < k} (1 + c_j c_k)(j, k) \in \mathcal{Y}_n \qquad (1 \le k \le n). \tag{13.22}$$

Lemma 13.2.5

(i) *For $2 \le k \le n$ we have $\varphi(M_k) = \frac{1}{2\sqrt{-2}}(1 - c_{k-1}c_k)L_k(c_{k-1} - c_k)$;*
(ii) *For $1 \le k \le n$ we have $\varphi(M_k^2) = \frac{1}{2}L_k^2$.*

Proof (i) is proved by induction on k using relations

$$M_{k+1} = -t_k M_k t_k + t_k,$$

which follow from (13.8), and

$$L_{k+1} = s_k L_k s_k + (1 + c_k c_{k+1}) s_k, \tag{13.23}$$

which are easy to check directly.

(ii) is checked using (i) and relations $c_k L_k = -L_k c_k$, $c_k L_j = L_j c_k$ for $k \neq j$. □

14
Affine Sergeev superalgebras

This chapter is parallel to Chapter 3. The role of the degenerate affine Hecke algebra is played by the affine Sergeev superalgebra $\mathcal{X}_n$, introduced and studied by Nazarov [N]. The superalgebra $\mathcal{X}_n$ is in the same relation to the Sergeev superalgebra $\mathcal{Y}_n$ as $\mathcal{H}_n$ to the group algebra $\mathcal{G}_n$ of the symmetric group.

We will sometimes suppress the word "super", as "everything is super" anyway. So we may speak of a *subalgebra* $\mathcal{C}_n$ of $\mathcal{X}_n$ rather than a *subsuperalgebra*, etc.

14.1 The superalgebras

Let $\mathcal{A}_n$ denote the superalgebra with even generators $x_1, \dots, x_n$ and odd generators $c_1, \dots, c_n$, where the x_i are subject to the polynomial relations (3.1), the c_i are subject to the Clifford superalgebra relations (12.8), (12.9), and there are the mixed relations

$$c_i x_j = x_j c_i, \quad c_i x_i = -x_i c_i \tag{14.1}$$

for all $1 \le i, j \le n$ with $i \ne j$. For $\alpha = (\alpha_1, \dots, \alpha_n) \in \mathbb{Z}^n$ and $\beta = (\beta_1, \dots, \beta_n) \in \mathbb{Z}_2^n$, we write x^α and c^β for the monomials $x_1^{\alpha_1} \dots x_n^{\alpha_n}$ and $c_1^{\beta_1} \dots c_n^{\beta_n}$, respectively. Then it is easy to see that the elements

$$\{x^\alpha c^\beta \mid \alpha \in \mathbb{Z}^n, \ \beta \in \mathbb{Z}_2^n\}$$

form a basis of $\mathcal{A}_n$. In particular, $\mathcal{A}_n \cong \mathcal{A}_1 \otimes \cdots \otimes \mathcal{A}_1$ (n times). Note that the polynomial algebra $\mathcal{P}_n$ can be identified with the subalgebra of $\mathcal{A}_n$ generated by the x_i, and the Clifford algebra $\mathcal{C}_n$ can be identified with the subalgebra of

$\mathcal{A}_n$ generated by the c_i. Thus, $\mathcal{A}_n$ is the twisted tensor product $\mathcal{P}_n \otimes \mathcal{C}_n$. We define a left action of S_n on $\mathcal{A}_n$ by algebra automorphisms so that

$$w \cdot x_i = x_{wi}, \quad w \cdot c_i = c_{wi} \tag{14.2}$$

for each $w \in S_n$, $i = 1, \dots, n$.

The *affine Sergeev superalgebra* $\mathcal{X}_n$ has even generators $s_1, \dots, s_{n-1}$, $x_1, \dots, x_n$ and odd generators $c_1, \dots, c_n$, subject to the relations (3.1), (12.8), (12.9), (14.1), (3.2), (3.3), together with the new relations:

$$s_i c_j = c_j s_i, \tag{14.3}$$

$$s_i c_i = c_{i+1} s_i, \quad s_i c_{i+1} = c_i s_i, \tag{14.4}$$

$$s_i x_j = x_j s_i, \tag{14.5}$$

$$s_i x_i = x_{i+1} s_i - 1 - c_i c_{i+1} \tag{14.6}$$

for all admissible i, j with $j \neq i, i+1$. We call $x_1, \dots, x_n$ *polynomial generators*, $c_1, \dots, c_n$ *Clifford generators*, and $s_1, \dots, s_{n-1}$ *Coxeter generators*. Note that $\mathcal{X}_1 \cong \mathcal{A}_1$. Also, *by agreement*, $\mathcal{X}_0 \cong F$. The relation (14.6) implies

$$s_i x_{i+1} = x_i s_i + 1 - c_i c_{i+1}. \tag{14.7}$$

By induction, we deduce more general formulas for $j \geq 1$:

$$s_i x_i^j = x_{i+1}^j s_i - \sum_{k=0}^{j-1} (x_i^k x_{i+1}^{j-1-k} + (-x_i)^k x_{i+1}^{j-1-k} c_i c_{i+1}), \tag{14.8}$$

$$s_i x_{i+1}^j = x_i^j s_i + \sum_{k=0}^{j-1} (x_i^k x_{i+1}^{j-1-k} - x_i^k (-x_{i+1})^{j-1-k} c_i c_{i+1}). \tag{14.9}$$

14.2 Basis Theorem for $\mathcal{X}_n$

There are obvious homomorphisms $\varphi : \mathcal{A}_n \to \mathcal{X}_n$, $\psi : \mathcal{G}_n \to \mathcal{X}_n$, and $\xi : \mathcal{Y}_n \to \mathcal{X}_n$ under which the elements x_i, c_k, and s_j map to the corresponding elements of $\mathcal{X}_n$. We write $\mathcal{A}_n$ for $\varphi(\mathcal{A}_n)$, $\mathcal{P}_n$ for $\varphi(\mathcal{P}_n)$, $\mathcal{G}_n$ for $\psi(\mathcal{G}_n)$, $\mathcal{Y}_n$ for $\xi(\mathcal{Y}_n)$, x^α for $\varphi(x^\alpha)$, c^β for $\varphi(c^\beta)$, and w for $\psi(w)$ (for $w \in S_n \subset \mathcal{G}_n$). This notation will be justified shortly, when we show that φ, ψ, and ξ are monomorphisms. The following is obvious from the defining relations:

Lemma 14.2.1 *Let $f \in \mathcal{A}_n$, $w \in S_n$. Then in $\mathcal{X}_n$ we have*

$$wf = (w \cdot f)w + \sum_{u<w} f_u u, \qquad fw = w(w^{-1} \cdot f) + \sum_{u<w} u f_u'$$

for some $f_y, f_y' \in \mathcal{A}_n$ of x-degrees less than the x-degree of f.

It follows easily from this lemma that that $\mathcal{X}_n$ is at least *spanned* by all $x^\alpha c^\beta w$, $\alpha \in \mathbb{Z}_+^n, \beta \in \mathbb{Z}_2^n, w \in S_n$.

Theorem 14.2.2 *The* $\{x^\alpha c^\beta w \mid \alpha \in \mathbb{Z}_+^n,\ \beta \in \mathbb{Z}_2^n,\ w \in S_n\}$ *form a basis for* $\mathcal{X}_n$.

Proof The proof using Bergman's Diamond Lemma is entirely similar to the first proof of Theorem 3.2.2. (The second proof seems to be harder to mimic). □

By Theorem 14.2.2, we have a right from now on to *identify* $\mathcal{P}_n$, $\mathcal{C}_n$, $\mathcal{A}_n$, and $\mathcal{G}_n$ with the corresponding subalgebras of $\mathcal{X}_n$. Then $\mathcal{X}_n$ is a free right $\mathcal{G}_n$-module on basis $\{x^\alpha c^\beta \mid \alpha \in \mathbb{Z}^n,\ \beta \in \mathbb{Z}_2^n\}$. As another consequence, if $m \leq n$, we can consider $\mathcal{X}_m$ as the subalgebra of $\mathcal{X}_n$ generated by $x_1, \dots, x_m$, $c_1, \dots, c_m$, and $s_1, \dots, s_{m-1}$. Sometimes, we will use another embedding of $\mathcal{X}_m$ into $\mathcal{X}_n$. In order to distinguish between the two, denote

$$\mathcal{X}'_m = \langle s_i, x_j, c_j \mid n+1-m \leq j \leq n,\ n+2-m \leq i < n \rangle \subset \mathcal{X}_n. \tag{14.10}$$

Finally, let us point out that there are obvious variants of Theorem 14.2.2. For example, $\mathcal{X}_n$ also has $\{w x^\alpha c^\beta \mid w \in S_n,\ \alpha \in \mathbb{Z}_+^n,\ \beta \in \mathbb{Z}_2^n\}$ as a basis. This follows using Lemma 14.2.1.

14.3 The center of $\mathcal{X}_n$

Theorem 14.3.1 *The center of* $\mathcal{X}_n$ *consists of all symmetric polynomials in* $x_1^2, \dots, x_n^2$.

Proof That the symmetric polynomials in $x_1^2, \dots, x_n^2$ are indeed central is easily verified. Indeed, it is clear that they commute with the Clifford generators. Moreover, as s_i commutes with all x_j for $j \neq i, i+1$, we just need to check that s_i commutes with a polynomial of the form $(x_i^2)^a (x_{i+1}^2)^b + (x_i^2)^b (x_{i+1}^2)^a$, which is done using (14.8) and (14.9).

Conversely, take a central element $z = \sum_{w \in S_n} f_w w \in \mathcal{X}_n$ where each $f_w \in \mathcal{A}_n$. Let w be maximal with respect to the Bruhat order such that $f_w \neq 0$. Assume $w \neq 1$. Then there exists $i \in \{1, \dots, n\}$ with $wi \neq i$. By Lemma 14.2.1, $x_i^2 z - z x_i^2$ looks like $f_w (x_i^2 - x_{wi}^2) w$ plus a linear combination of terms of the form $f'_u u$ for $f'_u \in \mathcal{A}_n$ and $u \in S_n$ with $u \not\geq w$ in the Bruhat order. So in view of Theorem 14.2.2, z is not central, giving a contradiction.

Hence, we must have that $z \in \mathcal{A}_n$. Commuting with polynomial generators, we see that z is actually in $\mathcal{P}_n$, and commuting with Clifford generators, we then deduce that $z \in F[x_1^2, \dots, x_n^2]$. To see that z is a symmetric polynomial, write $z = \sum_{i,j \geq 0} a_{i,j}(x_1^2)^i(x_2^2)^j$ where the coefficients $a_{i,j}$ lie in $F[x_3, \dots, x_n]$. Applying Lemma 14.2.1 to $s_1 z = z s_1$ now gives that $a_{i,j} = a_{j,i}$ for each i, j, hence z is symmetric in x_1 and x_2. Similar argument shows that z is symmetric in x_i and x_{i+1} for all $i = 1, \dots, n-1$. $\square$

14.4 Parabolic subalgebras of $\mathcal{X}_n$

Let $\mu = (\mu_1, \dots, \mu_r)$ be a composition of n. We define the *parabolic subalgebra* $\mathcal{X}_\mu$ of $\mathcal{X}_n$ as the subalgebra generated by $\mathcal{A}_n$ and all s_j for which $s_j \in S_\mu$. It follows from Theorem 14.2.2 that the elements

$$\{x^\alpha c^\beta w \mid \alpha \in \mathbb{Z}_+^n,\ \beta \in \mathbb{Z}_2^n,\ w \in S_\mu\}$$

form a basis for $\mathcal{X}_\mu$. In particular,

$$\mathcal{X}_\mu \cong \mathcal{X}_{\mu_1} \otimes \cdots \otimes \mathcal{X}_{\mu_r}$$

(tensor product of *superalgebras*). Note that the parabolic subalgebra $\mathcal{X}_{(1,1,\dots,1)}$ is precisely the subalgebra $\mathcal{A}_n$.

We will use the induction and restriction functors between $\mathcal{X}_n$ and $\mathcal{X}_\mu$. These will be denoted simply

$$\mathrm{ind}_\mu^n : \mathcal{X}_\mu\text{-smod} \to \mathcal{X}_n\text{-smod}, \qquad \mathrm{res}_\mu^n : \mathcal{X}_n\text{-smod} \to \mathcal{X}_\mu\text{-smod}, \tag{14.11}$$

the former being the tensor functor $\mathcal{X}_n \otimes_{\mathcal{X}_\mu} ?$ which is left adjoint to res_μ^n. More generally, we will consider induction and restriction between nested parabolic subalgebras, with obvious notation. We will also occasionally consider the restriction functor

$$\mathrm{res}_{n-1}^n : \mathcal{X}_n\text{-smod} \to \mathcal{X}_{n-1}\text{-smod}. \tag{14.12}$$

14.5 Mackey Theorem for $\mathcal{X}_n$

Recall the notation of Section 3.5. Fix some total order $\prec$ refining the Bruhat order $<$ on $D_{\mu,\nu}$. For $x \in D_{\mu,\nu}$, set

$$\mathcal{B}_{\preceq x} = \bigoplus_{y \in D_{\mu,\nu},\ y \preceq x} \mathcal{X}_\mu y \mathcal{G}_\nu, \tag{14.13}$$

$$\mathcal{B}_{\prec x} = \bigoplus_{y \in D_{\mu,\nu},\ y \prec x} \mathcal{X}_\mu y \mathcal{G}_\nu, \tag{14.14}$$

$$\mathcal{B}_x = \mathcal{B}_{\preceq x} / \mathcal{B}_{\prec x}. \tag{14.15}$$

It follows from Lemma 14.2.1 that $\mathcal{B}_{\leq x}$ and $\mathcal{B}_{<x}$ are invariant under right multiplication by $\mathcal{A}_n$. Hence, since $\mathcal{X}_\nu = \mathcal{G}_\nu \mathcal{A}_n$, we have a filtration of $\mathcal{X}_n$ as an $(\mathcal{X}_\mu, \mathcal{X}_\nu)$-bisupermodule.

Note using the property (2) from Section 3.5 that for each $y \in D_{\mu,\nu}$, there exists an algebra isomorphism

$$\varphi_{y^{-1}} : \mathcal{X}_{\mu \cap y\nu} \to \mathcal{X}_{y^{-1}\mu \cap \nu}$$

with $\varphi_{y^{-1}}(w) = y^{-1}wy$, $\varphi_{y^{-1}}(c_i) = c_{y^{-1}i}$, and $\varphi_{y^{-1}}(x_i) = x_{y^{-1}i}$ for $w \in S_{\mu \cap y\nu}$, $1 \leq i \leq n$. If N is a left $\mathcal{X}_{y^{-1}\mu \cap \nu}$-supermodule, then by twisting the action with the isomorphism $\varphi_{y^{-1}}$ we get a left $\mathcal{X}_{\mu \cap y\nu}$-supermodule, which will be denoted ${}^yN$.

Lemma 14.5.1 *Let us view $\mathcal{X}_\mu$ as an $(\mathcal{X}_\mu, \mathcal{X}_{\mu \cap x\nu})$-bisupermodule and $\mathcal{X}_\nu$ as an $(\mathcal{X}_{x^{-1}\mu \cap \nu}, \mathcal{X}_\nu)$-bisupermodule in the natural ways. Then ${}^x\mathcal{X}_\nu$ is an $(\mathcal{X}_{\mu \cap x\nu}, \mathcal{X}_\nu)$-bisupermodule, and*

$$\mathcal{B}_x \simeq \mathcal{X}_\mu \otimes_{\mathcal{X}_{\mu \cap x\nu}} {}^x\mathcal{X}_\nu$$

as $(\mathcal{X}_\mu, \mathcal{X}_\nu)$-bisupermodules.

Proof Similar to the proof of Lemma 3.5.1. □

Theorem 14.5.2 ("Mackey Theorem") *Let M be an $\mathcal{X}_\nu$-supermodule. Then* $\mathrm{res}^n_\mu \mathrm{ind}^n_\nu M$ *admits a filtration with subquotients $\simeq$ to*

$$\mathrm{ind}^\mu_{\mu \cap x\nu} {}^x(\mathrm{res}^\nu_{x^{-1}\mu \cap \nu} M),$$

one for each $x \in D_{\mu,\nu}$. Moreover, the subquotients can be taken in any order refining the Bruhat order on $D_{\mu,\nu}$, in particular $\mathrm{ind}^\mu_{\mu \cap \nu} \mathrm{res}^\nu_{\mu \cap \nu} M$ *appears as a subsupermodule.*

Proof Similar to the proof of Theorem 3.5.2, but using Lemma 14.5.1 instead of Lemma 3.5.1. □

14.6 Some (anti) automorphisms of $\mathcal{X}_n$

A check of relations shows that $\mathcal{X}_n$ possesses an automorphism σ and an antiautomorphism τ defined on the generators as follows:

$$\sigma : s_i \mapsto -s_{n-i}, \quad c_j \mapsto c_{n+1-j}, \quad x_j \mapsto x_{n+1-j}; \tag{14.16}$$

$$\tau : s_i \mapsto s_i, \quad c_j \mapsto c_j, \quad x_j \mapsto x_j, \tag{14.17}$$

for all $i = 1, \dots, n-1$, $j = 1, \dots, n$.

If $M \in \mathcal{X}_n$-smod, we can use τ to make the dual space M^* into an $\mathcal{X}_n$-module denoted M^τ, see Section 12.1. Note that τ leaves invariant every parabolic subalgebra of $\mathcal{X}_n$, so it also induces a duality on finite dimensional $\mathcal{X}_\mu$-supermodules for each composition μ of n.

Given $M \in \mathcal{X}_n$-smod, we can twist the action of $\mathcal{X}_n$ with σ to get a new $\mathcal{X}_n$-supermodule M^σ. The following result is proved similarly to Lemma 3.6.1.

Lemma 14.6.1 *Let* $M \in \mathcal{X}_m$-smod *and* $N \in \mathcal{X}_n$-smod. *Then*

$$(\mathrm{ind}_{m,n}^{m+n} M \boxtimes N)^\sigma \simeq \mathrm{ind}_{n,m}^{m+n} N^\sigma \boxtimes M^\sigma.$$

Moreover, if M and N are irreducible, the same holds for $\circledast$ *in place of* $\boxtimes$.

14.7 Duality for $\mathcal{X}_n$-supermodules

This section should be entirely parallel to Section 3.7. In particular, essentially the same arguments lead to:

Theorem 14.7.1 *For* $M \in \mathcal{X}_m$-smod *and* $N \in \mathcal{X}_n$-smod, *we have*

$$(\mathrm{ind}_{m,n}^{m+n} M \boxtimes N)^\tau \simeq \mathrm{ind}_{n,m}^{m+n} (N^\tau \boxtimes M^\tau).$$

Moreover, if M and N are irreducible, the same holds for $\circledast$ *in place of* $\boxtimes$.

14.8 Intertwining elements for $\mathcal{X}_n$

For $1 \leq i < n$ set

$$\Phi_i := s_i(x_i^2 - x_{i+1}^2) + (x_i + x_{i+1}) + c_i c_{i+1}(x_i - x_{i+1}). \tag{14.18}$$

A tedious calculation using (14.8) and (14.9) gives:

$$\Phi_i^2 = 2x_i^2 + 2x_{i+1}^2 - (x_i^2 - x_{i+1}^2)^2, \tag{14.19}$$

$$\Phi_i x_i = x_{i+1}\Phi_i, \quad \Phi_i x_{i+1} = x_i \Phi_i, \quad \Phi_i x_j = x_j \Phi_i, \tag{14.20}$$

$$\Phi_i c_i = c_{i+1}\Phi_i, \quad \Phi_i c_{i+1} = c_i \Phi_i, \quad \Phi_i c_j = c_j \Phi_i, \tag{14.21}$$

$$\Phi_i \Phi_k = \Phi_k \Phi_i, \quad \Phi_i \Phi_{i+1} \Phi_i = \Phi_{i+1} \Phi_i \Phi_{i+1} \tag{14.22}$$

for all admissible i, j, k with $j \neq i, i+1$ and $|i-k| > 1$. Property (14.22) means that for every $w \in S_n$ we have a well-defined element $\Phi_w \in \mathcal{X}_n$, namely,

$$\Phi_w := \Phi_{i_1} \dots \Phi_{i_m}$$

where $w = s_{i_1} \dots s_{i_m}$ is any reduced expression for w. According to (14.20), these elements have the property that

$$\Phi_w x_i = x_{wi} \Phi_w, \tag{14.23}$$

for all $w \in S_n$ and $1 \leq i \leq n$. We note that only the properties (14.19) and (14.20) will be essential in what follows:

Lemma 14.8.1 *Let V be an $\mathcal{A}_n$-module, $1 \leq j < n$, and $v \in V$ be a vector with $x_j^2 v = av$, $x_{j+1}^2 v = bv$ for some $a, b \in F$. Then $\Phi_j^2 v = 0$ if and only if $2a + 2b = (a - b)^2$, which in turn is equivalent to $a = c^2 - c$ and $b = c^2 + c$ for some $c \in F$.*

Proof By (14.19), Φ_j^2 acts on v with the scalar $2a + 2b - (a - b)^2$. The second part is an elementary exercise. □

15
Integral representations and cyclotomic Sergeev algebras

For notational reasons, when studying representation theory of $\mathcal{X}_n$, it is more convenient to start with the definition of *integral representations*, although in principle we could postpone doing this until after the chapter on character calculations, as was done in the classical case. So Chapter 15 is parallel to Chapter 7. Note that the definition of an *integral* representation is now less obvious. If you want to 'discover' this notion, you need to play with calculations made in Chapter 18 first.

15.1 Integral representations of $\mathcal{X}_n$

Recall that p stands for the characteristic of the ground field F, which is 0 or *odd* prime. Denote

$$\ell := \begin{cases} \infty & \text{if } p = 0, \\ (p-1)/2 & \text{if } p > 0. \end{cases} \tag{15.1}$$

We define the set $I \subset F$ as follows:

$$I := \begin{cases} \mathbb{Z}_{\geq 0} & \text{if } p = 0, \\ \{0, 1, \dots, \ell\} & \text{if } p > 0. \end{cases} \tag{15.2}$$

For $i \in I$ we set

$$q(i) = i(i+1). \tag{15.3}$$

Note that $q(i) = q(j)$ if and only if $i = j$ (for $i, j \in I$).

Definition 15.1.1 An $\mathcal{A}_n$-supermodule M is called *integral* if it is finite dimensional and all eigenvalues of $x_1^2, \dots, x_n^2$ on M are of the form $q(i)$

for $i \in I$. An $\mathcal{X}_n$-supermodule, or more generally an $\mathcal{X}_\mu$-supermodule for μ a composition of n, is called *integral* if it is integral on restriction to $\mathcal{A}_n$. We write $\operatorname{Rep}_I \mathcal{X}_n$ (resp. $\operatorname{Rep}_I \mathcal{A}_n$, $\operatorname{Rep}_I \mathcal{X}_\mu$) for the full subcategory of $\mathcal{X}_n$-smod (resp. $\mathcal{A}_n$-smod, $\mathcal{X}_\mu$-smod) consisting of all integral supermodules.

Lemma 15.1.2 *Let M be a finite dimensional $\mathcal{X}_n$-supermodule and fix j with $1 \le j \le n$. Assume that all eigenvalues of x_j on M belong to I. Then M is integral.*

Proof It suffices to show that the eigenvalues of x_k^2 are of the form $q(i)$ for $i \in I$ if and only if the eigenvalues of x_{k+1}^2 are of the same form, for an arbitrary k with $1 \le k < n$. Actually, by an argument involving conjugation with the automorphism σ, it suffices just to prove the 'if' part. So assume that all eigenvalues of x_{k+1}^2 on M are of the form $q(i)$ for $i \in I$. Let a be an eigenvalue of x_k on M. We have to prove that a^2 is of the form $q(i)$. As since $0 = q(0)$, we may assume that $a \neq 0$. Since x_k and x_{k+1} commute, we can pick v lying in the a-eigenspace of x_k so that v is also an eigenvector for x_{k+1}, of eigenvalue b say. By assumption we have $b^2 = q(i)$ for some $i \in I$. Now let Φ_k be the intertwining element (14.18). By (14.20), we have $x_{k+1}\Phi_k = \Phi_k x_k$. So if $\Phi_k v \neq 0$, we get that

$$a\Phi_k v = \Phi_k x_k v = x_{k+1}\Phi_k v,$$

whence a is an eigenvalue of x_{k+1}, and so $a = q(i')$ for some $i' \in I$ by assumption. Else, $\Phi_k v = 0$ so $\Phi_k^2 v = 0$. So applying Lemma 14.8.1, we again get that $a = q(i')$ for some $i' \in I$. □

Lemma 15.1.3 *Let μ be a composition of n and M be an integral $\mathcal{X}_\mu$-supermodule. Then $\operatorname{ind}_\mu^n M$ is an integral $\mathcal{X}_n$-supermodule.*

Proof The proof is similar to that of Lemma 7.1.3 but uses the element

$$Y_j = \prod_{i \in I}(x_j^2 - q(i)), \quad 1 \le j \le n.$$

instead of the Y_j there and Lemma 15.1.2 instead of Lemma 7.1.2. □

It follows that the functors ind_μ^n, res_μ^n restrict to well-defined functors

$$\operatorname{ind}_\mu^n : \operatorname{Rep}_I \mathcal{X}_\mu \to \operatorname{Rep}_I \mathcal{X}_n, \qquad \operatorname{res}_\mu^n : \operatorname{Rep}_I \mathcal{X}_n \to \operatorname{Rep}_I \mathcal{X}_\mu \tag{15.4}$$

on integral representations. Similar remarks apply to more general induction and restriction between nested parabolic subalgebras of $\mathcal{X}_n$.

15.2 Some Lie theoretic notation

If $p > 0$, let $\mathfrak{g}$ denote the twisted affine Kac–Moody algebra of type $A_{2\ell}^{(2)}$ (over $\mathbb{C}$), see [Kc, Ch. 4, Table Aff 2]. In particular we label the Dynkin diagram by the index set $I = \{0, 1, \dots, \ell\}$ as follows:

if $\ell \geq 2$, and
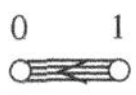
if $\ell = 1$.

The weight lattice is denoted P, the simple roots are $\{\alpha_i \mid i \in I\} \subset P$ and the corresponding simple coroots are $\{h_i \mid i \in I\} \subset P^*$. The Cartan matrix $(\langle h_i, \alpha_j \rangle)_{0 \leq i,j \leq \ell}$ is

$$\begin{pmatrix} 2 & -2 & 0 & \cdots & 0 & 0 & 0 \\ -1 & 2 & -1 & \cdots & 0 & 0 & 0 \\ 0 & -1 & 2 & \cdots & 0 & 0 & 0 \\ & & & \ddots & & & \\ 0 & 0 & 0 & \dots & 2 & -1 & 0 \\ 0 & 0 & 0 & \dots & -1 & 2 & -2 \\ 0 & 0 & 0 & \dots & 0 & -1 & 2 \end{pmatrix} \quad \text{if } \ell \geq 2, \text{ and}$$

$$\begin{pmatrix} 2 & -4 \\ -1 & 2 \end{pmatrix} \quad \text{if } \ell = 1.$$

Let $\{\Lambda_i \mid i \in I\} \subset P$ denote fundamental dominant weights, so that $\langle h_i, \Lambda_j \rangle = \delta_{i,j}$, and let $P_+ \subset P$ denote the set of all dominant integral weights. Set

$$c = h_0 + \sum_{i=1}^{\ell} 2h_i, \qquad \delta = \sum_{i=0}^{\ell-1} 2\alpha_i + \alpha_\ell. \tag{15.5}$$

Then the $\Lambda_0, \dots, \Lambda_\ell, \delta$ form a $\mathbb{Z}$-basis for P, and $\langle c, \alpha_i \rangle = \langle h_i, \delta \rangle = 0$ for all $i \in I$.

In the case $p = 0$, we make the following changes to these definitions. First, $\mathfrak{g}$ denotes the Kac–Moody algebra of type B_∞, see [Kc, Section 7.11]. So $I = \{0, 1, 2, \dots\}$, corresponding to the nodes of the Dynkin diagram

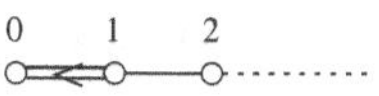

The Cartan matrix $\left(\langle h_i, \alpha_j \rangle\right)_{i,j\geq 0}$ is

$$\begin{pmatrix} 2 & -2 & 0 & & \\ -1 & 2 & -1 & 0 & \\ 0 & -1 & 2 & -1 & \\ & 0 & -1 & 2 & \ddots \\ & & & \ddots & \ddots \end{pmatrix}$$

Note certain notions, for example the element c from (15.5), only make sense if we pass to the completed algebra b_∞, see [Kc, Section 7.12], though the intended meaning whenever we make use of them should be obvious regardless.

The meaning of $U_\mathbb{Q}$, e_i, f_i, h_i, $U_\mathbb{Z}$, $U_\mathbb{Z}^-$, $U_\mathbb{Z}^0$, $e_i^{(n)}$, etc. is the same as in Section 7.2.

15.3 Cyclotomic Sergeev superalgebras

For $\lambda \in P_+$ we set

$$f_\lambda := x_1^{\langle h_0, \lambda\rangle} \prod_{i \in I \setminus \{0\}}^{\ell} (x_1^2 - q(i))^{\langle h_i, \lambda \rangle} \in \mathcal{X}_n. \tag{15.6}$$

Let $\mathcal{J}_\lambda$ denote the two-sided ideal of $\mathcal{X}_n$ generated by f_λ, and define the *cyclotomic Sergeev superalgebra* to be the quotient

$$\mathcal{X}_n^\lambda := \mathcal{X}_n / \mathcal{J}_\lambda.$$

Also, *by agreement*, $\mathcal{X}_0^\lambda \cong F$.

Lemma 15.3.1 *Let M be a finite dimensional $\mathcal{X}_n$-supermodule. Then M is integral if and only if $\mathcal{J}_\lambda M = 0$ for some $\lambda \in P_+$.*

Proof If $\mathcal{J}_\lambda M = 0$, then the eigenvalues of x_1^2 on M are all of the form $q(i)$ for $i \in I$, by definition of $\mathcal{J}_\lambda$. Hence M is integral in view of Lemma 15.1.2. Conversely, suppose that M is integral. Then the minimal polynomial of x_1^2 on M is of the form $\prod_{i\in I}(t - q(i))^{\lambda_i}$ for some $\lambda_i \geq 0$. So if we set $\lambda = 2\lambda_0\Lambda_0 + \lambda_1\Lambda_1 + \cdots + \lambda_\ell\Lambda_\ell \in P_+$, we certainly have that $\mathcal{J}_\lambda M = 0$. □

Lemma 15.3.1 allows us to introduce the functors

$$\mathrm{pr}^\lambda : \mathrm{Rep}_I \mathcal{X}_n \to \mathcal{X}_n^\lambda\text{-smod}, \qquad \mathrm{infl}^\lambda : \mathcal{X}_n^\lambda\text{-smod} \to \mathrm{Rep}_I \mathcal{X}_n. \tag{15.7}$$

Here, infl^λ is simply inflation along the canonical epimorphism $\mathcal{X}_n \to \mathcal{X}_n^\lambda$, while on a supermodule M, $\text{pr}^\lambda M = M/\mathcal{J}_\lambda M$ with the induced action of $\mathcal{X}_n^\lambda$. The functor infl^λ is right adjoint to pr^λ, that is there is a functorial isomorphism

$$\text{Hom}_{\mathcal{X}_n^\lambda}(\text{pr}^\lambda M, N) \cong \text{Hom}_{\mathcal{X}_n}(M, \text{infl}^\lambda N). \tag{15.8}$$

Note we will generally be sloppy and omit the functor infl^λ in our notation. In other words, we generally identify $\mathcal{X}_n^\lambda$-smod with the full subcategory of $\text{Rep}_I\, \mathcal{X}_n$ consisting of all supermodules M with $\mathcal{J}_\lambda M = 0$.

15.4 Basis Theorem for cyclotomic Sergeev superalgebras

Let $d = \langle c, \lambda \rangle$. Then f_λ is a monic polynomial of degree d. Write

$$f_\lambda = x_1^d + a_{d-1}x_1^{d-1} + \cdots + a_1 x_1 + a_0.$$

Set $f_1 = f_\lambda$ and for $i = 2, \ldots, n$, define inductively

$$f_i = s_{i-1} f_{i-1} s_{i-1}.$$

Note using (15.6):

$$c_j f_i = \pm f_i c_j \qquad (1 \le i, j \le n). \tag{15.9}$$

The first lemma follows easily by induction using relations in $\mathcal{X}_n$, especially (14.8).

Lemma 15.4.1 *For $i = 1, \ldots, n$, we have*

$$f_i = x_i^d + (\textit{terms lying in } \mathcal{P}_{i-1} x_i^e \mathcal{Y}_i \textit{ for } 0 \le e < d).$$

Given $Z = \{z_1 < \cdots < z_u\} \subseteq \{1, \ldots, n\}$, let

$$f_Z = f_{z_1} f_{z_2} \cdots f_{z_u} \in \mathcal{X}_n.$$

Also, define

$$\Pi_n = \{(\alpha, Z) \,|\, Z \subseteq \{1, \ldots, n\}, \alpha \in \mathbb{Z}_+^n \text{ with } \alpha_i < d \text{ whenever } i \notin Z\},$$
$$\Pi_n^+ = \{(\alpha, Z) \in \Pi_n \,|\, Z \neq \varnothing\}.$$

Lemma 15.4.2 *$\mathcal{X}_n$ is a free right $\mathcal{Y}_n$-module on basis*

$$\{x^\alpha f_Z \,|\, (\alpha, Z) \in \Pi_n\}.$$

Proof The proof is similar to that of Lemma 7.5.2 with $\mathcal{Y}_i$ in place of $\mathcal{G}_i$, and uses Lemma 15.4.1 instead of Lemma 7.5.1. □

Lemma 15.4.3 *For $n > 1$ we have $\mathcal{Y}_{n-1} f_n \mathcal{Y}_n = f_n \mathcal{Y}_n$.*

Proof It suffices to show that the left multiplication by the elements $s_1, \dots, s_{n-2}$ and $c_1, \dots, c_{n-1}$ leaves the space $f_n \mathcal{Y}_n$ invariant. But this follows from the definition of f_n and relations in $\mathcal{X}_n$. □

Lemma 15.4.4 *We have $\mathcal{I}_\lambda = \sum_{i=1}^{n} \mathcal{P}_n f_i \mathcal{Y}_n$.*

Proof We have

$$\begin{aligned}
\mathcal{I}_\lambda &= \mathcal{X}_n f_1 \mathcal{X}_n = \mathcal{X}_n f_1 \mathcal{P}_n \mathcal{Y}_n = \mathcal{X}_n f_1 \mathcal{Y}_n \\
&= \mathcal{P}_n \mathcal{Y}_n f_1 \mathcal{Y}_n = \sum_{i=1}^{n} \sum_{u \in S_{2\dots n}} \mathcal{A}_n s_{i-1} \dots s_1 u f_1 \mathcal{Y}_n \\
&= \sum_{i=1}^{n} \mathcal{A}_n s_{i-1} \dots s_1 f_1 \mathcal{Y}_n = \sum_{i=1}^{n} \mathcal{A}_n f_i \mathcal{Y}_n \\
&= \sum_{i=1}^{n} \mathcal{P}_n \mathcal{C}_n f_i \mathcal{Y}_n = \sum_{i=1}^{n} \mathcal{P}_n f_i \mathcal{Y}_n,
\end{aligned}$$

using (15.9) for the last equality. □

Lemma 15.4.5 *For $d > 0$ we have $\mathcal{I}_\lambda = \sum_{(\alpha, Z) \in \Pi_n^+} x^\alpha f_Z \mathcal{Y}_n$.*

Proof The proof is similar to that of Lemma 7.5.5 using Lemmas 15.4.4, 15.4.2, 15.4.3, 15.4.1 instead of Lemmas 7.5.4, 7.5.2, 7.5.3, 7.5.1, respectively. □

Theorem 15.4.6 *The canonical images of the elements*

$$\{x^\alpha c^\beta w \mid \alpha \in \mathbb{Z}_+^n \text{ with } \alpha_1, \dots, \alpha_n < d, \ \beta \in \mathbb{Z}_2^n, \ w \in S_n\}$$

form a basis for $\mathcal{X}_n^\lambda$.

Proof By Lemmas 15.4.2 and 15.4.5, the elements $\{x^\alpha f_Z \mid (\alpha, Z) \in \Pi_n^+\}$ form a basis for $\mathcal{I}_\lambda$ viewed as a right $\mathcal{Y}_n$-module. Hence Lemma 15.4.2 implies that the elements

$$\{x^\alpha \mid \alpha \in \mathbb{Z}_+^n \text{ with } \alpha_1, \dots, \alpha_n < d\}$$

form a basis for a complement to $\mathcal{J}_\lambda$ in $\mathcal{X}_n$ viewed as a right $\mathcal{Y}_n$-module. The theorem follows at once. □

Remark 15.4.7 It follows from Theorem 15.4.6 that $\mathcal{X}_n^{\Lambda_0}$ is isomorphic to the Sergeev superalgebra $\mathcal{Y}_n$. If we identify the two using the map $c^\beta w + \mathcal{J}_\lambda \mapsto c^\beta w$, then the projection

$$\mathcal{X}_n \twoheadrightarrow \mathcal{X}_n/\mathcal{J}_\lambda = \mathcal{Y}_n$$

can be described as follows: it is identity on $\mathcal{Y}_n$, i.e. $c^\beta w \in \mathcal{Y}_n \subset \mathcal{X}_n$ is mapped to $c^\beta w \in \mathcal{Y}_n$, and x_k is mapped to the kth Jucys–Murphy element $L_k \in \mathcal{Y}_n$, defined in (13.22). This is easy to see because $L_1 = 0$ by definition, and x_1 must go to 0. Moreover, the exression (13.23) of L_{k+1} in terms of L_k is the same as that of x_{k+1} in terms of x_k.

15.5 Cyclotomic Mackey Theorem

Given any $y \in \mathcal{X}_n$, we will denote its canonical image in $\mathcal{X}_n^\lambda$ by the same symbol. Thus, Theorem 15.4.6 says that

$$B_n := \{x^\alpha c^\beta w \mid \alpha \in \mathbb{Z}_+^n \text{ with } \alpha_1, \dots, \alpha_n < d,\ \beta \in \mathbb{Z}_2^n,\ w \in S_n\} \tag{15.10}$$

is a basis for $\mathcal{X}_n^\lambda$. Also Theorem 15.4.6 implies that the subalgebra of $\mathcal{X}_{n+1}^\lambda$ generated by $x_1, \dots, x_n, c_1, \dots, c_n$ and w, for $w \in S_n$, is isomorphic to $\mathcal{X}_n^\lambda$. We will write $\mathrm{ind}_{\mathcal{X}_n^\lambda}^{\mathcal{X}_{n+1}^\lambda}$ and $\mathrm{res}_{\mathcal{X}_n^\lambda}^{\mathcal{X}_{n+1}^\lambda}$ for the induction and restriction functors between $\mathcal{X}_n^\lambda$ and $\mathcal{X}_{n+1}^\lambda$. So,

$$\mathrm{ind}_{\mathcal{X}_n^\lambda}^{\mathcal{X}_{n+1}^\lambda} M = \mathcal{X}_{n+1}^\lambda \otimes_{\mathcal{X}_n^\lambda} M.$$

Lemma 15.5.1 (i) *$\mathcal{X}_{n+1}^\lambda$ is a free right $\mathcal{X}_n^\lambda$-supermodule on basis*

$$\{x_j^a c_j^b s_j \dots s_n \mid 0 \le a < d,\ b \in \mathbb{Z}_2,\ 1 \le j \le n+1\}.$$

(ii) *As $(\mathcal{X}_n^\lambda, \mathcal{X}_n^\lambda)$-bisupermodules,*

$$\mathcal{X}_{n+1}^\lambda = \mathcal{X}_n^\lambda s_n \mathcal{X}_n^\lambda \oplus \bigoplus_{0 \le a < d,\ b \in \mathbb{Z}_2} x_{n+1}^a c_{n+1}^b \mathcal{X}_n^\lambda.$$

(iii) *Let $0 \le a < d$. Then there are isomorphisms of $(\mathcal{X}_n^\lambda, \mathcal{X}_n^\lambda)$-bisupermodules*

$$\mathcal{X}_n^\lambda s_n \mathcal{X}_n^\lambda \simeq \mathcal{X}_n^\lambda \otimes_{\mathcal{X}_{n-1}^\lambda} \mathcal{X}_n^\lambda, \quad x_{n+1}^a \mathcal{X}_n^\lambda \simeq \mathcal{X}_n^\lambda, \quad x_{n+1}^a c_{n+1} \mathcal{X}_n^\lambda \simeq \Pi \mathcal{X}_n^\lambda.$$

Proof (i) By Theorem 15.4.6 and dimension considerations, we just need to check that $\mathcal{X}^\lambda_{n+1}$ is generated as a right $\mathcal{X}^\lambda_n$-module by the given elements. This follows from relations, especially (14.8).

(ii) It suffices to notice, using (i) and (14.8), that

$$\{x_j^a c_j^b s_j \dots s_n \mid 0 \le a < d,\ b \in \mathbb{Z}_2,\ 1 \le j \le n\}$$

is a basis of $\mathcal{X}^\lambda_n s_n \mathcal{X}^\lambda_n$ as a free right $\mathcal{X}^\lambda_n$-module.

(iii) The isomorphisms $x^a_{n+1}\mathcal{X}^\lambda_n \simeq \mathcal{X}^\lambda_n$ and $x^a_{n+1}c_{n+1}\mathcal{X}^\lambda_n \simeq \Pi\mathcal{X}^\lambda_n$ are clear from (i). Furthermore, the map

$$\mathcal{X}^\lambda_n \times \mathcal{X}^\lambda_n \to \mathcal{X}^\lambda_n s_n \mathcal{X}^\lambda_n,\ (u, v) \mapsto u s_n v$$

is $\mathcal{X}^\lambda_{n-1}$-balanced, so it induces an (even) homomorphism

$$\Phi : \mathcal{X}^\lambda_n \otimes_{\mathcal{X}^\lambda_{n-1}} \mathcal{X}^\lambda_n \to \mathcal{X}^\lambda_n s_n \mathcal{X}^\lambda_n$$

of $(\mathcal{X}^\lambda_n, \mathcal{X}^\lambda_n)$-bisupermodules. By (i), $\mathcal{X}^\lambda_n \otimes_{\mathcal{X}^\lambda_{n-1}} \mathcal{X}^\lambda_n$ is a free right $\mathcal{X}^\lambda_n$-supermodule on basis

$$\{x_j^a c_j^b s_j \dots s_{n-1} \otimes 1 \mid 1 \le j \le n,\ b \in \mathbb{Z}_2,\ 0 \le a < d\}.$$

But Φ maps these elements to a basis for $\mathcal{X}^\lambda_n s_n \mathcal{X}^\lambda_n$ as a free right $\mathcal{X}^\lambda_n$-supermodule, using a fact observed in the proof of (ii). This shows that Φ is an isomorphism. □

We have now decomposed $\mathcal{X}^\lambda_{n+1}$ as an $(\mathcal{X}^\lambda_n, \mathcal{X}^\lambda_n)$-bisupermodule. So the same argument as for Theorem 3.5.2 gives:

Theorem 15.5.2 *Let M be an $\mathcal{X}^\lambda_n$-supermodule. Then there is a natural isomorphism*

$$\mathrm{res}^{\mathcal{X}^\lambda_{n+1}}_{\mathcal{X}^\lambda_n} \mathrm{ind}^{\mathcal{X}^\lambda_{n+1}}_{\mathcal{X}^\lambda_n} M \simeq (M \oplus \Pi M)^{\oplus d} \oplus \mathrm{ind}^{\mathcal{X}^\lambda_n}_{\mathcal{X}^\lambda_{n-1}} \mathrm{res}^{\mathcal{X}^\lambda_n}_{\mathcal{X}^\lambda_{n-1}} M$$

of $\mathcal{X}^\lambda_n$-supermodules.

15.6 Duality for cyclotomic superalgebras

Lemma 15.6.1 *For $1 \le i \le n$ and $a \ge 0$, we have*

$$s_n \dots s_i x_i^a s_i \dots s_n = x^a_{n+1} + (*),$$

$$s_n \dots s_i x_i^a c_i s_i \dots s_n = x^a_{n+1} c_{n+1} + (**),$$

where $(*)$ and $(**)$ lie in $\mathcal{X}^\lambda_n s_n \mathcal{X}^\lambda_n + \sum_{k=0}^{a-2}(x^k_{n+1}\mathcal{X}^\lambda_n + x^k_{n+1}c_{n+1}\mathcal{X}^\lambda_n)$.

Proof We apply induction on $n = i, i+1, \ldots$. In case $n = i$, the result follows from a calculation using (14.8). The induction step is similar, noting that s_n centralizes $\mathcal{X}_{n-1}^{\lambda}$. □

Lemma 15.6.2 *There is an* $(\mathcal{X}_n^{\lambda}, \mathcal{X}_n^{\lambda})$*-bisupermodule homomorphism* $\theta : \mathcal{X}_{n+1}^{\lambda} \to \mathcal{X}_n^{\lambda}$ *such that* $\ker\theta$ *contains no non-zero left ideals of* $\mathcal{X}_{n+1}^{\lambda}$.

Proof By Lemma 15.5.1(ii), we know that

$$\mathcal{X}_{n+1}^{\lambda} = x_{n+1}^{d-1}\mathcal{X}_n^{\lambda} \oplus \bigoplus_{a=0}^{d-2} x_{n+1}^{a}\mathcal{X}_n^{\lambda} \oplus \bigoplus_{a=0}^{d-1} x_{n+1}^{a}c_{n+1}\mathcal{X}_n^{\lambda} \oplus \mathcal{X}_n^{\lambda}s_n\mathcal{X}_n^{\lambda}$$

as an $(\mathcal{X}_n^{\lambda}, \mathcal{X}_n^{\lambda})$-bisupermodule. Let

$$\theta : \mathcal{X}_{n+1}^{\lambda} \to x_{n+1}^{d-1}\mathcal{X}_n^{\lambda}$$

be the projection on to the first summand of this decomposition, which by Lemma 15.5.1(iii) is isomorphic to $\mathcal{X}_n^{\lambda}$ as an $(\mathcal{X}_n^{\lambda}, \mathcal{X}_n^{\lambda})$-bisupermodule. So we just need to show that if $y \in \mathcal{X}_{n+1}^{\lambda}$ has the property that $\theta(hy) = 0$ for all $h \in \mathcal{X}_{n+1}^{\lambda}$, then $y = 0$. Using Lemma 15.5.1(i), we may write y as

$$\sum_{a=0}^{d-1}(x_{n+1}^{a}\sigma_a + x_{n+1}^{a}c_{n+1}\tau_a) + \sum_{a=0}^{d-1}\sum_{j=1}^{n}(x_j^a s_j \ldots s_n\mu_{a,j} + x_j^a c_j s_j \ldots s_n \nu_{a,j})$$

for some elements $\sigma_a, \tau_a, \mu_{a,j}, \nu_{a,j} \in \mathcal{X}_n^{\lambda}$. As $\theta(y) = 0$, we must have $\sigma_{d-1} = 0$. Now $\theta(x_{n+1}y) = 0$ implies $\sigma_{d-2} = 0$, then $\theta(x_{n+1}^2 y) = 0$ implies $\sigma_{d-3} = 0$, etc., proving that all σ_a are zero. Now $\theta(c_{n+1}y) = 0$ implies that all τ_a are zero. Next, considering $\theta(s_n y)$, $\theta(x_{n+1}s_n y)$, $\theta(x_{n+1}^2 s_n y), \ldots$, and using Lemma 15.6.1, we get $\mu_{d-1,n} = \mu_{d-2,n} = \cdots = 0$ and $\nu_{d-1,n} = \nu_{d-2,n} = \cdots = 0$. Next, repeat the argument again, this time considering $s_n s_{n-1}, x_{n+1}s_n s_{n-1}, \ldots$, to get that all $\mu_{a,n-1} = 0$ and all $\nu_{a,n-1} = 0$. Continuing in this way we eventually arrive to the desired conclusion that $y = 0$. □

Theorem 15.6.3 *There is a natural isomorphism*

$$\mathcal{X}_{n+1}^{\lambda} \otimes_{\mathcal{X}_n^{\lambda}} M \cong \mathrm{Hom}_{\mathcal{X}_n^{\lambda}}(\mathcal{X}_{n+1}^{\lambda}, M)$$

for all $\mathcal{X}_n^{\lambda}$*-supermodules* M.

Proof It suffices to show that there exists an even isomorphism

$$\varphi : \mathcal{X}_{n+1}^{\lambda} \to \mathrm{Hom}_{\mathcal{X}_n^{\lambda}}(\mathcal{X}_{n+1}^{\lambda}, \mathcal{X}_n^{\lambda})$$

of $(\mathcal{X}_{n+1}^{\lambda}, \mathcal{X}_n^{\lambda})$-bisupermodules. This is done as in the proof of Theorem 7.7.3 using Lemmas 15.6.2 instead of Lemmas 7.7.2. □

Corollary 15.6.4 *$\mathcal{X}_n^\lambda$ is a Frobenius superalgebra, that is there is an even isomorphism of left $\mathcal{X}_n^\lambda$-supermodules $\mathcal{X}_n^\lambda \cong \mathrm{Hom}_F(\mathcal{X}_n^\lambda, F)$.*

Proof Similar to the proof of Corollary 7.7.4. □

For the next corollary, recall the duality induced by τ (14.17) on finite dimensional $\mathcal{X}_n$-supermodules. Since τ leaves the two-sided ideal $\mathcal{I}_\lambda$ invariant, it induces a duality also denoted τ on finite dimensional $\mathcal{X}_n^\lambda$-supermodules.

Corollary 15.6.5 *The exact functor $\mathrm{ind}_{\mathcal{X}_n^\lambda}^{\mathcal{X}_{n+1}^\lambda}$ is both left and right adjoint to $\mathrm{res}_{\mathcal{X}_n^\lambda}^{\mathcal{X}_{n+1}^\lambda}$. Moreover, it commutes with duality in the sense that there is a natural isomorphism*

$$\mathrm{ind}_{\mathcal{X}_n^\lambda}^{\mathcal{X}_{n+1}^\lambda}(M^\tau) \simeq (\mathrm{ind}_{\mathcal{X}_n^\lambda}^{\mathcal{X}_{n+1}^\lambda} M)^\tau$$

for all finite dimensional $\mathcal{X}_n^\lambda$-supermodules M.

Proof Similar to the proof of Corollary 7.7.5. □

16
First results on $\mathcal{X}_n$-modules

This chapter is parallel to Chapter 4. The role of the subalgebra $\mathcal{P}_n$ is played by $\mathcal{A}_n$. However, it will be more convenient to consider only *integral* modules everywhere, as defined in Chapter 15.

16.1 Formal characters of $\mathcal{X}_n$-modules

Let $i \in I$, see (15.2). Denote by $L(i)$ the vector superspace on basis $\{v_1, v_{-1}\}$, where $\bar{v}_1 = \bar{0}$, $\bar{v}_{-1} = \bar{1}$, made into an $\mathcal{A}_1$-supermodule via

$$c_1 v_1 = v_{-1}, \;\; c_1 v_{-1} = v_1, \;\; x_1 v_1 = \sqrt{q(i)} v, \;\; x_1 v_{-1} = -\sqrt{q(i)} v_{-1}.$$

To be precise, we need to first make a choice of $\sqrt{q(i)}$, as it is defined only up to $\pm$. But it is not a big deal, since changing from $\sqrt{q(i)}$ to $-\sqrt{q(i)}$ leads to an isomorphic supermodule (the isomorphism swaps v_1 and v_{-1}). Note that $L(q(i))$ is irreducible of type $\mathtt{M}$ if $i \neq 0$, and irreducible of type $\mathtt{Q}$ if $i = 0$. It is also easy to see that the modules $\{L(i) \mid i \in I\}$ form a complete set of pairwise non-isomorphic irreducible $\mathcal{A}_1$-supermodules in the category $\operatorname{Rep}_I \mathcal{A}_1$. Now, $\mathcal{A}_n \cong \mathcal{A}_1 \otimes \cdots \otimes \mathcal{A}_1$, and so the next result follows from Lemmas 12.2.13 and 12.2.15:

Lemma 16.1.1 *The $\mathcal{A}_n$-supermodules*

$$\{L(i_1) \circledast \cdots \circledast L(i_n) \mid (i_1, \dots, i_n) \in I^n\}$$

form a complete set of pairwise non-isomorphic irreducible supermodules in the category $\operatorname{Rep}_I \mathcal{A}_n$. *Moreover, let γ_0 denote the number of $j = 1, \dots, n$ such that $i_j = 0$. Then $L(i_1) \circledast \cdots \circledast L(i_n)$ is of type* $\mathtt{M}$ *if γ_0 is even and type* $\mathtt{Q}$ *if γ_0 is odd. Finally,* $\dim L(i_1) \circledast \cdots \circledast L(i_n) = 2^{n - \lfloor \gamma_0/2 \rfloor}$.

Now take any $M \in \operatorname{Rep}_I \mathcal{A}_n$. For any $\underline{i} = (i_1, \dots, i_n) \in I^n$, let $M_{\underline{i}}$ be the largest submodule of M all of whose composition factors are isomorphic to $L(i_1) \circledast \cdots \circledast L(i_n)$. Alternatively, $M_{\underline{i}}$ is the simultaneous generalized eigenspace for the commuting operators $x_1^2, \dots, x_n^2$ corresponding to the eigenvalues $q(i_1), \dots, q(i_n)$, respectively. Hence:

Lemma 16.1.2 *For any $M \in \operatorname{Rep}_I \mathcal{A}_n$ we have $M = \bigoplus_{\underline{i} \in I^n} M_{\underline{i}}$ as an $\mathcal{A}_n$-supermodule.*

Since x_j^2 acts as $q(i_j)$ on $L(i_1) \circledast \cdots \circledast L(i_n)$, and the dimension of each $L(i_1) \circledast \cdots \circledast L(i_n)$ is known from Lemma 16.1.1, knowledge of the dimensions of the spaces $M_{\underline{i}}$ for all $\underline{i}$ is equivalent to knowing the coefficients $r_{\underline{i}}$ when the class $[M]$ of M in the Grothendieck group $K(\operatorname{Rep}_I \mathcal{A}_n)$ is expanded as

$$[M] = \sum_{\underline{i} \in I^n} r_{\underline{i}} [L(i_1) \circledast \cdots \circledast L(i_n)]$$

in terms of the basis $\{[L(i_1) \circledast \cdots \circledast L(i_n)] \mid \underline{i} \in I^n\}$.

Now let $M \in \operatorname{Rep}_I \mathcal{X}_n$. Recall that $\mathcal{X}_{1,\dots,1} = \mathcal{A}_n$. We define the *formal character* of M by:

$$\operatorname{ch} M := [\operatorname{res}^n_{1,\dots,1} M] \in K(\operatorname{Rep}_I \mathcal{A}_n). \tag{16.1}$$

Since the functor $\operatorname{res}^n_{1,\dots,1}$ is exact, ch induces a homomorphism

$$\operatorname{ch} : K(\operatorname{Rep}_I \mathcal{X}_n) \to K(\operatorname{Rep}_I \mathcal{A}_n)$$

at the level of Grothendieck groups.

As in Section 4.1, we get as special cases of Mackey Theorem:

Lemma 16.1.3 *Let $\underline{i} = (i_1, \dots, i_n) \in I^n$. Then*

$$\operatorname{ch} \operatorname{ind}^n_{1,\dots,1} L(i_1) \circledast \cdots \circledast L(i_n) = \sum_{w \in S_n} [L(i_{w^{-1}1}) \circledast \cdots \circledast L(i_{w^{-1}n})]$$

Lemma 16.1.4 ("Shuffle Lemma") *Let $n = m + k$, and let $M \in \operatorname{Rep}_I \mathcal{X}_m$, $K \in \operatorname{Rep}_I \mathcal{X}_k$. Assume*

$$\operatorname{ch} M = \sum_{\underline{i} \in I^m} r_{\underline{i}} [L(i_1) \circledast \cdots \circledast L(i_m)],$$

$$\operatorname{ch} K = \sum_{\underline{j} \in I^k} s_{\underline{j}} [L(j_1) \circledast \cdots \circledast L(j_k))].$$

Then

$$\operatorname{ch} \operatorname{ind}^n_{m,k} M \circledast K = \sum_{\underline{i} \in I^m} \sum_{\underline{j} \in I^k} r_{\underline{i}} s_{\underline{j}} (\sum_{\underline{c}} L(h_1) \circledast \cdots \circledast L(h_n)),$$

where the last sum is over all $\underline{h} = (h_1, \ldots, h_n) \in I^n$, which are obtained by shuffling $\underline{i}$ and $\underline{j}$.

16.2 Central characters and blocks

Recall by Theorem 14.3.1 that every element z of the center $Z(\mathcal{X}_n)$ can be written as a symmetric polynomial $f(x_1^2, \ldots, x_n^2)$. Given $\underline{i} \in I^n$, we associate the *central character*

$$\chi_{\underline{i}} : Z(\mathcal{X}_n) \to F, \quad f(x_1^2, \ldots, x_n^2) \mapsto f(q(i_1), \ldots, q(i_n)).$$

If $\underline{i} \in I^n$, define its *content* $\operatorname{cont}(\underline{i}) \in P$ by

$$\operatorname{cont}(\underline{i}) = \sum_{i \in I} \gamma_i \alpha_i \qquad \text{where} \qquad \gamma_i = \sharp\{j = 1, \ldots, n \mid i_j = i\}. \tag{16.2}$$

So $\operatorname{cont}(\underline{i})$ is an element of the set Γ_n of non-negative integral linear combinations $\gamma = \sum_{i \in I} \gamma_i \alpha_i$ of the simple roots such that $\sum_{i \in I} \gamma_i = n$. Obviously, the S_n-orbit of $\underline{i}$ is uniquely determined by the content of $\underline{i}$, so we obtain a labeling of the orbits of S_n on I^n by the elements of Γ_n. We will also use the notation χ_γ for the central character $\chi_{\underline{i}}$ where $\underline{i}$ is any element of I^n with $\operatorname{cont}(\underline{i}) = \gamma$. It is clear that χ_γ is well defined.

Now let $M \in \operatorname{Rep}_I \mathcal{X}_n$, and $\gamma \in \Gamma_n$. Denote

$$M[\gamma] = \{v \in M \mid (z - \chi_\gamma(z))^k v = 0 \text{ for all } z \in Z(\mathcal{X}_n) \text{ and } k \gg 0\}.$$

Observe this is an $\mathcal{X}_n$-subsupermodule of M. Now, for any $\underline{i} \in I^n$ with $\operatorname{cont}(\underline{i}) = \gamma$, $Z(\mathcal{X}_n)$ acts on $L(i_1) \circledast \cdots \circledast L(i_n)$ via the central character χ_γ. So applying Lemma 4.2.1, we see that

$$M[\gamma] = \bigoplus_{\operatorname{cont}(\underline{i}) = \gamma} M_{\underline{i}}, \tag{16.3}$$

recalling the decomposition of M from Lemma 16.1.2. Therefore:

Lemma 16.2.1 *Any $M \in \operatorname{Rep}_I \mathcal{X}_n$ decomposes as*

$$M = \bigoplus_{\gamma \in \Gamma_n} M[\gamma]$$

as an $\mathcal{X}_n$-supermodule.

Thus the $\{\chi_\gamma \mid \gamma \in \Gamma_n\}$ exhaust the possible central characters that can arise in an integral $\mathcal{X}_n$-supermodule, while Lemma 16.1.3 shows that every such central character does arise in some integral $\mathcal{X}_n$-supermodule.

If $\gamma \in \Gamma_n$, let us denote by $\operatorname{Rep}_I \mathcal{X}_n[\gamma]$ the full subcategory of $\operatorname{Rep}_I \mathcal{X}_n$ consisting of all supermodules M with $M[\gamma] = M$. Then Lemma 16.2.1 implies that there is an equivalence of categories

$$\operatorname{Rep}_I \mathcal{X}_n \cong \bigoplus_{\gamma \in \Gamma_n} \operatorname{Rep}_I \mathcal{X}_n[\gamma]. \tag{16.4}$$

We say that $\operatorname{Rep}_I \mathcal{X}_n[\gamma]$ is the *block* of $\operatorname{Rep}_I \mathcal{X}_n$ corresponding to γ (or to the central character χ_γ). If $M \in \operatorname{Rep}_I \mathcal{X}_n[\gamma]$, we say that M *belongs* to the block corresponding to γ. If $M \neq 0$ is indecomposable then $M \in \operatorname{Rep}_I \mathcal{X}_n[\gamma]$ for a unique $\gamma \in \Gamma_n$.

We can extend some of these notions to $\mathcal{X}_n^\lambda$-supermodules, for $\lambda \in P_+$. In particular, if $M \in \mathcal{X}_n^\lambda$-smod, we also write $M[\gamma]$ for the summand $M[\gamma]$ of M defined by first viewing M as an $\mathcal{X}_n$-supermodule by inflation. Also write $\mathcal{X}_n^\lambda$-smod$[\gamma]$ for the full subcategory of $\mathcal{X}_n^\lambda$-smod consisting of the modules M with $M = M[\gamma]$. Then we have a decomposition

$$\mathcal{X}_n^\lambda\text{-smod} \cong \bigoplus_{\gamma \in \Gamma_n} \mathcal{X}_n^\lambda\text{-smod}[\gamma]. \tag{16.5}$$

Note though that we should not yet refer to $\mathcal{X}_n^\lambda$-smod$[\gamma]$ as a block of $\mathcal{X}_n^\lambda$-smod: the center of $\mathcal{X}_n^\lambda$ may be larger than the image of the center of $\mathcal{X}_n$, so we cannot yet assert that $Z(\mathcal{X}_n^\lambda)$ acts on $M[\gamma]$ by a single central character. Also we no longer know precisely which $\gamma \in \Gamma_n$ have the property that $\mathcal{X}_n^\lambda$-smod$[\gamma]$ is non-trivial. These questions will be settled in Section 20.6.

16.3 Kato's Theorem for $\mathcal{X}_n$

Let $i \in I$. Introduce the *Kato (super)module*

$$L(i^n) := \operatorname{ind}_{1,\dots,1}^n L(i) \circledast \cdots \circledast L(i). \tag{16.6}$$

By Lemma 16.1.3,

$$\operatorname{ch} L(i^n) = n![L(i) \circledast \cdots \circledast L(i)].$$

In particular, for each $k = 1, \dots, n$, the only eigenvalue of the element x_k^2 on $L(i^n)$ is $q(i)$. We have to work a little harder than in Section 4.3 in order to prove "Kato's Theorem".

Lemma 16.3.1 *Let $n \geq 2$, $1 \leq k < n$, $i \in I \setminus \{0\}$, and $v \in L(i)^{\boxtimes n} \setminus \{0\}$. Then* $\big(x_k(1 - c_k c_{k+1}) + (1 - c_k c_{k+1}) x_{k+1}\big) v \neq 0.$

Proof The elements of $\mathcal{A}_n$ which are involved in the inequality act only on the positions k and $k+1$ in the tensor product. So we may assume that $n=2$ and $k=1$. Let

$$w = a_1 v_1 \otimes v_1 + a_2 v_1 \otimes v_{-1} + a_3 v_{-1} \otimes v_1 + a_4 v_{-1} \otimes v_{-1}$$

for $a_1, a_2, a_3, a_4 \in F$, and denote $b := \sqrt{q(i)}$. Then the result is clear from:

$$\begin{aligned} x_1(1-c_1c_2)w &= (ba_1+ba_4)v_1 \otimes v_1 + (ba_2+ba_3)v_1 \otimes v_{-1} \\ &\quad + (-ba_3+ba_2)v_{-1} \otimes v_1 + (-ba_4+ba_1)v_{-1} \otimes v_{-1}, \\ (1-c_1c_2)x_2 w &= (ba_1-ba_4)v_1 \otimes v_1 + (-ba_2+ba_3)v_1 \otimes v_{-1} \\ &\quad + (ba_3+ba_2)v_{-1} \otimes v_1 + (-ba_4-ba_1)v_{-1} \otimes v_{-1}. \end{aligned}$$

□

Lemma 16.3.2 *Let $i \in I$. Set $L = L(i)^{\circledast n}$, so $L(i^n) = \mathcal{X}_n \otimes_{\mathcal{A}_n} L$.*

(i) *If $i \neq 0$, the common $q(i)$-eigenspace of $x_1^2, \dots, x_{n-1}^2$ on $L(i^n)$ is precisely $1 \otimes L$, which is contained in the $q(i)$-eigenspace of x_n^2 too. Moreover, all Jordan blocks of x_1^2 on $L(i^n)$ are of size n.*

(ii) *If $i = 0$, the common 0-eigenspace of $x_1, \dots, x_{n-1}$ on $L(0^n)$ is precisely $1 \otimes L$, which is contained in the 0-eigenspace of x_n too. Moreover, all Jordan blocks of x_1 on $L(0^n)$ are of size n.*

Proof We prove (i), (ii) being similar. Note $L(i^n) = \bigoplus_{w \in S_n} w \otimes L$, since by Theorem 14.2.2 we know that $\mathcal{X}_n$ is a free right $\mathcal{A}_n$-module on basis $\{w \mid w \in S_n\}$.

We first claim that the $q(i)$-eigenspace of x_1^2 is a sum of the subspaces of the form $y \otimes L$, where $y \in S'_{n-1} = \langle s_2, \dots, s_{n-1} \rangle$. Well, any w can be written as $ys_1s_2 \dots s_j$ for some $y \in S'_{n-1}$ and $0 \le j < n$. Note that $(x_{j+1}^2 - q(i))v = 0$ for any $v \in L$, by definition of L. So the defining relations of $\mathcal{X}_n$ imply

$$\begin{aligned} &(x_1^2 - q(i))ys_1s_2 \dots s_j \otimes v = \\ &\qquad - ys_1 \dots s_{j-1} \otimes (x_j(1-c_jc_{j+1}) - (1-c_jc_{j+1})x_{j+1})v + (*), \end{aligned}$$

where $(*)$ stands for terms in subspaces of the form $y's_1 \dots s_k \otimes L$ for $y' \in S'_{n-1}$ and $0 \le k < j-1$. Now assume that a linear combination

$$z := \sum_{y \in S'_{n-1}} \sum_{0 \le j < n} \sum_{v \in L} c_{y,j,v} ys_1s_2 \dots s_j \otimes v$$

is an eigenvector for x_1^2. Choose the maximal j for which the coefficient $c_{y,j,v}$ is non-zero, and for this j choose the maximal (with respect to the

Bruhat order) y such that $c_{y,j,v}$ is non-zero. Then the calculation above and Lemma 16.3.1 show that $(x_1^2 - q(i))z \neq 0$ unless $j = 0$. This proves our claim.

Now apply the same argument to see that the common eigenspace of x_1^2 and x_2^2 is spanned by $y \otimes L$ for $y \in \langle s_3, \dots, s_n \rangle$, and so on, yielding the first claim in (i). Finally, define

$$V(m) := \{w \in L(i^n) \mid (x_1^2 - q(i))^m w = 0\}.$$

It follows by induction from the calculation above and Lemma 16.3.1 that

$$V(m) = \text{span}\{ys_1 s_2 \dots s_j \otimes v \mid y \in S'_{n-1},\ j < m,\ v \in L\},$$

giving the second claim. □

The main theorem on the structure of the Kato module $L(i^n)$ is

Theorem 16.3.3 *Let $i \in I$ and $\mu = (\mu_1, \dots, \mu_r)$ be a composition of n.*

(i) *$L(i^n)$ is irreducible, and it is the only irreducible module in its block. Moreover, $L(i^n)$ is of type* Q *if $i = 0$ and n is odd, and of type* M *otherwise.*

(ii) *All composition factors of $\text{res}^n_\mu L(i^n)$ are isomorphic to*

$$L(i^{\mu_1}) \circledast \cdots \circledast L(i^{\mu_r}),$$

and $\text{soc}\,\text{res}^n_\mu L(i^n)$ is irreducible.

(iii) $\text{soc}\,\text{res}^n_{n-1} L(i^n) \cong \text{res}^{n-1,1}_{n-1}\big(L(i^{n-1}) \circledast L(i)\big)$.

Proof Denote $L(i) \circledast \cdots \circledast L(i)$ by L.

(i) Let M be a non-zero $\mathcal{X}_n$-submodule of $L(i^n)$. Then $\text{res}^n_{1,\dots,1} M$ must contain a $\mathcal{A}_n$-submodule N isomorphic to L. But the commuting operators $x_1^2, \dots, x_n^2$ (or $x_1, \dots, x_n$ if $i = 0$) act on L as scalars, giving that N is contained in their common eigenspace on $L(i^n)$. But by Lemma 16.3.2, this implies that $N = 1 \otimes L$. This shows that M contains $1 \otimes L$, but $1 \otimes L$ generates the whole of $L(i^n)$ over $\mathcal{X}_n$. So $M = L(i^n)$. To see that $L(i^n)$ is the only irreducible in its block use Frobenius reciprocity and the fact just proved that $L(i^n)$ is irreducible.

Finally, in view of Lemma 16.1.1, it remains to see that the type of the $\mathcal{X}_n$-supermodule $L(i^n)$ is the same as the type of the $\mathcal{A}_n$-supermodule L. The functor $\text{ind}^n_{1,\dots,1}$ determines a map

$$\text{End}_{\mathcal{A}_n}(L) \to \text{End}_{\mathcal{X}_n}(L(i^n)).$$

We just need to see that this is an isomorphism, which we do by constructing the inverse map. Let $f \in \mathrm{End}_{\mathcal{X}_n}(L(i^n))$. Then f leaves $1 \otimes L$ invariant by Lemma 16.3.2, so f restricts to an $\mathcal{A}_n$-endomorphism of L.

(ii) The fact that all composition factors of $\mathrm{res}^n_\mu L(i^n)$ are isomorphic to $L(i^{\mu_1}) \circledast \cdots \circledast L(i^{\mu_r})$ follows by formal characters and (i). To see that $\mathrm{soc}\,\mathrm{res}^n_\mu L(i^n)$ is irreducible, note that the submodule $\mathcal{X}_\mu \otimes L$ of $\mathrm{res}_{\mathcal{X}_\mu} L(i^n)$ is isomorphic to $L(i^{\mu_1}) \circledast \cdots \circledast L(i^{\mu_l})$. This module is irreducible, and so it is contained in the socle. Conversely, let M be an irreducible $\mathcal{X}_\mu$-submodule of $L(i^n)$. Then using Lemma 16.3.2 as in the proof of (i), we see that M must contain $1 \otimes L$, hence $\mathcal{X}_\mu \otimes L$.

(iii) By part (ii), $L(i^n)$ has a unique $\mathcal{X}_{n-1,1}$-submodule isomorphic to $L(i^{n-1}) \circledast L(i)$, namely, $\mathcal{X}_{n-1,1} \otimes L$, which contributes a copy of $\mathrm{res}^{n-1,1}_{n-1}\big(L(i^{n-1}) \circledast L(i)\big)$ to $\mathrm{soc}\,\mathrm{res}^n_{n-1} L(i^n)$, since

$$\mathrm{res}^{n-1,1}_{n-1}\big(L(i^{n-1}) \circledast L(i)\big)$$

is completely reducible.

Conversely, take an irreducible $\mathcal{X}_{n-1}$-submodule M of $L(i^n)$. The common $q(i)$-eigenspace of $x_1^2, \ldots, x_{n-1}^2$ (or $x_1, \ldots, x_{n-1}$ if $i = 0$) on M must lie in $1 \otimes L$ by Lemma 16.3.2. Hence, $M \subseteq \mathcal{X}_{n-1,1} \otimes L$ which completes the proof. □

16.4 Covering modules for $\mathcal{X}_n$

Fix $i \in I$ and $n \geq 1$ throughout the section. We will construct for each $m \geq 1$ an $\mathcal{X}_n$-supermodule $L_m(i^n)$ with irreducible head isomorphic to $L(i^n)$. Let $\mathcal{J}(i^n)$ denote the annihilator in $\mathcal{X}_n$ of $L(i^n)$. Introduce the quotient superalgebra

$$\mathcal{R}_m(i^n) := \mathcal{X}_n / \mathcal{J}(i^n)^m \tag{16.7}$$

for each $m \geq 1$. Obviously $\mathcal{J}(i^n)$ contains $(x_k^2 - q(i))^{n!}$ for each $k = 1, \ldots, n$, whence each superalgebra $\mathcal{R}_m(i^n)$ is finite dimensional. Moreover, by Theorem 16.3.3, $L(i^n)$ is the unique irreducible $\mathcal{R}_m(i^n)$-module up to isomorphism. Let $L_m(i^n)$ denote a projective cover of $L(i^n)$ in the category $\mathcal{R}_m(i^n)$-smod (for convenience, we also define $L_0(i^n) = \mathcal{R}_0(i^n) = 0$).

Lemma 16.4.1 *For each $m \geq 1$,*

$$\mathcal{R}_m(i^n) \simeq \begin{cases} (L_m(i^n) \oplus \Pi L_m(i^n))^{\oplus n! 2^{n-1}} & \text{if } i \neq 0, \\ (L_m(i^n) \oplus \Pi L_m(i^n))^{\oplus n! 2^{(n-2)/2}} & \text{if } i = 0 \text{ and } n \text{ is even}, \\ L_m(i^n)^{\oplus n! 2^{(n-1)/2}} & \text{if } i = 0 \text{ and } n \text{ is odd}, \end{cases}$$

as left $\mathcal{X}_n$-modules. Moreover, $L_m(i^n)$ admits an odd involution if and only if $i = 0$ and n is odd.

Proof The superdimension of $L(i^n)$ is known from Lemma 16.1.1 and the definition. Now use Lemma 12.2.16 and Proposition 12.2.12. □

The obvious surjections

$$\mathcal{R}_1(i^n) \twoheadleftarrow \mathcal{R}_2(i^n) \twoheadleftarrow \dots \tag{16.8}$$

of superalgebras and properties of projective covers lead to even surjections

$$L(i^n) = L_1(i^n) \twoheadleftarrow L_2(i^n) \twoheadleftarrow \dots, \tag{16.9}$$

where $L_m(i^n)$ are considered as $\mathcal{X}_n$-modules by inflation. Moreover, in case $i = 0$ and n is odd, we can choose the odd involutions

$$\theta_m : L_m(i^n) \to L_m(i^n) \tag{16.10}$$

given by Lemma 16.4.1 in such a way that they are compatible with the maps in (16.9) (see the argument in Lemma 12.2.16).

Lemma 16.4.2 *Let M be an $\mathcal{X}_n$-supermodule annihilated by $\mathcal{J}(i^n)^k$ for some k. Then for all $m \geq k$ there is a natural isomorphism of $\mathcal{X}_n$-supermodules*

$$\mathrm{Hom}_{\mathcal{X}_n}(\mathcal{R}_m(i^n), M) \cong M.$$

Proof The assumption implies that M is the inflation of an $\mathcal{R}_m(i^n)$-supermodule. So

$$\mathrm{Hom}_{\mathcal{X}_n}(\mathcal{R}_m(i^n), M) \cong \mathrm{Hom}_{\mathcal{R}_m(i^n)}(\mathcal{R}_m(i^n), M) \cong M,$$

all isomorphisms being natural. □

We can also obtain an analogue of Lemma 4.4.3 for $\mathcal{X}_n$, but we will not need this result.

In the important special case $n = 1$ we easily check that the ideal $\mathcal{J}(i)^m$ is generated by $(x_1^2 - q(i))^m$ if $i \neq 0$ or x_1^m if $i = 0$. It follows that

$$\dim \mathcal{R}_m(i) = \begin{cases} 4m & \text{if } i \neq 0, \\ 2m & \text{if } i = 0. \end{cases} \tag{16.11}$$

Moreover, Lemma 16.4.1 shows in this case that

$$\mathcal{R}_m(i) \simeq \begin{cases} L_m(i) \oplus \Pi L_m(i) & \text{if } i \neq 0, \\ L_m(i) & \text{if } i = 0. \end{cases} \tag{16.12}$$

Hence, $\dim L_m(i) = 2m$ in either case. Using this, it follows easily that $L_m(i)$ can be described alternatively as the vector superspace on basis $w_1, \ldots, w_m$, $w'_1, \ldots, w'_m$, where each w_k is even and each w'_k is odd, with $\mathcal{X}_1$-module structure uniquely determined by

$$x_1 w_k = \sqrt{q(i)} w_k + w_{k+1}, \qquad c_1 w_k = w'_k$$

for each $k = 1, \ldots, m$, interpreting w_{m+1} as 0. Using this explicit description, we check that $L_m(i)$ is uniserial with m composition factors all $\simeq L(i)$.

We can also describe the map $L_m(i) \twoheadleftarrow L_{m+1}(i)$ from (16.9) explicitly: it is the identity on $w_1, \ldots, w_m, w'_1, \ldots, w'_m$ but maps w_{m+1} and w'_{m+1} to zero. Also, the map θ_m from (16.10) can be chosen so that

$$w_k \mapsto \sqrt{-1} w'_k, \qquad w'_k \mapsto -\sqrt{-1} w_k$$

for each $k = 1, \ldots, m$.

17
Crystal operators for $\mathcal{X}_n$

This chapter is parallel to Chapter 5.

17.1 Multiplicity-free socles

For $M \in \operatorname{Rep}_I \mathcal{X}_n$ and $i \in I$, define $\Delta_i M$ to be the generalized $q(i)$-eigenspace of x_n^2 on M. Equivalently,

$$\Delta_i M = \bigoplus_{\underline{i} \in I^n,\ i_n = i} M_{\underline{i}}, \tag{17.1}$$

recalling the decomposition from Lemma 16.1.2. Note since x_n^2 is central in the parabolic subalgebra $\mathcal{X}_{n-1,1}$ of $\mathcal{X}_n$, $\Delta_i M$ is invariant under this subalgebra. So, in fact, Δ_i can be viewed as an exact functor

$$\Delta_i : \operatorname{Rep}_I \mathcal{X}_n \to \operatorname{Rep}_I \mathcal{X}_{n-1,1}. \tag{17.2}$$

Slightly more generally, given $m \geq 0$, define

$$\Delta_{i^m} : \operatorname{Rep}_I \mathcal{X}_n \to \operatorname{Rep}_I \mathcal{X}_{n-m,m} \tag{17.3}$$

so that $\Delta_{i^m} M$ is the simultaneous generalized $q(i)$-eigenspace of the commuting operators x_k^2 for $k = n-m+1, \ldots, n$. In view of Theorem 16.3.3(i), $\Delta_{i^m} M$ can also be characterized as the largest submodule of $\operatorname{res}^n_{n-m,m} M$ all of whose composition factors are of the form $N \circledast L(i^m)$ for irreducible $N \in \operatorname{Rep}_I \mathcal{X}_{n-m}$. The definition of Δ_{i^m} implies a functorial isomorphism

$$\operatorname{Hom}_{\mathcal{X}_{n-m,m}}(N \boxtimes L(i^m), \Delta_{i^m} M) \cong \operatorname{Hom}_{\mathcal{X}_n}(\operatorname{ind}^n_{n-m,m} N \boxtimes L(i^m), M)$$

for $N \in \operatorname{Rep}_I \mathcal{X}_{n-m}$, $M \in \operatorname{Rep}_I \mathcal{X}_n$. For irreducible N this implies

$$\begin{aligned} &\operatorname{Hom}_{\mathcal{X}_{n-m,m}}(N \circledast L(i^m), \Delta_{i^m} M) \\ &\cong \operatorname{Hom}_{\mathcal{X}_n}(\operatorname{ind}^n_{n-m,m} N \circledast L(i^m), M) \end{aligned} \tag{17.4}$$

Also from definitions we get:

Lemma 17.1.1 *Let* $M \in \mathrm{Rep}_I \mathcal{X}_n$ *with*

$$\mathrm{ch}\, M = \sum_{\underline{i} \in I^n} r_{\underline{i}} [L(i_1) \boxtimes \cdots \boxtimes L(i_n)].$$

Then we have

$$\mathrm{ch}\, \Delta_{i^m} M = \sum_{\underline{j}} r_{\underline{j}} [L(j_1) \boxtimes \cdots \boxtimes L(j_n)],$$

summing over all $\underline{j} \in I^n$ *with* $j_{n-m+1} = \cdots = j_n = i$.

Now for $i \in I$ and $M \in \mathrm{Rep}_I \mathcal{X}_n$, define

$$\varepsilon_i(M) = \max\{m \geq 0 \mid \Delta_{i^m} M \neq 0\}. \tag{17.5}$$

The following results and their proofs are completely analogous to the corresponding results of Section 5.1. The only change we usually need to make is to go from "$\boxtimes$" to "$\circledast$" and apply the $\mathcal{X}_n$-analogues of the lemmas used in Section 5.1.

Lemma 17.1.2 *Let* $M \in \mathrm{Rep}_I \mathcal{X}_n$ *be irreducible,* $i \in I$, $\varepsilon = \varepsilon_i(M)$. *If* $N \circledast L(i^m)$ *is an irreducible submodule of* $\Delta_{i^m} M$ *for some* $0 \leq m \leq \varepsilon$, *then* $\varepsilon_i(N) = \varepsilon - m$.

Proof Analogous to the proof of Lemma 5.1.2. □

Lemma 17.1.3 *Let* $m \geq 0$, $i \in I$ *and* $N \in \mathrm{Rep}_I \mathcal{X}_n$ *be irreducible with* $\varepsilon_i(N) = 0$. *Set* $M = \mathrm{ind}_{n,m}^{n+m} N \circledast L(i^m)$. *Then:*

(i) $\Delta_{i^m} M \cong N \circledast L(i^m)$;
(ii) $\mathrm{hd}\, M$ *is irreducible with* $\varepsilon_i(\mathrm{hd}\, M) = m$;
(iii) *all other composition factors* L *of* M *have* $\varepsilon_i(L) < m$.

Proof Analogous to the proof of Lemma 5.1.3. □

Lemma 17.1.4 *Let* $M \in \mathrm{Rep}_I \mathcal{X}_n$ *be irreducible,* $i \in I$, *and* $\varepsilon = \varepsilon_i(M)$. *Then* $\Delta_{i^\varepsilon} M$ *is isomorphic to* $N \circledast L(i^\varepsilon)$ *for some irreducible* $\mathcal{X}_{n-\varepsilon}$*-supermodule* N *with* $\varepsilon_i(N) = 0$.

Proof Analogous to the proof of Lemma 5.1.4. □

Lemma 17.1.5 *Let* $m \geq 0$, $i \in I$ *and* $N \in \mathrm{Rep}_I \mathcal{X}_n$ *be irreducible. Set*

$$M = \mathrm{ind}_{n,m}^{n+m} (N \circledast L(i^m)).$$

Then hd M *is irreducible with* $\varepsilon_i(\mathrm{hd}\, M) = \varepsilon_i(N) + m$*, and all other composition factors* L *of* M *have* $\varepsilon_i(L) < \varepsilon_i(N) + m$*.*

Proof Analogous to the proof of Lemma 5.1.5. □

Theorem 17.1.6 *Let* $M \in \mathrm{Rep}_I\, \mathcal{X}_n$ *be irreducible and* $i \in I$*. Then, for any* $0 \leq m \leq \varepsilon_i(M)$*,* $\mathrm{soc}\, \Delta_{i^m} M$ *is an irreducible* $\mathcal{X}_{n-m,m}$*-supermodule of the same type as* M*, and is isomorphic to* $L \circledast L(i^m)$*, for some irreducible* $\mathcal{X}_{n-m}$*-supermodule* L *with* $\varepsilon_i(L) = \varepsilon_i(M) - m$*.*

Proof An argument analogous to the one given in the proof of Theorem 5.1.6 shows that

$$\mathrm{soc}\, \Delta_{i^m} M \cong L \circledast L(i^m)$$

for some irreducible $\mathcal{X}_{n-m}$-supermodule L with $\varepsilon_i(L) = \varepsilon_i(M) - m$. We just need to show that $L \circledast L(i^m)$ has the same type as M. Note by Lemma 17.1.5,

$$\mathrm{ind}^n_{n-m,m} L \circledast L(i^m)$$

has irreducible head, necessarily isomorphic to M by Frobenius reciprocity. So applying (17.4) we have that

$$\begin{aligned}
\mathrm{End}_{\mathcal{X}_{n-m,m}}(L \circledast L(i^m)) &\cong \mathrm{Hom}_{\mathcal{X}_{n-m,m}}(L \circledast L(i^m), \Delta_{i^m} M) \\
&\cong \mathrm{Hom}_{\mathcal{X}_n}(\mathrm{ind}^n_{n-m,m} L \circledast L(i^m), M) \\
&\cong \mathrm{End}_{\mathcal{X}_n}(M),
\end{aligned}$$

which implies the statement concerning types. □

Corollary 17.1.7 *For an irreducible supermodule* $M \in \mathrm{Rep}_I\, \mathcal{X}_n$*, the socle of* $\mathrm{res}^n_{n-1,1} M$ *is multiplicity free.*

Define the functor

$$\mathrm{res}_i := \mathrm{res}^{n-1,1}_{n-1} \circ \Delta_i : \mathrm{Rep}_I\, \mathcal{X}_n \to \mathrm{Rep}_I\, \mathcal{X}_{n-1}. \tag{17.6}$$

Record the following obvious equalities for $M \in \mathrm{Rep}_I(\mathcal{X}_n)$:

$$\varepsilon_i(M) = \max\{m \geq 0 \,|\, \mathrm{res}_i^m M \neq 0\}, \tag{17.7}$$

$$\mathrm{res}^n_{n-1} M = \bigoplus_{i \in I} \mathrm{res}_i M. \tag{17.8}$$

Corollary 17.1.8 *For an irreducible $M \in \mathrm{Rep}_I\, \mathcal{X}_n$ with $\varepsilon_i(M) > 0$,*

$$\mathrm{soc\,res}_i M \simeq \begin{cases} L & \text{if } M \text{ is of type } \mathtt{M} \text{ and } i = 0, \\ L \oplus \Pi L & \text{otherwise} \end{cases}$$

for some irreducible $\mathcal{X}_{n-1}$-supermodule L of the same type as M if $i \neq 0$ and of the opposite type to M if $i = 0$.

Proof Set $\delta := 1$ if M is of type $\mathtt{M}$ and $i = 0$, $\delta := 2$ otherwise. By Theorem 17.1.6, the socle of $\Delta_i M$ is isomorphic to $L \circledast L(i)$ for some irreducible $\mathcal{X}_{n-1}$-supermodule L, and

$$\mathrm{res}^{n-1,1}_{n-1} L \circledast L(i) \cong L^{\oplus \delta},$$

indeed it is exactly as in the statement of the corollary, including the statement about the type.

Now take any irreducible subsupermodule K of $\mathrm{res}_i M$. Consider the $\mathcal{X}_{n-1,1}$-subsupermodule $\mathcal{X}'_1 K \subseteq \mathrm{res}_i M$, where $\mathcal{X}'_1$ is the subalgebra generated by c_n, x_n. All composition factors of $\mathcal{X}'_1 K$ are isomorphic to $K \circledast L(i)$. As the socle of $\mathcal{X}'_1 K$ must be isomorphic to $L \circledast L(i)$, this implies $K \cong L$. We deduce that $\mathrm{soc\,res}_i M \cong L^{\oplus \delta'}$ for some $\delta' \geq \delta$.

By Lemma 17.1.2, $\varepsilon_i(L) = \varepsilon - 1$ where $\varepsilon := \varepsilon_i(M)$, and $\Delta_{i^{\varepsilon-1}} L$ is irreducible by Lemma 17.1.4. So at least δ' copies of $\Delta_{i^{\varepsilon-1}} L$ appear in $\mathrm{soc\,res}^{n-\varepsilon,\varepsilon}_{n-\varepsilon,\varepsilon-1} \Delta_{i^\varepsilon}(M)$. But $\Delta_{i^\varepsilon} M \cong N \circledast L(i^\varepsilon)$ for some irreducible N, thus applying Theorem 16.3.3 (iii) and the facts about type in Theorem 17.1.6,

$$\begin{aligned} \mathrm{soc\,res}^{n-\varepsilon,\varepsilon}_{n-\varepsilon,\varepsilon-1} \Delta_{i^\varepsilon}(M) &\cong \mathrm{soc\,res}^{n-\varepsilon,\varepsilon}_{n-\varepsilon,\varepsilon-1} N \circledast L(i^\varepsilon) \\ &\cong \mathrm{res}^{n-\varepsilon,\varepsilon-1,1}_{n-\varepsilon,\varepsilon-1} N \circledast L(i^{\varepsilon-1}) \circledast L(i) \\ &\cong (N \circledast L(i^{\varepsilon-1}))^{\oplus \delta}. \end{aligned}$$

Hence $\delta' \leq \delta$ also. □

17.2 Operators $\tilde{e}_i$ and $\tilde{f}_i$

Let M be an irreducible $\mathcal{X}_n$-supermodule. Define

$$\tilde{f}_i M := \mathrm{hd\,ind}^{n+1}_{n,1} M \circledast L(i), \tag{17.9}$$

Note $\tilde{f}_i M$ is irreducible by Lemma 17.1.5. To define $\tilde{e}_i M$, note by Theorem 17.1.6 that either $\Delta_i M = 0$ or $\mathrm{soc}\, \Delta_i M$ has form $N \circledast L(i)$ for an irreducible $\mathcal{X}_{n-1}$-supermodule N. In the former case we define $\tilde{e}_i M = 0$, and in the latter case we define $\tilde{e}_i M = N$. Thus, $\tilde{e}_i M$ is defined from

$$\operatorname{soc} \Delta_i M \cong (\tilde{e}_i M) \circledast L(i). \tag{17.10}$$

Note right away from Lemma 17.1.2 that

$$\varepsilon_i(M) = \max\{m \geq 0 \mid \tilde{e}_i^m M \neq 0\}, \tag{17.11}$$

while a special case of Lemma 17.1.5 shows that

$$\varepsilon_i(\tilde{f}_i M) = \varepsilon_i(M) + 1. \tag{17.12}$$

Lemma 17.2.1 *Let $M \in \operatorname{Rep}_I \mathcal{X}_n$ be irreducible, $i \in I$ and $m \geq 0$.*

(i) $\operatorname{soc} \Delta_{i^m} M \cong (\tilde{e}_i^m M) \circledast L(i^m)$;
(ii) $\operatorname{hd} \operatorname{ind}_{n,m}^{n+m} M \circledast L(i^m) \cong \tilde{f}_i^m M$.

Proof Analogous to the proof of Lemma 5.2.1. □

Lemma 17.2.2 *Let $M \in \operatorname{Rep}_I \mathcal{X}_n$ and $N \in \operatorname{Rep}_I \mathcal{X}_{n+1}$ be irreducible supermodules, and $i \in I$. Then $\tilde{f}_i M \cong N$ if and only if $\tilde{e}_i N \cong M$.*

Proof Analogous to the proof of Lemma 5.2.3. □

From Lemma 17.2.2 we immediately deduce the following:

Corollary 17.2.3 *Let $M, N \in \operatorname{Rep}_I \mathcal{X}_n$ be irreducible. Then $\tilde{f}_i M \cong \tilde{f}_i N$ if and only if $M \cong N$. Similarly, providing $\varepsilon_i(M), \varepsilon_i(N) > 0$, $\tilde{e}_i M \cong \tilde{e}_i N$ if and only if $M \cong N$.*

17.3 Independence of irreducible characters

Theorem 17.3.1 *The map* $\operatorname{ch} : K(\operatorname{Rep}_I \mathcal{X}_n) \to K(\operatorname{Rep}_I \mathcal{A}_n)$ *is injective.*

Proof Analogous to the proof of Theorem 5.3.1. □

Corollary 17.3.2 *If L is an irreducible $\mathcal{X}_n$-supermodule, then $L \cong L^\tau$.*

Proof Analogous to the proof of Corollary 5.3.2 □

Lemma 17.3.3 *Let $M \in \operatorname{Rep}_I \mathcal{X}_m$ and $N \in \operatorname{Rep}_I \mathcal{X}_n$ be irreducible supermodules. Suppose:*

(i) $\operatorname{ind}_{m,n}^{m+n} M \circledast N \cong \operatorname{ind}_{n,m}^{n+m} N \circledast M$;
(ii) *$M \circledast N$ appears in $\operatorname{res}_{m,n}^{m+n} \operatorname{ind}_{m,n}^{m+n} M \circledast N$ with multiplicity one.*

Then $\operatorname{ind}_{m,n}^{m+n} M \circledast N$ is irreducible.

Proof Analogous to the proof of Lemma 5.3.3. □

We can also show at this point that the type of an irreducible supermodule is determined by the type of its central character:

Lemma 17.3.4 *Suppose $L \in \operatorname{Rep}_I \mathcal{X}_n$ is irreducible with central character χ_γ, where $\gamma = \sum_{i \in I} \gamma_i \alpha_i \in \Gamma_n$. Then, L is of type* Q *if γ_0 is odd, type* M *if γ_0 is even.*

Proof Proceed by induction on n, the case $n = 0$ being trivial. If $n > 1$, choose $i \in I$ so that $\tilde{e}_i L \neq 0$. By definition, $\tilde{e}_i L$ has central character $\gamma - \alpha_i$. So by the induction hypothesis, $\tilde{e}_i L$ is of type Q if $\gamma_0 - \delta_{i,0}$ is odd, type M otherwise. But by Lemmas 12.2.13 and 16.1.1, $(\tilde{e}_i L) \circledast L(i)$ is of the opposite type to $\tilde{e}_i L$ if $i = 0$, of the same type if $i \neq 0$. Hence, $(\tilde{e}_i L) \circledast L(i)$ is of type Q if γ_0 is odd, type M otherwise. Finally, the proof is completed by Theorem 17.1.6, since this shows that L has the same type as $\operatorname{soc} \Delta_i L = (\tilde{e}_i L) \circledast L(i)$. □

17.4 Labels for irreducibles

Write $\mathbf{1}$ for the trivial irreducible supermodule over $\mathcal{X}_0 \cong F$. If L is an irreducible $\mathcal{X}_n$-supermodule, it follows from Lemma 17.2.2 that

$$L \cong \tilde{f}_{i_n} \dots \tilde{f}_{i_2} \tilde{f}_{i_1} \mathbf{1}$$

for at least one tuple $\underline{i} = (i_1, i_2, \dots, i_n) \in I^n$. So if we define

$$L(\underline{i}) = L(i_1, \dots, i_n) := \tilde{f}_{i_n} \dots \tilde{f}_{i_2} \tilde{f}_{i_1} \mathbf{1}, \tag{17.13}$$

we obtain a labeling of all irreducibles by tuples in I^n. For example, $L(i, i, \dots, i)$ (n times) is precisely the Kato module $L(i^n)$ introduced in (16.6). A given irreducible L will in general be parametrized by several *different* tuples $\underline{i} \in I^n$. Some basic properties of $L(\underline{i})$ are easy to read off from the notation: for instance the central character of $L(\underline{i})$ is $\chi_{\underline{i}}$.

18

Character calculations for $\mathcal{X}_n$

This chapter is parallel to Chapter 6. The explicit character calculations in Section 18.2 determine the type of the Kac–Moody algebra acting on the Grothendieck group of $\mathcal{X}_n$-supermodules.

18.1 Some irreducible induced supermodules

Given $\underline{i} = (i_1, \dots, i_n) \in I^n$, let

$$\mathrm{ind}(\underline{i}) = \mathrm{ind}(i_1, \dots, i_n) := \mathrm{ind}^n_{1,\dots,1} L(i_1) \circledast \cdots \circledast L(i_n).$$

By Lemma 16.1.3 and (16.3), every irreducible constituent of $\mathrm{ind}(\underline{i})$ belongs to the block corresponding to the orbit $S_n \cdot \underline{i}$.

Lemma 18.1.1 *Let $\underline{i} \in I^n$. Then:*

(i) $\mathrm{res}^n_{1,\dots,1} L(\underline{i})$ *has a submodule isomorphic to* $L(i_1) \circledast \cdots \circledast L(i_n)$*;*

(ii) $\mathrm{ind}(\underline{i})$ *contains a copy of* $L(\underline{i})$ *in its head;*

(iii) *every irreducible module in the block corresponding to the orbit* $S_n \cdot \underline{i}$ *appears at least once as a constituent of* $\mathrm{ind}(\underline{i})$.

Proof Analogous to the proof of Lemma 6.1.1. □

Lemma 18.1.2 *Let $\underline{i} \in I^m$, $\underline{j} \in I^n$ be tuples such that $i_r - j_s \notin \{0, \pm 1\}$ for all $1 \le r \le m$ and $1 \le s \le n$. Then*

$$\mathrm{ind}^{m+n}_{m,n} L(\underline{i}) \circledast L(\underline{j}) \cong \mathrm{ind}^{m+n}_{n,m} L(\underline{j}) \circledast L(\underline{i})$$

is irreducible.

Proof By the Shuffle Lemma, $L(\underline{i}) \circledast L(\underline{j})$ appears in

$$\mathrm{res}^{m+n}_{m,n}\mathrm{ind}^{m+n}_{m,n} L(\underline{i}) \circledast L(\underline{j})$$

with multiplicity 1. So, in view of Lemma 17.3.3, it suffices to show that

$$\mathrm{ind}^{m+n}_{m,n} L(\underline{i}) \circledast L(\underline{j}) \cong \mathrm{ind}^{m+n}_{n,m} L(\underline{j}) \circledast L(\underline{i}).$$

By the Mackey Theorem and central characters argument,

$$\mathrm{res}^{m+n}_{m,n}\mathrm{ind}^{m+n}_{n,m} L(\underline{j}) \circledast L(\underline{i})$$

contains $L(\underline{i}) \circledast L(\underline{j})$ as a summand with multiplicity one, all other constituents lying in different blocks. Hence by Frobenius reciprocity, there exists a non-zero homomorphism

$$f: \mathrm{ind}^{m+n}_{m,n} L(\underline{i}) \circledast L(\underline{j}) \to \mathrm{ind}^{m+n}_{n,m} L(\underline{j}) \circledast L(\underline{i}).$$

Every homomorphic image of $\mathrm{ind}^{m+n}_{m,n} L(\underline{i}) \circledast L(\underline{j})$ contains an $\mathcal{X}_{m,n}$-submodule isomorphic to $L(\underline{i}) \circledast L(\underline{j})$. So, by Lemma 18.1.1(i), we see that the image of f contains a $\mathcal{A}_{m+n}$-submodule V isomorphic to

$$L(i_1) \circledast \cdots \circledast L(i_m) \circledast L(j_1) \circledast \cdots \circledast L(j_n).$$

Next we claim that the image of f also contains an $\mathcal{A}_{m+n}$-submodule isomorphic to

$$L(j_1) \circledast \cdots \circledast L(j_n) \circledast L(i_1) \circledast \cdots \circledast L(i_m). \tag{18.1}$$

Indeed, by Lemma 14.8.1 and the assumption that $i_m - j_1 \notin \{0, \pm 1\}$, Φ_m^2 acts on v by a non-zero scalar. So by (14.20), (14.21), $\Phi_m V \neq 0$ is a $\mathcal{A}_{m+n}$-submodule isomorphic to

$$L(i_1) \circledast \cdots \circledast L(i_{m-1}) \circledast L(j_1) \circledast L(i_m) \circledast L(j_2) \circledast \cdots \circledast L(j_n).$$

Next apply $\Phi_{m-1}, \ldots, \Phi_1$ to move $L(j_1)$ to the first position, and continue in this way to complete the proof of the claim.

Now, by the Shuffle Lemma, all composition factors of

$$\mathrm{res}^{m+n}_{1,\ldots,1}\mathrm{ind}^{m+n}_{n,m} L(\underline{j}) \circledast L(\underline{i})$$

isomorphic to (18.1) necessarily lie in the irreducible $\mathcal{X}_{n,m}$-submodule $1 \otimes L(\underline{j}) \circledast L(\underline{i})$ of the induced module. Since this generates all of $\mathrm{ind}^{m+n}_{n,m} L(\underline{j}) \circledast L(\underline{i})$ as an $\mathcal{X}_{m+n}$-module, this shows that f is surjective. Hence f is an isomorphism by dimension. □

Remark 18.1.3 Keep assumptions of the lemma. Set

$$\gamma = S_m \cdot \underline{i},\ \delta := S_n \cdot \underline{j},\ \gamma \cup \delta := S_{m+n} \cdot (\underline{i} \cup \underline{j}),\ (\gamma, \delta) := S_m \times S_n \cdot (\underline{i}, \underline{j}).$$

Then the argument as above shows that the functor $\mathrm{ind}_{m,n}^{m+n}$ induces an equivalence of categories

$$\mathrm{Rep}_I\, \mathcal{X}_m \otimes \mathcal{X}_n[(\gamma, \delta)] \simeq \mathrm{Rep}_I\, \mathcal{X}_{m+n}[\gamma \cup \delta].$$

Theorem 18.1.4 *Let $\underline{i} \in I^m$, $\underline{j} \in I^n$ be tuples such that $i_r - j_s \notin \{\pm 1\}$ for all $1 \le r \le m$ and $1 \le s \le n$. Then*

$$\mathrm{ind}_{m,n}^{m+n} L(\underline{i}) \circledast L(\underline{j}) \cong \mathrm{ind}_{n,m}^{m+n} L(\underline{j}) \circledast L(\underline{i})$$

is irreducible. Moreover, every other irreducible $\mathcal{X}_{m+n}$-supermodule lying in the same block as $\mathrm{ind}_{m,n}^{m+n} L(\underline{i}) \circledast L(\underline{j})$ is of the form $\mathrm{ind}_{m,n}^{m+n} L(\underline{i}') \circledast L(\underline{j}')$ for permutations $\underline{i}'$ of $\underline{i}$ and $\underline{j}'$ of $\underline{j}$.

Proof Analogous to the proof of Theorem 6.1.4. □

18.2 Calculations for small rank

Recall the notation $\langle h_i, \alpha_j \rangle$ from Section 15.2.

Theorem 18.2.1 *Let $i, j \in I$, $i - j \in \{\pm 1\}$, and $k = -\langle h_i, \alpha_j \rangle$.*

(i) *For all $r, s \ge 0$ with $r + s \le k$ we have*

$$\mathrm{ch}\, L(i^r j i^s) = (r!)(s!)[L(i)^{\circledast r} \circledast L(j) \circledast L(i)^{\circledast s}].$$

Moreover, if additionally we have $r > 0$, then there is a non-split short exact sequence

$$0 \longrightarrow L(i^r j i^s) \longrightarrow \mathrm{ind}_{r+s,1}^{r+s+1} L(i^{r-1} j i^s) \circledast L(i) \longrightarrow L(i^{r-1} j i^{s+1}) \longrightarrow 0.$$

(ii) *We have $L(i^{k+1} j) \cong L(i^k j i)$. Moreover, for all $r \ge 0$ and $s > 0$ with $r + s = k + 1$ we have*

$$\mathrm{ch}\, L(i^r j i^s) = (r!)(s!)[L(i)^{\circledast r} \circledast L(j) \circledast L(i)^{\circledast s}] + (r+1)!(s-1)![L(i)^{\circledast r+1} \circledast L(j) \circledast L(i)^{\circledast s-1}],$$

and

$$L(i^r j i^s) \cong \mathrm{ind}_{k,1}^{k+1} L(i^r j i^{s-1}) \circledast L(i) \cong \mathrm{ind}_{1,k}^{k+1} L(i) \circledast L(i^r j i^{s-1}).$$

Proof The general scheme here is similar to the classical case, see Lemmas 6.2.1 and 6.2.2.

(i) We first proceed by induction on $n = 0, 1, \dots, k$ to show that

$$\operatorname{ch} L(i^n j) = n![L(i)^{\circledast n} \circledast L(j)],$$

the induction base being clear. For $n > 0$, let

$$M_n = \operatorname{ind}_{n,1}^{n+1} L(i^{n-1} j) \circledast L(j). \tag{18.2}$$

We know by the inductive hypothesis and the Shuffle Lemma that

$$\operatorname{ch} M_n = n![L(i)^{\circledast n} \circledast L(j)] + (n-1)![L(i)^{\circledast n-1} \circledast L(j) \circledast L(i)].$$

Now consider the $\mathcal{X}_{n,1}$-submodule

$$N_n := (x_{n+1}^2 - q(i))M \cong L(i^n) \circledast L(j) \tag{18.3}$$

of M_n. We claim that N_n is stable under the action of s_n, hence all of $\mathcal{X}_{n+1}$. This is a brutal calculation outlined in Lemmas 18.2.2–18.2.5. So there exists an irreducible $\mathcal{X}_{n+1}$-supermodule N_n with character $n![L(i)^{\circledast n} \circledast L(j)]$. This must be $L(i^n j)$, by Lemma 18.1.1(i), completing the proof of the induction step.

Now we explain how to deduce the characters of the remaining irreducibles in the block. The quotient module M_n/N_n has character

$$\operatorname{ch} M_n/N_n = (n-1)![L(i)^{\circledast(n-1)} \circledast L(j) \circledast L(i)],$$

so $M_n/N_n \cong L(i^{n-1} j i)$. Twisting with the automorphism σ proves that there exist irreducibles with characters

$$n![L(j) \circledast L(i)^{\circledast n}] \quad \text{and} \quad (n-1)![L(i) \circledast L(j) \circledast L(i)^{\circledast(n-1)}],$$

which must be $L(ji^n)$ and $L(iji^{n-1})$ respectively by Lemma 18.1.1(i) once more. This covers everything unless $n = 4$, when we necessarily have that $i = 0$, $j = 1$ and $p = 3$. In this case, we have shown already that there exist four irreducibles with characters

$$\operatorname{ch} L(0^4 1) = 24[L(0)^{\circledast 4} \circledast L(1)],\ \operatorname{ch} L(0^3 10) = 6[L(0)^{\circledast 3} \circledast L(1) \circledast L(0)],$$
$$\operatorname{ch} L(10^4) = 24[L(1) \circledast L(0)^{\circledast 4}],\ \operatorname{ch} L(010^3) = 6[L(0) \circledast L(1) \circledast L(0)^{\circledast 3}].$$

So by Lemma 18.1.1, there must be exactly one more irreducible module in the block, namely $L(00100)$, since none of the above involves the character $[L(0)^{\circledast 2} \circledast L(1) \circledast L(0)^{\circledast 2}]$. Considering the character of $\operatorname{ind}_{4,1}^{5} L(0010) \circledast L(0)$ shows that $\operatorname{ch} L(00100)$ is either $4[L(0)^{\circledast 2} \circledast L(1) \circledast L(0)^{\circledast 2}]$ or $4[L(0)^{\circledast 2} \circledast$

$L(1) \circledast L(0)^{\circledast 2}] + 6[L(0)^{\circledast 3} \circledast L(1) \circledast L(0)]$, but the latter is not σ-invariant so cannot occur as there would then be too many irreducibles.

Now that the characters are known, it is a routine matter using the Shuffle Lemma and Lemma 17.1.5 to prove the existence of the required non-split sequence.

(ii) Let $n = k+1$, and $M_n = \mathrm{ind}_{n,1}^{n+1} L(i^{n-1}j) \circledast L(i)$. We first claim that M_n is irreducible. To prove this, arguing as in (i), it suffices to show that the $\mathcal{X}_{n,1}$-submodule

$$(x_{n+1}^2 - q(i))M_n \cong L(i^n) \circledast L(j)$$

of M_n is *not* invariant under s_n. Again, this is a brutal calculation, see Lemmas 18.2.2–18.2.5. Hence, in view of (i), there is an irreducible $\mathcal{X}_{n+1}$-module M_n with character

$$n![L(i)^{\circledast n} \circledast L(j)] + (n-1)![L(i)^{\circledast (n-1)} \circledast L(j) \circledast L(i)].$$

Therefore $\tilde{e}_i M_n \cong L(i^{n-1}j)$ and $\tilde{e}_j M_n \cong L(i^n)$. So we deduce that $M_n \cong L(i^{n-1}ji) \cong L(i^n j)$ thanks to Lemma 17.2.2.

Now consider the remaining irreducibles in the block. There are at most k remaining, namely $L(i^r j i^s)$ for $r \geq 0, s \geq 2$ with $r+s = k+1$. Considering the known characters of $\mathrm{ind}_{n,1}^{n+1} L(i^r j i^{s-1}) \circledast L(i)$ and arguing in a similar way to (i), the remainder of the lemma follows without further calculation. □

We now sketch the calculations skipped in the proof of Theorem 18.2.1. From now on till the end of this section i, j are as in the assumptions of Theorem 18.2.1, and M_n, N_n are as in (18.2), (18.3), respectively. Until the end of Section 18.2 we write a for $q(i)$ and b for $q(j)$.

Lemma 18.2.2 *N_1 is s_1-invariant.*

Proof By assumption i, j are not both zero, so $L(j) \circledast L(i) = L(j) \boxtimes L(i)$, see Lemma 16.1.1. Recall the basis $\{v_1, v_{-1}\}$ of $L(i)$ from Section 16.1. We use the same notation for a similar basis of $L(j)$. Then M_1 has basis

$$\{1 \otimes v_\alpha \otimes v_\beta,\ s_1 \otimes v_\alpha \otimes v_\beta \mid \alpha, \beta \in \{\pm 1\}\}.$$

Note that $(x_2^2 - a)(1 \otimes v_\alpha \otimes v_\beta) = 0$, and also, using (14.8), we have

$$\begin{aligned}(x_2^2 - a)(s_1 \otimes v_\alpha \otimes v_\beta) &= (b-a)(s_1 \otimes v_\alpha \otimes v_\beta) + (\alpha\sqrt{b} + \beta\sqrt{a})1 \otimes v_\alpha \otimes v_\beta \\ &\quad + (\sqrt{b} - \alpha\beta\sqrt{a})1 \otimes v_{-\alpha} \otimes v_{-\beta}. \qquad (18.4)\end{aligned}$$

Denote

$$v_{\alpha,\beta} := (x_2^2 - a)(s_1 \otimes v_\alpha \otimes v_\beta) \qquad (\alpha, \beta \in \{\pm 1\}).$$

Then $\{v_{\alpha,\beta} \mid \alpha, \beta \in \{\pm 1\}\}$ is a basis of N_1. Note for future reference:

$$\begin{aligned} x_1 v_{\alpha,\beta} &= \beta\sqrt{a} v_{\alpha,\beta}, \quad x_2 v_{\alpha,\beta} = \alpha\sqrt{b} v_{\alpha,\beta}, \\ c_1 v_{\alpha,\beta} &= -\alpha v_{\alpha,-\beta}, \quad c_2 v_{\alpha,\beta} = -v_{-\alpha,\beta}. \end{aligned} \tag{18.5}$$

Using (18.4) and the assumption that $i - j \in \{\pm 1\}$, we check that

$$s_1 v_{\alpha,\beta} = \frac{\alpha\sqrt{b} + \beta\sqrt{a}}{b-a} v_{\alpha,\beta} + \frac{\sqrt{b} - \alpha\beta\sqrt{a}}{b-a} v_{-\alpha,-\beta}. \tag{18.6}$$

In particular, N_1 is s_1-invariant. □

Lemma 18.2.3 *N_2 is s_2-invariant if and only if*

$$(i, j) \in \{(0, 1), (\ell - 1, \ell)\}.$$

Proof We have found in the proof of Lemma 18.2.2 that $L(i, j)$ has basis $\{v_{\alpha,\beta} \mid \alpha, \beta \in \{\pm 1\}\}$ with the action given by (18.5) and (18.6). So

$$N_2' := (x_3^2 - a)\mathrm{ind}_{2,1}^3 L(i, j) \boxtimes L(i)$$

is spanned by the vectors

$$z_{\alpha,\beta,\gamma} := (x_3^2 - a)s_2 \otimes v_{\alpha,\beta} \otimes v_\gamma, \; y_{\alpha,\beta,\gamma} := (x_3^2 - a)s_1 s_2 \otimes v_{\alpha,\beta} \otimes v_\gamma$$

for $\alpha, \beta, \gamma \in \{\pm 1\}$. If $i \neq 0$, we have $N_2' = N_2$. If $i = 0$, then $L(i, j) \boxtimes L(i) \cong (L(i, j) \circledast L(i))^{\oplus 2}$, so $N_2' \cong N_2^{\oplus 2}$. In any case, it is enough to check the lemma for N_2' instead of N_2. Using relations we get:

$$\begin{aligned} z_{\alpha,\beta,\gamma} &= (b-a)s_2 \otimes v_{\alpha,\beta} \otimes v_\gamma + (\alpha\sqrt{b} + \gamma\sqrt{a})(1 \otimes v_{\alpha,\beta} \otimes v_\gamma) \\ &\quad + (\alpha\beta\gamma\sqrt{a} - \beta\sqrt{b})(1 \otimes v_{-\alpha,\beta} \otimes v_{-\gamma}), \end{aligned} \tag{18.7}$$

$$\begin{aligned} y_{\alpha,\beta,\gamma} &= (b-a)s_1 s_2 \otimes v_{\alpha,\beta} \otimes v_\gamma \\ &\quad + \frac{(\alpha\sqrt{b} + \gamma\sqrt{a})(\alpha\sqrt{b} + \beta\sqrt{a})}{b-a}(1 \otimes v_{\alpha,\beta} \otimes v_\gamma) \\ &\quad + \frac{(\alpha\sqrt{b} + \gamma\sqrt{a})(\sqrt{b} - \alpha\beta\sqrt{a})}{b-a}(1 \otimes v_{-\alpha,-\beta} \otimes v_\gamma) \\ &\quad + \frac{(\alpha\beta\gamma\sqrt{a} - \beta\sqrt{b})(-\alpha\sqrt{b} + \beta\sqrt{a})}{b-a}(1 \otimes v_{-\alpha,\beta} \otimes v_{-\gamma}) \\ &\quad + \frac{(\alpha\beta\gamma\sqrt{a} - \beta\sqrt{b})(\sqrt{b} + \alpha\beta\sqrt{a})}{b-a}(1 \otimes v_{\alpha,-\beta} \otimes v_{-\gamma}). \end{aligned} \tag{18.8}$$

Now, a calculation entirely similar to the $n = 1$ case shows that

$$s_2 z_{\alpha,\beta,\gamma} = \frac{\alpha\sqrt{b}+\gamma\sqrt{a}}{b-a} z_{\alpha,\beta,\gamma} + \frac{\alpha\beta\gamma\sqrt{a}-\beta\sqrt{b}}{b-a} z_{-\alpha,\beta,-\gamma}. \tag{18.9}$$

Next, we check that

$$\begin{aligned} s_2 y_{\alpha,\beta,\gamma} = &(\alpha\sqrt{b}+\beta\sqrt{a})s_1 s_2 \otimes v_{\alpha,\beta} \otimes v_\gamma \\ &+ (\sqrt{b}-\alpha\beta\sqrt{a})s_1 s_2 \otimes v_{-\alpha,-\beta} \otimes v_\gamma \\ &+ \frac{(\alpha\sqrt{b}+\gamma\sqrt{a})(\alpha\sqrt{b}+\beta\sqrt{a})}{b-a} s_2 \otimes v_{\alpha,\beta} \otimes v_\gamma \\ &+ \frac{(\alpha\sqrt{b}+\gamma\sqrt{a})(\sqrt{b}-\alpha\beta\sqrt{a})}{b-a} s_2 \otimes v_{-\alpha,-\beta} \otimes v_\gamma \\ &+ \frac{(\alpha\beta\gamma\sqrt{a}-\beta\sqrt{b})(-\alpha\sqrt{b}+\beta\sqrt{a})}{b-a} s_2 \otimes v_{-\alpha,\beta} \otimes v_{-\gamma} \\ &+ \frac{(\alpha\beta\gamma\sqrt{a}-\beta\sqrt{b})(\sqrt{b}+\alpha\beta\sqrt{a})}{b-a} s_2 \otimes v_{\alpha,-\beta} \otimes v_{-\gamma}. \end{aligned} \tag{18.10}$$

So, N_2' is s_2-invariant if and only if

$$\begin{aligned} s_2 y_{\alpha,\beta,\gamma} = &\frac{\alpha\sqrt{b}+\beta\sqrt{a}}{b-a} y_{\alpha,\beta,\gamma} + \frac{\sqrt{b}-\alpha\beta\sqrt{a}}{b-a} y_{-\alpha,-\beta,\gamma} \\ &+ \frac{(\alpha\sqrt{b}+\gamma\sqrt{a})(\alpha\sqrt{b}+\beta\sqrt{a})}{(b-a)^2} z_{\alpha,\beta,\gamma} \\ &+ \frac{(\alpha\sqrt{b}+\gamma\sqrt{a})(\sqrt{b}-\alpha\beta\sqrt{a})}{(b-a)^2} z_{-\alpha,-\beta,\gamma} \\ &+ \frac{(\alpha\beta\gamma\sqrt{a}-\beta\sqrt{b})(-\alpha\sqrt{b}+\beta\sqrt{a})}{(b-a)^2} z_{-\alpha,\beta,-\gamma} \\ &+ \frac{(\alpha\beta\gamma\sqrt{a}-\beta\sqrt{b})(\sqrt{b}+\alpha\beta\sqrt{a})}{(b-a)^2} z_{\alpha,-\beta,-\gamma}. \end{aligned} \tag{18.11}$$

Now, a calculation, using (18.10), (18.7), and (18.8), shows that (18.11) is equivalent to $(i, j) \in \{(0, 1), (\ell - 1, \ell)\}$. □

Lemma 18.2.4 *Let $(i, j) = (0, 1)$ or $(\ell - 1, \ell)$. Then N_3 is s_3-invariant if and only if $p = 3$.*

Proof As in the proof of Lemma 18.2.3, we can work with the module

$$N_3' := (x_4^2 - a)\mathrm{ind}_{3,1}^4 N_2' \boxtimes L(i) \subset M_3' := \mathrm{ind}_{3,1}^4 N_2' \boxtimes L(i).$$

Recall the basis $\{z_{\alpha\beta\gamma}, y_{\alpha\beta\gamma}\}$ of N_2' found in the proof of Lemma 18.2.3. The action of the generators of $\mathcal{X}_3$ on the basis elements is given by (18.11) together with

$$s_1 z_{\alpha\beta\gamma} = y_{\alpha\beta\gamma}, \quad s_1 y_{\alpha\beta\gamma} = z_{\alpha\beta\gamma},$$

$$s_2 z_{\alpha\beta\gamma} = \frac{\alpha\sqrt{b}+\gamma\sqrt{a}}{b-a} z_{\alpha\beta\gamma} + \frac{\alpha\beta\gamma\sqrt{a}-\beta\sqrt{b}}{b-a} z_{-\alpha,\beta,-\gamma},$$

$$c_1 z_{\alpha\beta\gamma} = -\alpha z_{\alpha,-\beta,\gamma}, \quad c_1 y_{\alpha\beta\gamma} = \alpha\beta y_{\alpha,\beta,-\gamma},$$

$$c_2 z_{\alpha\beta\gamma} = \alpha\beta z_{\alpha,\beta,-\gamma}, \quad c_2 y_{\alpha\beta\gamma} = -\alpha y_{\alpha,-\beta,\gamma},$$

$$c_3 z_{\alpha\beta\gamma} = -z_{-\alpha,\beta,\gamma}, \quad c_3 y_{\alpha\beta\gamma} = -y_{-\alpha,\beta,\gamma},$$

$$x_1 z_{\alpha\beta\gamma} = \beta\sqrt{a} z_{\alpha\beta\gamma}, \quad x_1 y_{\alpha\beta\gamma} = \gamma\sqrt{a} y_{\alpha\beta\gamma} - z_{\alpha\beta\gamma} + \alpha\beta z_{-\alpha,\beta,-\gamma},$$

$$x_2 z_{\alpha\beta\gamma} = \gamma\sqrt{a} z_{\alpha\beta\gamma}, \quad x_2 y_{\alpha\beta\gamma} = \beta\sqrt{a} y_{\alpha\beta\gamma} + z_{\alpha\beta\gamma} + \alpha\beta z_{-\alpha,\beta,-\gamma},$$

$$x_3 z_{\alpha\beta\gamma} = \alpha\sqrt{b} z_{\alpha\beta\gamma}, \quad x_3 y_{\alpha\beta\gamma} = \alpha\sqrt{b} y_{\alpha\beta\gamma}.$$

Let us write $z^r_{\alpha\beta\gamma\delta}$ and $y^r_{\alpha\beta\gamma\delta}$ for the elements $(x_4^2-a)s_r s_{r+1}\dots s_3 \otimes z_{\alpha\beta\gamma} \otimes v_\delta$ and $(x_4^2-a)s_r s_{r+1}\dots s_3 \otimes y_{\alpha\beta\gamma} \otimes v_\delta$ of N_3', respectively. We can easily see that N_3' is spanned by the vectors

$$y^1_{\alpha\beta\gamma\delta},\ y^2_{\alpha\beta\gamma\delta},\ y^3_{\alpha\beta\gamma\delta},\ z^1_{\alpha\beta\gamma\delta},\ z^2_{\alpha\beta\gamma\delta},\ z^3_{\alpha\beta\gamma\delta}.$$

Let us also denote by $y_{\alpha\beta\gamma\delta}$ and $z_{\alpha\beta\gamma\delta}$ the elements $1\otimes y_{\alpha\beta\gamma}\otimes v_\delta$ and $1\otimes z_{\alpha\beta\gamma}\otimes v_\delta$ of M_3', respectively. A calculation using the action formulas on N_2' above shows that

$$\begin{aligned} z^3_{\alpha\beta\gamma\delta} &= (b-a)s_3 \otimes z_{\alpha\beta\gamma} \otimes v_\delta + (\delta\sqrt{a}+\alpha\sqrt{b}) z_{\alpha\beta\gamma\delta} \\ &\quad + (\alpha\beta\gamma\delta\sqrt{a} - \beta\gamma\sqrt{b}) z_{-\alpha,\beta,\gamma,-\delta}, \\ y^3_{\alpha\beta\gamma\delta} &= (b-a)s_3 \otimes y_{\alpha\beta\gamma} \otimes v_\delta + (\delta\sqrt{a}+\alpha\sqrt{b}) y_{\alpha\beta\gamma\delta} \\ &\quad + (\alpha\beta\gamma\delta\sqrt{a} - \beta\gamma\sqrt{b}) y_{-\alpha,\beta,\gamma,-\delta}, \end{aligned}$$

$$\begin{aligned} z^2_{\alpha\beta\gamma\delta} &= (b-a)s_2 s_3 \otimes z_{\alpha\beta\gamma} \otimes v_\delta \\ &\quad + \frac{1}{b-a}\Big((\delta\sqrt{a}+\alpha\sqrt{b})(\alpha\sqrt{b}+\gamma\sqrt{a}) z_{\alpha\beta\gamma\delta} \\ &\quad + (\delta\sqrt{a}+\alpha\sqrt{b})(\alpha\beta\gamma\sqrt{a}-\beta\sqrt{b}) z_{-\alpha,\beta,-\gamma,\delta} \\ &\quad + (\alpha\beta\gamma\delta\sqrt{a}-\beta\gamma\sqrt{b})(-\alpha\sqrt{b}+\gamma\sqrt{a}) z_{-\alpha,\beta,\gamma,-\delta} \\ &\quad + (\alpha\beta\gamma\delta\sqrt{a}-\beta\gamma\sqrt{b})(-\alpha\beta\gamma\sqrt{a}-\beta\sqrt{b}) z_{\alpha,\beta,-\gamma,-\delta}\Big), \end{aligned}$$

$$
\begin{aligned}
y^2_{\alpha\beta\gamma\delta} &= (b-a)s_2s_3 \otimes z_{\alpha\beta\gamma} \otimes v_\delta \\
&+ \frac{1}{b-a}\Big((\delta\sqrt{a}+\alpha\sqrt{b})(\alpha\sqrt{b}+\beta\sqrt{a})y_{\alpha\beta\gamma\delta} \\
&+ (\delta\sqrt{a}+\alpha\sqrt{b})(\sqrt{b}-\alpha\beta\sqrt{a})y_{-\alpha,-\beta,\gamma,\delta} \\
&+ (\alpha\beta\gamma\delta\sqrt{a}-\beta\gamma\sqrt{b})(-\alpha\sqrt{b}+\beta\sqrt{a})y_{-\alpha,\beta,\gamma,-\delta} \\
&+ (\alpha\beta\gamma\delta\sqrt{a}-\beta\gamma\sqrt{b})(\sqrt{b}+\alpha\beta\sqrt{a})y_{\alpha,-\beta,\gamma,-\delta}\Big) \\
&+ \frac{1}{(b-a)^2}\Big((\delta\sqrt{a}+\alpha\sqrt{b})(\alpha\sqrt{b}+\gamma\sqrt{a})(\alpha\sqrt{b}+\beta\sqrt{a})z_{\alpha\beta\gamma\delta} \\
&+ (\delta\sqrt{a}+\alpha\sqrt{b})(\alpha\sqrt{b}+\gamma\sqrt{a})(\sqrt{b}-\alpha\beta\sqrt{a})z_{-\alpha,-\beta,\gamma,\delta} \\
&+ (\delta\sqrt{a}+\alpha\sqrt{b})(\alpha\beta\gamma\sqrt{a}-\beta\sqrt{b})(-\alpha\sqrt{b}+\beta\sqrt{a})z_{-\alpha,\beta,-\gamma,\delta} \\
&+ (\delta\sqrt{a}+\alpha\sqrt{b})(\alpha\beta\gamma\sqrt{a}-\beta\sqrt{b})(\sqrt{b}+\alpha\beta\sqrt{a})z_{\alpha,-\beta,-\gamma,\delta} \\
&+ (\alpha\beta\gamma\delta\sqrt{a}-\beta\gamma\sqrt{b})(-\alpha\sqrt{b}+\gamma\sqrt{a})(-\alpha\sqrt{b}+\beta\sqrt{a})z_{-\alpha,\beta,\gamma,-\delta} \\
&+ (\alpha\beta\gamma\delta\sqrt{a}-\beta\gamma\sqrt{b})(-\alpha\sqrt{b}+\gamma\sqrt{a})(\sqrt{b}+\alpha\beta\sqrt{a})z_{\alpha,-\beta,\gamma,-\delta} \\
&+ (\alpha\beta\gamma\delta\sqrt{a}-\beta\gamma\sqrt{b})(-\alpha\beta\gamma\sqrt{a}-\beta\sqrt{b})(\alpha\sqrt{b}+\beta\sqrt{a})z_{\alpha,\beta,-\gamma,-\delta} \\
&+ (\alpha\beta\gamma\delta\sqrt{a}-\beta\gamma\sqrt{b})(-\alpha\beta\gamma\sqrt{a}-\beta\sqrt{b})(\sqrt{b}-\alpha\beta\sqrt{a})z_{-\alpha,-\beta,-\gamma,-\delta}\Big).
\end{aligned}
$$

Next we calculate $s_3y^2_{\alpha\beta\gamma\delta}$, and note that for $s_3y^2_{\alpha\beta\gamma\delta}$ to belong to N'_3 it must equal

$$
\begin{aligned}
&\frac{1}{b-a}\Big((\alpha\sqrt{b}+\beta\sqrt{a})y^2_{\alpha\beta\gamma\delta}+(\sqrt{b}-\alpha\beta\sqrt{a})y^2_{-\alpha,-\beta,\gamma,\delta}\Big) \\
&+ \frac{1}{(b-a)^2}\Big((\alpha\sqrt{b}+\gamma\sqrt{a})(\alpha\sqrt{b}+\beta\sqrt{a})z^2_{\alpha\beta\gamma\delta} \\
&+ (\alpha\sqrt{b}+\gamma\sqrt{a})(\sqrt{b}-\alpha\beta\sqrt{a})z^2_{-\alpha,-\beta,\gamma,\delta} \\
&+ (\alpha\beta\gamma\sqrt{a}-\beta\sqrt{b})(-\alpha\sqrt{b}+\beta\sqrt{a})z^2_{-\alpha,\beta,-\gamma,\delta} \\
&+ (\alpha\beta\gamma\sqrt{a}-\beta\sqrt{b})(\sqrt{b}+\alpha\beta\sqrt{a})z^2_{\alpha,-\beta,-\gamma,\delta} \\
&+ (\delta\sqrt{a}+\alpha\sqrt{b})(\alpha\sqrt{b}+\beta\sqrt{a})y^3_{\alpha\beta\gamma\delta} \\
&+ (\delta\sqrt{a}+\alpha\sqrt{b})(\sqrt{b}-\alpha\beta\sqrt{a})y^3_{-\alpha,-\beta,\gamma,\delta} \\
&+ (\alpha\beta\gamma\delta\sqrt{a}-\beta\gamma\sqrt{b})(-\alpha\sqrt{b}+\beta\sqrt{a})y^3_{-\alpha,\beta,\gamma,-\delta} \\
&+ (\alpha\beta\gamma\delta\sqrt{a}-\beta\gamma\sqrt{b})(\sqrt{b}+\alpha\beta\sqrt{a})y^3_{\alpha,-\beta,\gamma,-\delta}\Big)
\end{aligned}
$$

$$+\frac{1}{(b-a)^3}\Big((\delta\sqrt{a}+\alpha\sqrt{b})(\alpha\sqrt{b}+\gamma\sqrt{a})(\alpha\sqrt{b}+\beta\sqrt{a})z^3_{\alpha\beta\gamma\delta}$$
$$+(\delta\sqrt{a}+\alpha\sqrt{b})(\alpha\sqrt{b}+\gamma\sqrt{a})(\sqrt{b}-\alpha\beta\sqrt{a})z^3_{-\alpha,-\beta,\gamma,\delta}$$
$$+(\delta\sqrt{a}+\alpha\sqrt{b})(\alpha\beta\gamma\sqrt{a}-\beta\sqrt{b})(-\alpha\sqrt{b}+\beta\sqrt{a})z^3_{-\alpha,\beta,-\gamma,\delta}$$
$$+(\delta\sqrt{a}+\alpha\sqrt{b})(\alpha\beta\gamma\sqrt{a}-\beta\sqrt{b})(\sqrt{b}+\alpha\beta\sqrt{a})z^3_{\alpha,-\beta,-\gamma,\delta}$$
$$+(\alpha\beta\gamma\delta\sqrt{a}-\beta\gamma\sqrt{b})(-\alpha\sqrt{b}+\gamma\sqrt{a})(-\alpha\sqrt{b}+\beta\sqrt{a})z^3_{-\alpha,\beta,\gamma,-\delta}$$
$$+(\alpha\beta\gamma\delta\sqrt{a}-\beta\gamma\sqrt{b})(-\alpha\sqrt{b}+\gamma\sqrt{a})(\sqrt{b}+\alpha\beta\sqrt{a})z^3_{\alpha,-\beta,\gamma,-\delta}$$
$$+(\alpha\beta\gamma\delta\sqrt{a}-\beta\gamma\sqrt{b})(-\alpha\beta\gamma\sqrt{a}-\beta\sqrt{b})(\alpha\sqrt{b}+\beta\sqrt{a})z^3_{\alpha,\beta,-\gamma,-\delta}$$
$$+(\alpha\beta\gamma\delta\sqrt{a}-\beta\gamma\sqrt{b})(-\alpha\beta\gamma\sqrt{a}-\beta\sqrt{b})(\sqrt{b}-\alpha\beta\sqrt{a})z^3_{-\alpha,-\beta,-\gamma,-\delta}\Big).$$

Using the formulas for $z^r_{\alpha\beta\gamma\delta}$ and $y^r_{\alpha\beta\gamma\delta}$ given above and considering the coefficient of $z_{\alpha\beta\gamma\delta}$ in the last expression we conclude that $p=3$. Now it is easy to check that in the case $p=3$ the equality above does hold, and, moreover, all other vectors $z^r_{\alpha\beta\gamma\delta}$ and $y^r_{\alpha\beta\gamma\delta}$ are s_3-invariant. In fact, for $p=3$ we get the following formulas:

$$s_1y^1_{\alpha\beta\gamma\delta}=y^2_{\alpha\beta\gamma\delta},\ s_1z^1_{\alpha\beta\gamma\delta}=z^2_{\alpha\beta\gamma\delta},\ s_1y^2_{\alpha\beta\gamma\delta}=y^1_{\alpha\beta\gamma\delta},\ s_1z^2_{\alpha\beta\gamma\delta}=z^1_{\alpha\beta\gamma\delta},$$
$$s_1y^3_{\alpha\beta\gamma\delta}=z^3_{\alpha\beta\gamma\delta},\ s_1z^3_{\alpha\beta\gamma\delta}=y^3_{\alpha\beta\gamma\delta},\ s_2y^1_{\alpha\beta\gamma\delta}=z^1_{\alpha\beta\gamma\delta},\ s_2z^1_{\alpha\beta\gamma\delta}=y^1_{\alpha\beta\gamma\delta},$$
$$s_2y^2_{\alpha\beta\gamma\delta}=y^3_{\alpha\beta\gamma\delta},\ s_2z^2_{\alpha\beta\gamma\delta}=z^3_{\alpha\beta\gamma\delta},\ s_2y^3_{\alpha\beta\gamma\delta}=y^2_{\alpha\beta\gamma\delta},\ s_2z^3_{\alpha\beta\gamma\delta}=z^2_{\alpha\beta\gamma\delta},$$

$$s_3y^1_{\alpha\beta\gamma\delta}=-\alpha\sqrt{2}y^1_{\alpha\beta\gamma\delta}-\sqrt{2}y^1_{-\alpha,-\beta,\gamma,\delta}-z^1_{\alpha\beta\gamma\delta}-\alpha z^1_{-\alpha,-\beta,\gamma,\delta}$$
$$-\alpha\beta z^1_{-\alpha,\beta,-\gamma,\delta}+\beta z^1_{\alpha,-\beta,-\gamma,\delta}-z^3_{\alpha\beta\gamma\delta}-\alpha z^3_{-\alpha,-\beta,\gamma,\delta}$$
$$-\alpha\beta\gamma z^3_{-\alpha,\beta,\gamma,-\delta}+\beta\gamma z^3_{\alpha,-\beta,\gamma,-\delta}+\alpha\sqrt{2}y^3_{\alpha\beta\gamma\delta}+\sqrt{2}y^3_{-\alpha,-\beta,\gamma,\delta}$$
$$+\beta\sqrt{2}y^3_{-\alpha,\beta,-\gamma,\delta}-\alpha\beta\sqrt{2}y^3_{\alpha,-\beta,-\gamma,\delta}-\beta\gamma\sqrt{2}y^3_{-\alpha,\beta,\gamma,-\delta}$$
$$+\alpha\beta\gamma\sqrt{2}y^3_{\alpha,-\beta,\gamma,-\delta}+\alpha\gamma\sqrt{2}y^3_{\alpha,\beta,-\gamma,-\delta}+\gamma\sqrt{2}y^3_{-\alpha,-\beta,-\gamma,-\delta},$$
$$s_3z^1_{\alpha\beta\gamma\delta}=-\alpha\sqrt{2}z^1_{\alpha\beta\gamma\delta}+\beta\sqrt{2}z^1_{-\alpha,\beta,-\gamma,\delta}-y^3_{\alpha\beta\gamma\delta}+\alpha\beta y^3_{-\alpha,\beta,-\gamma,\delta}$$
$$-\alpha\beta\gamma y^3_{-\alpha,\beta,\gamma,-\delta}-\gamma y^3_{\alpha,\beta,-\gamma,-\delta},$$
$$s_3y^2_{\alpha\beta\gamma\delta}=-\alpha\sqrt{2}y^2_{\alpha\beta\gamma\delta}-\sqrt{2}y^2_{-\alpha,-\beta,\gamma,\delta}-z^2_{\alpha\beta\gamma\delta}-\alpha z^2_{-\alpha,-\beta,\gamma,\delta}$$
$$-\alpha\beta z^2_{-\alpha,\beta,-\gamma,\delta}+\beta z^2_{\alpha,-\beta,-\gamma,\delta}-y^3_{\alpha\beta\gamma\delta}-\alpha y^3_{-\alpha,-\beta,\gamma,\delta}$$
$$-\alpha\beta\gamma y^3_{-\alpha,\beta,\gamma,-\delta}+\beta\gamma y^3_{\alpha,-\beta,\gamma,-\delta}+\alpha\sqrt{2}z^3_{\alpha\beta\gamma\delta}+\sqrt{2}z^3_{-\alpha,-\beta,\gamma,\delta}$$

$$+\beta\sqrt{2}z^3_{-\alpha,\beta,-\gamma,\delta} - \alpha\beta\sqrt{2}z^3_{\alpha,-\beta,-\gamma,\delta} - \beta\gamma\sqrt{2}z^3_{-\alpha,\beta,\gamma,-\delta}$$

$$+\alpha\beta\gamma\sqrt{2}z^3_{\alpha,-\beta,\gamma,-\delta} + \alpha\gamma\sqrt{2}z^3_{\alpha,\beta,-\gamma,-\delta} + \gamma\sqrt{2}z^3_{-\alpha,-\beta,-\gamma,-\delta},$$

$$s_3 z^2_{\alpha\beta\gamma\delta} = -\alpha\sqrt{2}z^2_{\alpha\beta\gamma\delta} + \beta\sqrt{2}z^2_{-\alpha,\beta,-\gamma,\delta} - z^3_{\alpha\beta\gamma\delta} + \alpha\beta z^3_{-\alpha,\beta,-\gamma,\delta}$$

$$-\alpha\beta\gamma z^3_{-\alpha,\beta,\gamma,-\delta} - \gamma z^3_{\alpha,\beta,-\gamma,-\delta},$$

$$s_3 y^3_{\alpha\beta\gamma\delta} = -\alpha\sqrt{2}y^3_{\alpha\beta\gamma\delta} + \beta\gamma\sqrt{2}y^3_{-\alpha,\beta,\gamma,-\delta},$$

$$s_3 z^3_{\alpha\beta\gamma\delta} = -\alpha\sqrt{2}z^3_{\alpha\beta\gamma\delta} + \beta\gamma\sqrt{2}z^3_{-\alpha,\beta,\gamma,-\delta}.$$

□

Lemma 18.2.5 *Let $p = 3$ and $(i, j) = (0, 1)$. Then N_4 is s_4-invariant and N_5 is not s_5-invariant.*

Proof By now you have either got the point or decided to skip these calculations altogether. Anyway, we proceed as above using formulas at the end of the proof of Lemma 18.2.4. □

18.3 Higher crystal operators

Lemma 18.3.1 *Let $i, j \in I$ with $i \neq j$. For any $r, s \geq 0$ with $r + s = -\langle h_i, \alpha_j \rangle$, and $m \geq 0$,*

$$\operatorname{ind} L(i^r, j, i^s) \circledast L(i^m) \cong \operatorname{ind} L(i^m) \circledast L(i^r, j, i^s)$$

is irreducible.

Proof We may assume that $i - j = \pm 1$, as otherwise the result is immediate from Theorem 18.1.4. Now, we claim that

$$\operatorname{ind} L(i^r, j, i^s) \circledast L(i^m) \cong \operatorname{ind} L(i^m) \circledast L(i^r, j, i^s).$$

Indeed, transitivity of induction and Theorem 18.2.1(ii) give that

$$\begin{aligned}\operatorname{ind} L(i^r, j, i^s) \circledast L(i^m) &\cong \operatorname{ind} L(i^r, j, i^s) \circledast L(i) \circledast L(i^{m-1}) \\ &\cong \operatorname{ind} L(i) \circledast L(i^r, j, i^s) \circledast L(i^{m-1}),\end{aligned}$$

and now repeating this argument $(m-1)$ more times gives the claim.

Hence, by Corollary 17.3.2 and Theorem 14.7.1,

$$K := \operatorname{ind} L(i^r, j, i^s) \circledast L(i^m)$$

is self-dual. Now suppose for a contradiction that K is reducible. Then we can pick a proper irreducible submodule S of K, and set $Q := K/S$. By Theorem 18.2.1(i) and the Shuffle Lemma, $\operatorname{ch} K$ equals

$$\sum_{t=0}^{m} \binom{m}{t} (r+t)!(s+m-t)![L(i)^{\circledast(r+t)} \circledast L(j) \circledast L(i)^{\circledast(s+m-t)}].$$

By Frobenius reciprocity, Q contains an $\mathcal{X}_{r+s+1,m}$-submodule isomorphic to $L(i^r, j, i^s) \circledast L(i^m)$. Hence by Lemma 18.1.1(i), the irreducible $\mathcal{A}_{r+s+m+1}$-supermodule

$$V := L(i)^{\circledast r} \circledast L(j) \circledast L(i)^{\circledast(s+m)}$$

appears in Q as a submodule. It now follows from Theorem 16.3.3(i) and the above formula for $\operatorname{ch} K$ that V must appear with multiplicity $r!(s+m)!$ (viewing Q as a module over $\mathcal{X}_{r,1,s+m}$). It follows that V is not a composition factor of S. But this is a contradiction, since, as K is self-dual, $S \cong S^\tau$ is a quotient module of K and hence must contain V by the argument above applied to the quotient $Q = S$. □

Lemma 18.3.2 *Let $i, j \in I$ with $i \neq j$, and m be a non-negative integer. For any $r \geq 1$, $s \geq 0$ with $r + s = -\langle h_i, \alpha_j \rangle$ and any irreducible supermodule $M \in \operatorname{Rep}_I \mathcal{X}_n$,*

$$\operatorname{hd} \operatorname{ind} M \circledast L(i^m) \circledast L(i^r, j, i^s)$$

is irreducible.

Proof By the argument as in the proof of Lemma 5.1.5, it suffices to prove this in the special case that $\varepsilon_i(M) = 0$. Let $t = m + r + s + 1$. Recall from the previous lemma that

$$N := \operatorname{ind} L(i^m) \circledast L(i^r, j, i^s)$$

is an irreducible $\mathcal{X}_t$-supermodule. Moreover by Theorem 18.2.1(i),

$$\operatorname{ch} L(i^r, j, i^s) = (r!)(s!)[L(i)^{\circledast r} \circledast L(j) \circledast L(i)^{\circledast s}].$$

So, since $\varepsilon_i(M) = 0$ and $r > 0$, the Mackey Theorem and a block argument imply that

$$\operatorname{res}^{n+t}_{n,t} \operatorname{ind} M \circledast L(i^m) \circledast L(i^r, j, i^s) \cong (M \circledast N) \oplus U$$

for some $\mathcal{X}_{n,t}$-module U all of whose composition factors lie in different blocks to those of $M \circledast N$. Now let

$$H := \operatorname{hd} \operatorname{ind} M \circledast L(i^m) \circledast L(i^r, j, i^s).$$

It follows from above that

$$\mathrm{res}^{n+t}_{n,t} H \cong (M \circledast N) \oplus \bar{U}$$

where $\bar{U}$ is some quotient module of U. Then:

$$\begin{aligned}\mathrm{Hom}_{\mathcal{X}_{n+t}}(H, H) &\cong \mathrm{Hom}_{\mathcal{X}_{n+t}}(\mathrm{ind}\, M \circledast L(i^m) \circledast L(i^r, j, i^s), H) \\ &\cong \mathrm{Hom}_{\mathcal{X}_{n,t}}(M \circledast N, \mathrm{res}^{n+t}_{n,t} H) \\ &\cong \mathrm{Hom}_{\mathcal{X}_{n,t}}(M \circledast N, M \circledast N \oplus \bar{U}) \\ &\cong \mathrm{Hom}_{\mathcal{X}_{n,t}}(M \circledast N, M \circledast N).\end{aligned}$$

Since H is completely reducible and $M \circledast N$ is irreducible, this implies that H is irreducible, as required. □

Let $i, j \in I$ with $i \neq j$, and $r \geq 1$, $s \geq 0$ satisfy $r + s = -\langle h_i, \alpha_j \rangle$. Then the special case $m = 0$ of Lemma 18.3.2 shows that

$$\tilde{f}_{i^r j i^s} M := \mathrm{hd}\,\mathrm{ind}\, M \circledast L(i^r, j, i^s)$$

is irreducible for every irreducible $M \in \mathrm{Rep}_I\, \mathcal{X}_n$.

Lemma 18.3.3 *Take $i, j \in I$ with $i \neq j$ and set $k = -\langle h_i, \alpha_j \rangle$. Let $M \in \mathrm{Rep}_I\, \mathcal{X}_n$ be irreducible.*

(i) *There exists a unique integer r with $0 \leq r \leq k$ such that for every $m \geq 0$ we have*

$$\varepsilon_i(\tilde{f}_i^m \tilde{f}_j M) = m + \varepsilon_i(M) - r.$$

(ii) *Assume $m \geq k$. Then a copy of $\tilde{f}_i^m \tilde{f}_j M$ appears in the head of*

$$\mathrm{ind}\, (\tilde{f}_i^{m-k} M) \circledast L(i^r, j, i^{k-r}),$$

where r is as in (i). *In particular, if $r \geq 1$, then*

$$\tilde{f}_i^m \tilde{f}_j M \cong \tilde{f}_{i^r j i^{k-r}} \tilde{f}_i^{m-k} M.$$

Proof The proof is similar to that of Lemma 6.3.3, using (17.12), Theorem 18.2.1(i), Lemma 18.3.1, Lemma 17.1.5, and Lemma 18.3.2 instead of (5.12), Lemma 6.2.1, Lemma 6.3.1, Lemma 5.1.5, and Lemma 6.3.2, respectively. □

19
Operators e_i^λ and f_i^λ

This chapter is more or less parallel to Chapter 8. Some additional complications arise because the dimension of the $\mathcal{A}_1$-supermodule $L(i)$ is 2 rather than 1, and so sometimes we need to "halve" res_i^λ and ind_i^λ. Whether we need to halve them or not depends on the properties of the operation $\circledast$. Note also that our "halving" procedure is *not functorial* and so e_i^λ and f_i^λ end up being not functors, but rather functions from irreducible $\mathcal{X}_n^\lambda$-supermodules to *isomorphism classes* of supermodules over $\mathcal{X}_{n-1}^\lambda$ and $\mathcal{X}_{n+1}^\lambda$, respectively.

19.1 i-induction and i-restriction

Fix $\lambda \in P_+$. Recall the functors Δ_i from (17.1) and res_i from (17.6) for $i \in I$. Note if M is an $\mathcal{X}_n^\lambda$-supermodule, then $\text{res}_i M$ is automatically an $\mathcal{X}_{n-1}^\lambda$-supermodule. So the restriction of the functor res_i gives a functor

$$\text{res}_i^\lambda : \mathcal{X}_n^\lambda\text{-smod} \to \mathcal{X}_{n-1}^\lambda\text{-smod}. \tag{19.1}$$

There is an alternative definition of res_i^λ: if M is a supermodule in $\mathcal{X}_n^\lambda\text{-smod}[\gamma]$ for some fixed $\gamma = \sum_{j \in I} \gamma_j \alpha_j \in \Gamma_n$ then

$$\text{res}_i^\lambda M = \begin{cases} (\text{res}_{\mathcal{X}_{n-1}^\lambda}^{\mathcal{X}_n^\lambda} M)[\gamma - \alpha_i] & \text{if } \gamma_i > 0, \\ 0 & \text{if } \gamma_i = 0. \end{cases} \tag{19.2}$$

This description makes it clear how to define an analogous (additive) functor

$$\text{ind}_i^\lambda : \mathcal{X}_n^\lambda\text{-smod} \to \mathcal{X}_{n+1}^\lambda\text{-smod}. \tag{19.3}$$

Using (16.5) and additivity, it suffices to define this on an object M belonging to $\mathcal{X}_n^\lambda\text{-smod}[\gamma]$ for fixed $\gamma = \sum_{j \in I} \gamma_j \alpha_j \in \Gamma_n$. Then we set

$$\text{ind}_i^\lambda M = (\text{ind}_{\mathcal{X}_n^\lambda}^{\mathcal{X}_{n+1}^\lambda} M)[\gamma + \alpha_i]. \tag{19.4}$$

To complete the definition of the functor ind_i^λ, it is defined on a morphism f simply by restriction of the corresponding morphism $\mathrm{ind}_{\mathcal{X}_n^\lambda}^{\mathcal{X}_{n+1}^\lambda} f$. We stress that the functor ind_i^λ depends fundamentally on the fixed choice of λ, unlike res_i^λ, which is just the restriction of its affine counterpart res_i.

Lemma 19.1.1 *For $\lambda \in P_+$ and each $i \in I$:*

(i) ind_i^λ *and* res_i^λ *are both left and right adjoint to each other, hence they are exact and send projectives to projectives;*
(ii) ind_i^λ *and* res_i^λ *commute with duality, that is there are natural isomorphisms*

$$\mathrm{ind}_i^\lambda(M^\tau) \simeq (\mathrm{ind}_i^\lambda M)^\tau, \qquad \mathrm{res}_i^\lambda(M^\tau) \simeq (\mathrm{res}_i^\lambda M)^\tau$$

for each finite dimensional $\mathcal{X}_n^\lambda$-supermodule M.

Proof Similar to the proof of Lemma 8.2.2 using Corollary 15.6.5 instead of Corollary 7.7.5. □

By (17.8) and Lemma 19.1.1, we have that

$$\mathrm{res}_{\mathcal{X}_{n-1}^\lambda}^{\mathcal{X}_n^\lambda} M = \bigoplus_{i\in I} \mathrm{res}_i^\lambda M, \qquad \mathrm{ind}_{\mathcal{X}_n^\lambda}^{\mathcal{X}_{n+1}^\lambda} M = \bigoplus_{i\in I} \mathrm{ind}_i^\lambda M. \tag{19.5}$$

In order to give an alternative description of res_i^λ and ind_i^λ recall the $\mathcal{X}_1$-supermodules $\mathcal{R}_m(i)$ from (16.7). Let $\mathcal{X}_1' \cong \mathcal{X}_1$ denote the subalgebra of $\mathcal{X}_n$ generated by c_n, x_n, see (14.10). The limits in the next lemma are taken with respect to the systems induced by the maps (16.8).

Lemma 19.1.2 *For every finite dimensional $\mathcal{X}_n^\lambda$-supermodule M and $i \in I$, there are natural isomorphisms*

$$\mathrm{ind}_i^\lambda M \simeq \varprojlim \mathrm{pr}^\lambda \mathrm{ind}_{n,1}^{n+1} M \boxtimes \mathcal{R}_m(i),$$

$$\mathrm{res}_i^\lambda M \simeq \varinjlim \mathrm{pr}^\lambda \mathrm{Hom}_{\mathcal{X}_1'}(\mathcal{R}_m(i), M)$$

(in the second case, the $\mathcal{X}_{n-1}$-module structure is defined by $(hf)(r) = h(f(r))$ for $f \in \mathrm{Hom}_{\mathcal{X}_1'}(\mathcal{R}_m(i), M)$ and $r \in \mathcal{R}_m(i)$).

Proof For res_i^λ, it suffices to consider the effect on $M \in \mathcal{X}_n^\lambda\text{-smod}[\gamma]$ for $\gamma = \sum_{j\in I} \gamma_j \alpha_j \in \Gamma_n$ with $\gamma_i > 0$, both sides of what we are trying to prove clearly being zero if $\gamma_i = 0$. Then, for all sufficiently large m, Lemma 16.4.2 (in the special case where $n = 1$) implies that

$$\mathrm{Hom}_{\mathcal{X}_1'}(\mathcal{R}_m(i), M) \simeq (\mathrm{res}_{\mathcal{X}_{n-1}^\lambda}^{\mathcal{X}_n^\lambda} M)[\gamma - \alpha_i].$$

Hence,

$$\varinjlim \mathrm{pr}^\lambda \mathrm{Hom}_{\mathcal{X}_1'}(\mathcal{R}_m(i), M) \simeq (\mathrm{res}^{\mathcal{X}_n^\lambda}_{\mathcal{X}_{n-1}^\lambda} M)[\gamma - \alpha_i] = \mathrm{res}_i^\lambda M.$$

To deduce the statement about induction, it now suffices by uniqueness of adjoint functors to show that $\varprojlim \mathrm{pr}^\lambda \mathrm{ind}_{n,1}^{n+1} ? \boxtimes \mathcal{R}_m(i)$ is left adjoint to $\varinjlim \mathrm{pr}^\lambda \mathrm{Hom}_{\mathcal{X}_1'}(\mathcal{R}_m(i), ?)$. Let $N \in \mathcal{X}_{n-1}^\lambda$-smod and $M \in \mathcal{X}_n^\lambda$-smod. First observe as explained in the previous paragraph that the direct system

$$\mathrm{pr}^\lambda \mathrm{Hom}_{\mathcal{X}_1'}(\mathcal{R}_1(i), M) \hookrightarrow \mathrm{pr}^\lambda \mathrm{Hom}_{\mathcal{X}_1'}(\mathcal{R}_2(i), M) \hookrightarrow \dots$$

stabilizes after finitely many terms. We claim that the inverse system

$$\mathrm{pr}^\lambda \mathrm{ind}_{n,1}^{n+1} N \boxtimes \mathcal{R}_1(i) \twoheadleftarrow \mathrm{pr}^\lambda \mathrm{ind}_{n,1}^{n+1} N \boxtimes \mathcal{R}_2(i) \twoheadleftarrow \dots$$

also stabilizes after finitely many terms. To see this, it suffices to show that the dimension of $\mathrm{pr}^\lambda \mathrm{ind}_{n,1}^{n+1} N \boxtimes \mathcal{R}_m(i)$ is bounded above independently of m. Well, each $\mathcal{R}_m(i)$ is generated as an $\mathcal{X}_1'$-module by a subspace W isomorphic (as a vector space) to the head $\mathcal{R}_1(i)$ of $\mathcal{R}_m(i)$. Then $\mathrm{ind}_{n,1}^{n+1} N \boxtimes \mathcal{R}_m(i)$ is generated as an $\mathcal{X}_{n+1}$-supermodule by the subspace $W' = 1 \otimes (N \otimes W)$, also of dimension independent of m. Finally, $\mathrm{pr}^\lambda \mathrm{ind}_{n,1}^{n+1} N \boxtimes \mathcal{R}_m(i)$ is a quotient of the vector space $\mathcal{X}_{n+1}^\lambda \otimes_F W'$, whose dimension is independent of m.

Now we can complete the proof of adjointness. Using the fact from the previous paragraph that the direct and inverse systems stabilize after finitely many terms, we have natural isomorphisms

$$\begin{aligned}
&\mathrm{Hom}_{\mathcal{X}_n^\lambda}(\varprojlim \mathrm{pr}^\lambda \mathrm{ind}_{n-1,1}^n N \boxtimes \mathcal{R}_m(i), M) \\
&\simeq \varinjlim \mathrm{Hom}_{\mathcal{X}_n^\lambda}(\mathrm{pr}^\lambda \mathrm{ind}_{n-1,1}^n N \boxtimes \mathcal{R}_m(i), M) \\
&\simeq \varinjlim \mathrm{Hom}_{\mathcal{X}_n}(\mathrm{ind}_{n-1,1}^n N \boxtimes \mathcal{R}_m(i), M) \\
&\simeq \varinjlim \mathrm{Hom}_{\mathcal{X}_{n-1,1}}(N \boxtimes \mathcal{R}_m(i), \mathrm{res}_{n-1,1}^n M) \\
&\simeq \varinjlim \mathrm{Hom}_{\mathcal{X}_{n-1}}(N, \mathrm{Hom}_{\mathcal{X}_1'}(\mathcal{R}_m(i), M)) \\
&\simeq \varinjlim \mathrm{Hom}_{\mathcal{X}_{n-1}^\lambda}(N, \mathrm{pr}^\lambda \mathrm{Hom}_{\mathcal{X}_1'}(\mathcal{R}_m(i), M)) \\
&\simeq \mathrm{Hom}_{\mathcal{X}_{n-1}^\lambda}(N, \varinjlim \mathrm{pr}^\lambda \mathrm{Hom}_{\mathcal{X}_1'}(\mathcal{R}_m(i), M)).
\end{aligned}$$

This completes the argument. □

19.2 Operators e_i^λ and f_i^λ

We wish to refine the functors res_i^λ (resp. ind_i^λ) to give operators, denoted e_i^λ (resp. f_i^λ) from *irreducible* $\mathcal{X}_n^\lambda$-supermodules to *isomorphism classes* of $\mathcal{X}_{n-1}^\lambda$- (resp. $\mathcal{X}_{n+1}^\lambda$-) supermodules.

Actually, e_i^λ is the restriction of an operator denoted e_i on the irreducible supermodules in $\mathrm{Rep}_I \mathcal{X}_n$, which we define first. Recall the definition of $L_m(i)$ from Section 16.4. Let $M \in \mathrm{Rep}_I \mathcal{X}_n$ be irreducible. For each $m \geq 1$, we define an $\mathcal{X}_{n-1}$-supermodule

$$\overline{\mathrm{Hom}}_{\mathcal{X}_1'}(L_m(i), M) \tag{19.6}$$

as follows. If M is of type $\mathtt{M}$ or $i \neq 0$, this is simply $\mathrm{Hom}_{\mathcal{X}_1'}(L_m(i), M)$ viewed as an $\mathcal{X}_{n-1}$-module in the same way as in Lemma 19.1.2. But if M is of type $\mathtt{Q}$ and $i = 0$, we can pick an odd involution $\theta_M : M \to M$ and also have the odd involutions $\theta_m : L_m(i) \to L_m(i)$ from (16.10). Let

$$\theta_M \otimes \theta_m : \mathrm{Hom}_{\mathcal{X}_1'}(L_m(i), M) \to \mathrm{Hom}_{\mathcal{X}_1'}(L_m(i), M)$$

denote the map defined by $((\theta_M \otimes \theta_m)f)(v) = (-1)^{\bar{f}}\theta_M(f(\theta_m v))$. It is easy to check that $(\theta_M \otimes \theta_m)^2 = 1$, hence the ± 1-eigenspaces of $\theta_M \otimes \theta_m$ split $\mathrm{Hom}_{\mathcal{X}_1'}(L_m(i), M)$ into a direct sum of two isomorphic $\mathcal{X}_{n-1}$-supermodules (there is an obvious odd automorphism swapping the two eigenspaces). Now in this case, we define $\overline{\mathrm{Hom}}_{\mathcal{X}_1'}(L_m(i), M)$ to be the 1-eigenspace (say).

In either case, we have a direct system

$$\overline{\mathrm{Hom}}_{\mathcal{X}_1'}(L_1(i), M) \hookrightarrow \overline{\mathrm{Hom}}_{\mathcal{X}_1'}(L_2(i), M) \hookrightarrow \dots$$

induced by the inverse system (16.9). Now define

$$e_i M := \varinjlim \overline{\mathrm{Hom}}_{\mathcal{X}_1'}(L_m(i), M). \tag{19.7}$$

Note if M is an $\mathcal{X}_n^\lambda$-supermodule then each $\overline{\mathrm{Hom}}_{\mathcal{X}_1'}(L_m(i), M)$ is an $\mathcal{X}_{n-1}^\lambda$-supermodule, so

$$e_i M = \varinjlim \mathrm{pr}^\lambda \overline{\mathrm{Hom}}_{\mathcal{X}_1'}(L_m(i), M). \tag{19.8}$$

We take (19.8) as our definition of the operator e_i^λ in the cyclotomic case. Comparing (19.7) and (19.8) with Lemma 19.1.2 and using (16.12), we get:

Lemma 19.2.1 *Let $i \in I$ and $M \in \mathrm{Rep}_I \mathcal{X}_n$ or $M \in \mathcal{X}_n^\lambda$-smod be an irreducible supermodule. Then*

$$\mathrm{res}_i^{(\lambda)} M \simeq \begin{cases} e_i^{(\lambda)} M & \text{if } i = 0 \text{ and } M \text{ is of type } \mathtt{M}, \\ e_i^{(\lambda)} M \oplus \Pi e_i^{(\lambda)} M & \text{otherwise.} \end{cases}$$

The action of e_i and e_i^λ on the level of characters is similar to the classical case (cf. (5.9)):

Corollary 19.2.2 *Let $M \in \mathrm{Rep}_I\, \mathcal{X}_n$ or $M \in \mathcal{X}_n^\lambda$-smod be an irreducible supermodule, and*

$$\mathrm{ch}\, M = \sum_{\underline{i} \in I^n} c_{\underline{i}} [L(i_1) \circledast \cdots \circledast L(i_n)].$$

Then

$$\mathrm{ch}\,(e_i^{(\lambda)} M) = \sum_{\underline{i} \in I^{n-1}} c_{(i_1,\dots,i_{n-1},i)} [L(i_1) \circledast \cdots \circledast L(i_{n-1})].$$

Now we turn to the definition of $f_i^\lambda M$ which, just like $\mathrm{ind}_i^\lambda M$, only makes sense in the cyclotomic case. Let M be an irreducible $\mathcal{X}_n^\lambda$-supermodule. We need to extend the definition of the operation $\circledast$ to give meaning to the notation $M \circledast L_m(i)$, for each $m \geq 1$. If either M is of type $\mathtt{M}$ or $i \neq 0$, then $M \circledast L_m(i) := M \boxtimes L_m(i)$. But if M is of type $\mathtt{Q}$ and $i = 0$, pick an odd involution $\theta_M : M \to M$. Then the $\pm\sqrt{-1}$-eigenspaces of $\theta_M \otimes \theta_m$ acting on the left of $M \boxtimes L_m(i)$ split it into a direct sum of two isomorphic $\mathcal{X}_{n,1}$-supermodules. Let $M \circledast L_m(i)$ denote the $\sqrt{-1}$-eigenspace (say) for each m.

We then have an inverse system

$$M \circledast L_1(i) \twoheadleftarrow M \circledast L_2(i) \twoheadleftarrow \dots$$

of $\mathcal{X}_{n,1}$-supermodules induced by the maps from (16.9). Now we can define

$$f_i^\lambda M = \varprojlim \mathrm{pr}^\lambda \mathrm{ind}_{n,1}^{n+1} M \circledast L_m(i). \tag{19.9}$$

Comparing the definition with the proof of Lemma 19.1.2, we see that the inverse limit stabilizes after finitely many steps, hence that $f_i^\lambda M$ is a well-defined finite dimensional $\mathcal{X}_{n+1}^\lambda$-supermodule. In fact, Lemma 19.1.2 and (16.12) imply:

Lemma 19.2.3 *Let $i \in I$ and M be an irreducible $\mathcal{X}_n^\lambda$-supermodule. Then*

$$\mathrm{ind}_i^\lambda M \simeq \begin{cases} f_i^\lambda M & \text{if } i = 0 \text{ and } M \text{ is of type } \mathtt{M}, \\ f_i^\lambda M \oplus \Pi f_i^\lambda M & \text{otherwise.} \end{cases}$$

Lemma 19.2.4 *Let $i \in I$ and $M \in \mathrm{Rep}_I\, \mathcal{X}_n$ be irreducible. Then $e_i M$ is non-zero if and only if $\tilde{e}_i M \neq 0$, in which case it is a self-dual indecomposable supermodule with irreducible socle and head isomorphic to $\tilde{e}_i M$.*

Proof To see that $e_i M$ has irreducible socle $\tilde{e}_i M$ whenever it is non-zero, combine Lemma 19.2.1 with Corollary 17.1.8. The remaining facts follow since M is self-dual by Lemma 17.3.2, and res_i commutes with duality by Lemma 19.1.1(ii). □

Now we define the *cyclotomic crystal operators* on irreducible supermodules

$$\tilde{e}_i^\lambda = \mathrm{pr}^\lambda \circ \tilde{e}_i \circ \mathrm{infl}^\lambda, \tag{19.10}$$

$$\tilde{f}_i^\lambda = \mathrm{pr}^\lambda \circ \tilde{f}_i \circ \mathrm{infl}^\lambda \tag{19.11}$$

for each $i \in I$ and $\lambda \in P_+$. Let $B(\infty)$ and $B(\lambda)$ denote the set of isomorphism classes of irreducible supermodules in $\mathrm{Rep}_I \mathcal{X}_n$ and $\mathcal{X}_n^\lambda$-smod for all $n \geq 0$, respectively. Note that we may consider $B(\lambda)$ as a subset of $B(\infty)$ via inflation. We have maps

$$\tilde{e}_i : B(\infty) \to B(\infty) \cup \{0\},$$

$$\tilde{f}_i : B(\infty) \to B(\infty),$$

$$\tilde{e}_i^\lambda, \tilde{f}_i^\lambda : B(\lambda) \to B(\lambda) \cup \{0\}.$$

The reader is advised to consult with Remark 8.2.4 at this point.

Theorem 19.2.5 *Let $\lambda \in P_+$ and $i \in I$. Then for any irreducible $\mathcal{X}_n^\lambda$-supermodule M:*

(i) *$e_i^\lambda M$ is non-zero if and only if $\tilde{e}_i^\lambda M \neq 0$, in which case it is a self-dual indecomposable supermodule with irreducible socle and head isomorphic to $\tilde{e}_i^\lambda M$;*
(ii) *$f_i^\lambda M$ is non-zero if and only if $\tilde{f}_i^\lambda M \neq 0$, in which case it is a self-dual indecomposable supermodule with irreducible socle and head isomorphic to $\tilde{f}_i^\lambda M$.*

Proof (i) This is immediate from Lemma 19.2.4.

(ii) Let N be an irreducible $\mathcal{X}_{n+1}^\lambda$-supermodule. Let δ_M equal 1 if $i = 0$ and M is of type M, 2 otherwise, and define δ_N similarly. Then, by Lemmas 19.1.1(i), 19.2.1, and 19.2.3,

$$\begin{aligned}\dim \mathrm{Hom}_{\mathcal{X}_{n+1}^\lambda}(f_i^\lambda M, N) &= \frac{1}{\delta_M} \dim \mathrm{Hom}_{\mathcal{X}_{n+1}^\lambda}(\mathrm{ind}_i^\lambda M, N) \\ &= \frac{1}{\delta_M} \dim \mathrm{Hom}_{\mathcal{X}_n^\lambda}(M, \mathrm{res}_i^\lambda N) \\ &= \frac{\delta_N}{\delta_M} \dim \mathrm{Hom}_{\mathcal{X}_n^\lambda}(M, e_i^\lambda N).\end{aligned}$$

By (i), the latter is zero unless $M = \tilde{e}_i N$, or equivalently $N = \tilde{f}_i^\lambda M$ by Lemma 17.2.2. Taking into account Lemmas 12.2.3 and 17.3.4, we deduce that hd $f_i^\lambda M \cong \tilde{f}_i^\lambda M$. Finally, note $f_i^\lambda M$ is self-dual by Lemma 19.1.1(ii) so everything else follows. □

Remark 19.2.6 Let us also point out, as follows easily from the definitions, that $e_i^\lambda M$ and $f_i^\lambda M$ admit odd involutions if either $i \neq 0$ and M is of type $\mathtt{Q}$, or $i = 0$ and M is of type $\mathtt{M}$.

19.3 Divided powers

We recall the covering modules $L_m(i^r)$ from Section 16.4 and the embedding (14.10). Let M be an irreducible supermodule in $\mathrm{Rep}_I\,\mathcal{X}_n$. If $r > n$, we set $e_i^{(r)} M = 0$. Otherwise, we have a direct system

$$\overline{\mathrm{Hom}}_{\mathcal{X}'_r}(L_1(i^r), M) \hookrightarrow \overline{\mathrm{Hom}}_{\mathcal{X}'_r}(L_2(i^r), M) \hookrightarrow \dots$$

induced by the inverse system (16.9), where $\overline{\mathrm{Hom}}$ is defined in exactly the same way as in Section 19.2 using the generalized maps θ_m from (16.10) in case $i = 0$. Now define

$$e_i^{(r)} M = \varinjlim \overline{\mathrm{Hom}}_{\mathcal{X}'_r}(L_m(i^r), M). \tag{19.12}$$

As in Section 19.2, if M is an irreducible $\mathcal{X}_n^\lambda$-supermodule then $e_i^{(r)} M$ is an $\mathcal{X}_{n-r}^\lambda$-supermodule, so that

$$e_i^{(r)} M = \varinjlim \mathrm{pr}^\lambda \overline{\mathrm{Hom}}_{\mathcal{X}'_1}(L_m(i^r), M). \tag{19.13}$$

We take (19.13) as our definition of $(e_i^\lambda)^{(r)}$ in the cyclotomic case.

To define $(f_i^\lambda)^{(r)}$, which as usual only makes sense in the cyclotomic case, let M be an irreducible $\mathcal{X}_n^\lambda$-supermodule. We have an inverse system

$$M \circledast L_1(i^r) \twoheadleftarrow M \circledast L_2(i^r) \twoheadleftarrow \dots$$

of $\mathcal{X}_{n,r}$-supermodules induced by the maps from (16.9), again interpreting $\circledast$ as in Section 19.2. Now define

$$(f_i^\lambda)^{(r)} M = \varprojlim \mathrm{pr}^\lambda \mathrm{ind}_{n,r}^{n+r} M \circledast L_m(i^r). \tag{19.14}$$

Lemma 19.3.1 *Let $i \in I$, $r \geq 1$ and M be an irreducible supermodule in $\mathcal{X}_n^\lambda$-smod. Then $(\mathrm{res}_i^\lambda)^r M$ is evenly isomorphic to*

$$\begin{cases} ((e_i^\lambda)^{(r)} M \oplus \Pi (e_i^\lambda)^{(r)} M)^{\oplus 2^{r-1}(r!)} & \text{if } i \neq 0, \\ (e_i^\lambda)^{(r)} M^{\oplus 2^{(r-1)/2}(r!)} & \text{if } i = 0,\ r \text{ is odd, and} \\ & \qquad M \text{ is of type } \mathtt{M}, \\ ((e_i^\lambda)^{(r)} M \oplus \Pi (e_i^\lambda)^{(r)} M)^{\oplus 2^{\lfloor (r-1)/2 \rfloor}(r!)} & \text{otherwise;} \end{cases}$$

and $(\mathrm{ind}_i^\lambda)^r M$ *is evenly isomorphic to*

$$\begin{cases} ((f_i^\lambda)^{(r)} M \oplus \Pi (f_i^\lambda)^{(r)} M)^{\oplus 2^{r-1}(r!)} & \text{if } i \neq 0, \\ (f_i^\lambda)^{(r)} M^{\oplus 2^{(r-1)/2}(r!)} & \text{if } i = 0,\ r \text{ is odd, and} \\ & \quad M \text{ is of type } \mathtt{M}, \\ ((f_i^\lambda)^{(r)} M \oplus \Pi (f_i^\lambda)^{(r)} M)^{\oplus 2^{\lfloor (r-1)/2 \rfloor}(r!)} & \text{otherwise.} \end{cases}$$

Proof Using Lemma 16.4.1 and the definitions, it suffices to show that

$$(\mathrm{res}_i^\lambda)^r M \simeq \varinjlim \mathrm{pr}^\lambda \mathrm{Hom}_{\mathcal{X}_r'}(\mathcal{R}_m(i^r), M),$$
$$(\mathrm{ind}_i^\lambda)^r M \simeq \varprojlim \mathrm{pr}^\lambda \mathrm{ind}_{n,r}^{n+r} M \boxtimes \mathcal{R}_m(i^r).$$

For $(\mathrm{res}_i^\lambda)^r$, this follows from Lemma 16.4.2 in exactly the same way as in the proof of Lemma 19.1.2. Now $(\mathrm{ind}_i^\lambda)^r$ is left adjoint to $(\mathrm{res}_i^\lambda)^r$, so the statement for induction follows from uniqueness of adjoint functors on checking that the functor $\varprojlim \mathrm{pr}^\lambda \mathrm{ind}_{n,r}^{n+r} ? \boxtimes \mathcal{R}_m(i^r)$ is left adjoint to $\varinjlim \mathrm{pr}^\lambda \mathrm{Hom}_{\mathcal{X}_r'}(\mathcal{R}_m(i^r), ?)$. The latter follows as in the proof of Lemma 19.1.2. □

Since we have defined the operators $e_i^{(r)}$ and $(e_i^{(\lambda)})^{(r)}$ on irreducible supermodules we get induced operators also denoted $e_i^{(r)}$ and $(e_i^{(\lambda)})^{(r)}$ at the level of Grothendieck groups, namely,

$$e_i^{(r)} : K(\mathrm{Rep}_I \mathcal{X}_n) \to K(\mathrm{Rep}_I \mathcal{X}_{n-r}), \text{ and}$$
$$(e_i^\lambda)^{(r)} : K(\mathcal{X}_n^\lambda\text{-smod}) \to K(\mathcal{X}_{n-r}^\lambda\text{-smod}),$$

in the affine and cyclotomic cases respectively. Similarly $(f_i^\lambda)^{(r)}$ induces an operator

$$(f_i^\lambda)^{(r)} : K(\mathcal{X}_n^\lambda\text{-smod}) \to K(\mathcal{X}_{n+r}^\lambda\text{-smod})$$

on Grothendieck groups.

Lemma 19.3.2 *As operators on the Grothendieck groups, we have that* $e_i^r = (r!)e_i^{(r)}$, $(e_i^\lambda)^r = (r!)(e_i^\lambda)^{(r)}$ *and* $(f_i^\lambda)^r = (r!)(f_i^\lambda)^{(r)}$.

Proof If $i \neq 0$, by Lemmas 19.2.1 and 19.2.3, we have that $(\mathrm{res}_i^\lambda)^r = 2^r (e_i^\lambda)^r$, and by Lemma 19.3.1, we have that $(\mathrm{res}_i^\lambda)^r = 2^r (r!)(e_i^\lambda)^{(r)}$, as operators on the Grothendieck group. The result for e follows, the proof for f being similar.

Now, let $i = 0$. The idea is the same here as for $i \neq 0$, but we have to check several cases. We explain one of them. Let r be odd and $M \in \mathcal{X}_n^\lambda$-smod be irreducible of type M. By Lemmas 19.2.1, 19.2.3, 19.3.1, and 17.3.4 we get

$$[(\mathrm{res}_i^\lambda)^r M] = 2^{(r-1)/2}[(e_i^\lambda)^r M] = 2^{(r-1)/2}(r!)[(e_i^\lambda)^{(r)} M],$$

and the result for e follows in this case. □

Let us finally note that we have only defined the operators $(e_i^\lambda)^{(r)}$ and $(f_i^\lambda)^{(r)}$ on irreducible supermodules. However, the definitions could be made more generally on pairs (M, θ_M), where M is an $\mathcal{X}_n^\lambda$-supermodule (or an integral $\mathcal{X}_n$-supermodule in the case of e_i) and $\theta_M : M \to M$ is either the identity map or else an odd involution of M. In case $\theta_M = \mathrm{id}_M$, the definitions of $(e_i^\lambda)^{(r)} M$ and $(f_i^\lambda)^{(r)} M$ are exactly the same as in the case where M is irreducible of type M above. In the case where θ_M is an odd involution, the definitions of $(e_i^\lambda)^{(r)} M$ and $(f_i^\lambda)^{(r)} M$ are exactly the same as in the case where M is irreducible of type Q above, substituting the given map θ_M for the canonical odd involution of M in the situation above.

This remark applies especially to give us $(e_i^\lambda)^{(r)} P_M$, $(f_i^\lambda)^{(r)} P_M$, where P_M is the projective cover of an irreducible $\mathcal{X}_n^\lambda$-supermodule: in this case, if M is of type Q, the odd involution θ_M of M lifts to a unique odd involution also denoted θ_M of the projective cover, see Lemma 12.2.16. On doing this, we have that

$$[(e_i^\lambda)^{(r)} P_M] = (e_i^\lambda)^{(r)}[P_M], \quad [(f_i^\lambda)^{(r)} P_M] = (f_i^\lambda)^{(r)}[P_M] \tag{19.15}$$

where the equalities are written in $K(\mathcal{X}_{n-r}^\lambda\text{-smod})$ and $K(\mathcal{X}_{n+r}^\lambda\text{-smod})$ respectively. To prove this, we need to observe that all composition factors of P_M are of the same type as M by Lemma 17.3.4.

Note Lemma 19.3.1 is also true if M is replaced by its projective cover P_M, the proof being the same as above. In particular, this shows that $(e_i^\lambda)^{(r)} P_M$ is a summand of $(\mathrm{res}_i^\lambda)^r P_M$, and similarly for (f_i^λ). So Lemma 19.1.1(i) gives that $(e_i^\lambda)^{(r)} P_M$ and $(f_i^\lambda)^{(r)} P_M$ are also projective modules. Hence $(e_i^\lambda)^{(r)}$ and $(f_i^\lambda)^{(r)}$ induce operators with the same names on the Grothendieck groups of *projective* modules too:

$$(e_i^\lambda)^{(r)} : K(\mathcal{X}_n^\lambda\text{-proj}) \to K(\mathcal{X}_{n-r}^\lambda\text{-proj}),$$

$$(f_i^\lambda)^{(r)} : K(\mathcal{X}_n^\lambda\text{-proj}) \to K(\mathcal{X}_{n+r}^\lambda\text{-proj}).$$

Moreover, by the same argument as in the proof of Lemma 19.3.2, we have:

Lemma 19.3.3 *As operators on the Grothendieck group $K(\mathcal{X}_n^\lambda\text{-proj})$, we have that*

$$(e_i^\lambda)^r[P_M] = (r!)[(e_i^\lambda)^{(r)} P_M], \qquad (f_i^\lambda)^r[P_M] = (r!)[(f_i^\lambda)^{(r)} P_M],$$

for all irreducible $\mathcal{X}_n^\lambda$-modules M.

19.4 Alternative descriptions of ε_i

Theorem 19.4.1 *Let $i \in I$ and M be an irreducible supermodule in* $\mathrm{Rep}_I\, \mathcal{X}_n$, *Then:*

(i) $[e_iM] = \varepsilon_i(M)[\tilde{e}_iM] + \sum c_a[N_a]$, *where the N_a are irreducibles with* $\varepsilon_i(N_a) < \varepsilon_i(\tilde{e}_iM)$;

(ii) $\varepsilon_i(M)$ *is the maximal size of a Jordan block of x_n^2 (resp. x_n if $i = 0$) on M with eigenvalue $q(i)$;*

(iii) $\mathrm{End}_{\mathcal{X}_{n-1}}(e_iM) \simeq \mathrm{End}_{\mathcal{X}_{n-1}}(\tilde{e}_iM)^{\oplus \varepsilon_i(M)}$ *as vector superspaces.*

Proof Let $\varepsilon = \varepsilon_i(M)$.

(i) The proof is similar to that of Theorem 5.5.1(i), except that we use $\circledast$ and Lemmas 17.1.4, and 17.1.3 instead of $\boxtimes$ and Lemmas 5.1.4, and 5.1.3, respectively, to deduce

$$[\Delta_iM] = \varepsilon[\tilde{e}_iM \circledast L(i)] + \sum c_a[N_a \circledast L(i)]$$

for irreducibles N_a with $\varepsilon_i(N_a) < \varepsilon_i(\tilde{e}_iM)$. The conclusion follows on applying Lemma 19.2.1.

(ii) Similar to the proof of Theorem 5.5.1(ii) but using Lemma 16.3.2, instead of Lemma 4.3.1.

(iii) Similar to the proof of Theorem 5.5.1(iii), but more delicate, so we provide details. Let $z = (x_n^2 - q(i))$ if $i \neq 0$ and $z = x_n$ if $i = 0$. Consider the effect of left multiplication by z on the $\mathcal{X}_{n-1}$-supermodule $R := \mathrm{res}_i(M)$. Note R is equal to either e_iM or $e_iM \oplus \Pi e_iM$, by Lemma 19.2.1. In the latter case, x_n^2 (resp. x_n) acts as a scalar on $\mathrm{soc}\, R \simeq \tilde{e}_iM \oplus \Pi\tilde{e}_iM$, hence it leaves the two indecomposable summands invariant. This shows that in any case left multiplication by z (which centralizes the subalgebra $\mathcal{X}_{n-1}$ of $\mathcal{X}_n$) induces an $\mathcal{X}_{n-1}$-endomorphism

$$\theta : e_iM \to e_iM.$$

But by (ii), $\theta^{\varepsilon-1} \neq 0$ and $\theta^{\varepsilon} = 0$. Hence, $1, \theta, \ldots, \theta^{\varepsilon-1}$ give ε linearly independent even $\mathcal{X}_{n-1}$-endomorphisms of e_iM. In view of Remark 19.2.6, we automatically get from these ε linearly independent odd endomorphisms in the case where $\tilde{e}_iM$ is of type Q, so we have now shown that

$$\dim \mathrm{End}_{\mathcal{X}_{n-1}}(e_iM) \geq \varepsilon \dim \mathrm{End}_{\mathcal{X}_{n-1}}(\tilde{e}_iM).$$

However, e_iM has irreducible head $\tilde{e}_iM$, and this appears in e_iM with multiplicity ε by (i), so the reverse inequality also holds. ☐

Corollary 19.4.2 *Let M, N be irreducible $\mathcal{X}_n$-supermodules with $M \not\cong N$. Then, for every $i \in I$,* $\mathrm{Hom}_{\mathcal{X}_{n-1}}(e_i M, e_i N) = 0$.

Proof Similar to the proof of Corollary 5.5.2. □

19.5 The $*$-operation

Suppose M is an irreducible supermodule in $\mathrm{Rep}_I\, \mathcal{X}_n$ and $0 \le m \le n$. Using Lemma 14.6.1 for the second equality in (19.17), define

$$\tilde{e}_i^* M = (\tilde{e}_i(M^\sigma))^\sigma, \tag{19.16}$$

$$\tilde{f}_i^* M = (\tilde{f}_i(M^\sigma))^\sigma = \mathrm{hd}\,\mathrm{ind}_{1,n}^{n+1} L(i) \circledast M, \tag{19.17}$$

$$\varepsilon_i^*(M) = \varepsilon_i(M^\sigma) = \max\{m \ge 0 \mid (\tilde{e}_i^*)^m M \ne 0\}. \tag{19.18}$$

We may think of the "starred" notions as left-hand versions of the original notions which are right hand. For example, $\varepsilon_i^*(M)$ can be worked out just from knowledge of the character of M as the maximal k such that $[L(i)^{\circledast k} \circledast \dots]$ appears in $\mathrm{ch}\, M$, while for $\varepsilon_i(M)$ we would take $[\dots \circledast L(i)^{\circledast k}]$ here.

Recalling the definition (15.7) of pr^λ, Theorem 19.4.1(ii) has the following important corollary:

Corollary 19.5.1 *Let M be an irreducible supermodule in $\mathrm{Rep}_I\, \mathcal{X}_n$ and $\lambda \in P_+$. Then $\mathrm{pr}^\lambda M = M$ if and only if $\varepsilon_i^*(M) \le \langle h_i, \lambda\rangle$ for all $i \in I$.*

Proof In view of Theorem 19.4.1(ii), $\varepsilon_i^*(M)$ is the maximal size of a Jordan block of x_1^2 (resp. x_1) on M with eigenvalue $q(i)$ (resp. 0) if $i \ne 0$ (resp. $i = 0$). The result follows immediately. □

19.6 Functions φ_i^λ

Let M be an irreducible $\mathcal{X}_n^\lambda$-supermodule. Define

$$\varepsilon_i^\lambda(M) = \max\{m \ge 0 \mid (\tilde{e}_i^\lambda)^m M \ne 0\}, \tag{19.19}$$

$$\varphi_i^\lambda(M) = \max\{m \ge 0 \mid (\tilde{f}_i^\lambda)^m M \ne 0\}. \tag{19.20}$$

As $\tilde{e}_i^\lambda$ is simply the restriction of $\tilde{e}_i$, we have $\varepsilon_i^\lambda(M) = \varepsilon_i(M)$, see (17.11). However, the integer $\varphi_i^\lambda(M)$ depends on the fixed choice of λ.

Recall that $\mathcal{X}_0^\lambda$ is interpreted as F (concentrated in degree $\bar{0}$), and let $\mathbf{1}_\lambda \cong F$ denote the irreducible $\mathcal{X}_0^\lambda$-supermodule.

Lemma 19.6.1 $\varepsilon_i^\lambda(\mathbf{1}_\lambda) = 0$ *and* $\varphi_i^\lambda(\mathbf{1}_\lambda) = \langle h_i, \lambda\rangle$.

Proof Similar to the proof of Lemma 8.4.1 but uses Corollary 19.5.1, instead of Corollary 7.4.1. □

Lemma 19.6.2 *Let* $i, j \in I$ *with* $i \neq j$ *and* M *be an irreducible supermodule in* $\operatorname{Rep}_I \mathcal{X}_n$. *Then* $\varepsilon_j^*(\tilde{f}_i^m M) \leq \varepsilon_j^*(M)$ *for every* $m \geq 0$.

Proof Follows from the Shuffle Lemma. □

Lemma 19.6.3 *Let* $i, j \in I$ *with* $i \neq j$. *Let* $M \in \mathcal{X}_n^\lambda$-smod *be irreducible with* $\varphi_j^\lambda(M) > 0$. *Then*

$$\varphi_i^\lambda(\tilde{f}_j^\lambda M) - \varepsilon_i^\lambda(\tilde{f}_j^\lambda M) \leq \varphi_i^\lambda(M) - \varepsilon_i^\lambda(M) - \langle h_i, \alpha_j\rangle.$$

Proof Similar to the proof of Lemma 8.4.3 but uses Lemma 18.3.3, Corollary 19.5.1, Lemma 19.6.2, Lemma 18.3.3(ii), and Theorem 18.2.1(i) instead of Lemma 6.3.3, Corollary 7.4.1, Lemma 8.4.2, Lemma 6.3.3(ii), and Lemma 6.2.1, respectively. □

Corollary 19.6.4 *Let* $\lambda \in P_+$ *and* M *be an irreducible* $\mathcal{X}_n^\lambda$*-supermodule with central character* χ_γ *for some* $\gamma \in \Gamma_n$. *Then*

$$\varphi_i^\lambda(M) - \varepsilon_i^\lambda(M) \leq \langle h_i, \lambda - \gamma\rangle.$$

Proof Similar to the proof of Corollary 8.4.4. □

19.7 Alternative descriptions of φ_i^λ

Let M be an irreducible $\mathcal{X}_n^\lambda$-supermodule. Recall that

$$f_i^\lambda M = \varprojlim \operatorname{pr}^\lambda \operatorname{ind}_{n,1}^{n+1} M \circledast L_m(i), \quad \operatorname{ind}_i^\lambda M = \varprojlim \operatorname{pr}^\lambda \operatorname{ind}_{n,1}^{n+1} M \boxtimes \mathcal{R}_m(i),$$

and that the inverse limits stabilize after finitely many terms. Define $\tilde{\varphi}_i^\lambda(M)$ to be the stabilization point of the limit, i.e. the least $m \geq 0$ such that

$$f_i^\lambda M = \operatorname{pr}^\lambda \operatorname{ind}_{n,1}^{n+1} M \circledast L_m(i),$$

or equivalently

$$\operatorname{ind}_i^\lambda M = \operatorname{pr}^\lambda \operatorname{ind}_{n,1}^{n+1} M \boxtimes \mathcal{R}_m(i).$$

Lemma 19.7.1 *Let M be an irreducible $\mathcal{X}_n^\lambda$-supermodule and $i \in I$. Then:*

(i) $[f_i^\lambda M] = \tilde{\varphi}_i^\lambda(M)[\tilde{f}_i^\lambda M] + \sum c_a[N_a]$ *where the N_a are irreducible $\mathcal{X}_{n+1}^\lambda$-supermodules with $\varepsilon_i^\lambda(N_a) < \varepsilon_i^\lambda(\tilde{f}_i^\lambda M)$;*

(ii) $\mathrm{End}_{\mathcal{X}_{n+1}^\lambda}(f_i^\lambda M) \simeq \mathrm{End}_{\mathcal{X}_{n+1}^\lambda}(\tilde{f}_i^\lambda M)^{\oplus \tilde{\varphi}_i^\lambda(M)}$ *as vector superspaces.*

Proof (i) Similar to the proof of Lemma 8.5.1(i).

(ii) Take $m = \tilde{\varphi}_i^\lambda(M)$. We easily show using the explicit construction of $L_m(i)$ in Section 16.4 that there is an even endomorphism

$$\theta : L_m(i) \to L_m(i)$$

of $\mathcal{X}_1$-modules, such that the image of θ^k is $\simeq L_{m-k}(i)$ for each $0 \le k \le m$. Frobenius reciprocity induces superalgebra homomorphisms

$$\mathrm{End}_{\mathcal{X}_1}(L_m(i)) \hookrightarrow \mathrm{End}_{\mathcal{X}_{n,1}}(M \circledast L_m(i)) \hookrightarrow \mathrm{End}_{\mathcal{X}_{n+1}}(\mathrm{ind}_{n,1}^{n+1} M \circledast L_m(i)).$$

So θ induces an even $\mathcal{X}_{n+1}$-endomorphism $\tilde{\theta}$ of $\mathrm{ind}_{n,1}^{n+1} M \circledast L_m(i)$, such that the image of $\tilde{\theta}^k$ is $\simeq \mathrm{ind}_{n,1}^{n+1} M \circledast L_{m-k}(i)$ for $0 \le k \le m$. Now apply the right exact functor pr^λ to get an even $\mathcal{X}_{n+1}^\lambda$-endomorphism

$$\hat{\theta} : \mathrm{pr}^\lambda \mathrm{ind}_{n,1}^{n+1} M \circledast L_m(i) \to \mathrm{pr}^\lambda \mathrm{ind}_{n,1}^{n+1} M \circledast L_m(i)$$

induced by $\tilde{\theta}$. Note $\hat{\theta}^m = 0$ and $\hat{\theta}^{m-1} \neq 0$ because its image coincides with the image of the non-zero map α_m in the proof of (i). Hence, $1, \hat{\theta}, \dots, \hat{\theta}^{m-1}$ are linearly independent, even endomorphisms of $f_i^\lambda M$. Now the proof of (ii) is completed in the same way as in the proof of Theorem 19.4.1(iii). □

Corollary 19.7.2 *Let M, N be irreducible $\mathcal{X}_n^\lambda$-supermodules with $M \not\simeq N$. Then, for every $i \in I$, $\mathrm{Hom}_{\mathcal{X}_{n+1}^\lambda}(f_i^\lambda M, f_i^\lambda N) = 0$.*

Proof Similar to the proof of Corollary 8.5.2. □

As in the classical case we want to prove that $\tilde{\varphi}_i^\lambda(M) = \varphi_i^\lambda(M)$. Note right away from the definitions that for any irreducible $\mathcal{X}_n^\lambda$-supermodule M we have $\tilde{\varphi}_i^\lambda(M) = 0$ if and only if $\varphi_i^\lambda(M) = 0$.

Lemma 19.7.3 *If M is an irreducible $\mathcal{X}_n^\lambda$-supermodule then*

$$(\tilde{\varphi}_0(M) - \varepsilon_0(M)) + 2\sum_{i=1}^{\ell}(\tilde{\varphi}_i^\lambda(M) - \varepsilon_i(M)) = \langle c, \lambda \rangle.$$

Proof By Frobenius reciprocity and Theorem 15.5.2, we have that

$$\begin{aligned}\mathrm{End}_{\mathcal{X}_{n+1}^\lambda}(\mathrm{ind}_{\mathcal{X}_n^\lambda}^{\mathcal{X}_{n+1}^\lambda} M) &\simeq \mathrm{Hom}_{\mathcal{X}_n^\lambda}(M, \mathrm{res}_{\mathcal{X}_n^\lambda}^{\mathcal{X}_{n+1}^\lambda} \mathrm{ind}_{\mathcal{X}_n^\lambda}^{\mathcal{X}_{n+1}^\lambda} M) \\ &\simeq \mathrm{Hom}_{\mathcal{X}_n^\lambda}(M, M \oplus \Pi M)^{\oplus \langle c, \lambda\rangle} \\ &\quad \oplus \mathrm{Hom}_{\mathcal{X}_n^\lambda}(M, \mathrm{ind}_{\mathcal{X}_{n-1}^\lambda}^{\mathcal{X}_n^\lambda} \mathrm{res}_{\mathcal{X}_{n-1}^\lambda}^{\mathcal{X}_n^\lambda} M) \\ &\simeq \mathrm{Hom}_{\mathcal{X}_n^\lambda}(M, M \oplus \Pi M)^{\oplus \langle c, \lambda\rangle} \\ &\quad \oplus \mathrm{End}_{\mathcal{X}_{n-1}^\lambda}(\mathrm{res}_{\mathcal{X}_{n-1}^\lambda}^{\mathcal{X}_n^\lambda} M).\end{aligned}$$

Hence, by Schur's Lemma,

$$\dim \mathrm{End}_{\mathcal{X}_{n+1}^\lambda}(\mathrm{ind}_{\mathcal{X}_n^\lambda}^{\mathcal{X}_{n+1}^\lambda} M) - \dim \mathrm{End}_{\mathcal{X}_{n-1}^\lambda}(\mathrm{res}_{\mathcal{X}_{n-1}^\lambda}^{\mathcal{X}_n^\lambda} M)$$
$$= \begin{cases} 2\langle c, \lambda\rangle & \text{if } M \text{ is of type } \mathtt{M}, \\ 4\langle c, \lambda\rangle & \text{if } M \text{ is of type } \mathtt{Q}. \end{cases}$$

Now if M is of type $\mathtt{M}$, then

$$\mathrm{ind}_{\mathcal{X}_n^\lambda}^{\mathcal{X}_{n+1}^\lambda} M \simeq f_0^\lambda M \oplus \bigoplus_{i=1}^{\ell} (f_i^\lambda M \oplus \Pi f_i^\lambda M),$$

$$\mathrm{res}_{\mathcal{X}_{n-1}^\lambda}^{\mathcal{X}_n^\lambda} M \simeq e_0^\lambda M \oplus \bigoplus_{i=1}^{\ell} (e_i^\lambda M \oplus \Pi e_i^\lambda M),$$

by Lemmas 19.2.1 and 19.2.3 and (19.5). Hence by Lemma 19.7.1 (ii) and Theorem 19.4.1(iii),

$$\dim \mathrm{End}_{\mathcal{X}_{n+1}^\lambda}(\mathrm{ind}_{\mathcal{X}_n^\lambda}^{\mathcal{X}_{n+1}^\lambda} M) = 2\tilde{\varphi}_0^\lambda(M) + 4\sum_{i=1}^{\ell} \tilde{\varphi}_i^\lambda(M),$$

$$\dim \mathrm{End}_{\mathcal{X}_{n-1}^\lambda}(\mathrm{res}_{\mathcal{X}_{n-1}^\lambda}^{\mathcal{X}_n^\lambda} M) = 2\varepsilon_0^\lambda(M) + 4\sum_{i=1}^{\ell} \varepsilon_i^\lambda(M),$$

and the conclusion follows in this case. The argument for M of type $\mathtt{Q}$ is similar. □

Lemma 19.7.4 *Let M be an irreducible $\mathcal{X}_n^\lambda$-supermodule and $i \in I$. Let $c_i = 1$ if $i = 0$, $c_i = 2$ otherwise. Then*

$$[\mathrm{res}_i^\lambda \mathrm{ind}_i^\lambda M : M] = 2c_i \varepsilon_i^\lambda(\tilde{f}_i^\lambda M)\tilde{\varphi}_i^\lambda(M),$$
$$[\mathrm{ind}_i^\lambda \mathrm{res}_i^\lambda M : M] = 2c_i \varepsilon_i^\lambda(M)\tilde{\varphi}_i^\lambda(\tilde{e}_i M),$$
$$\mathrm{soc}\,\mathrm{res}_i^\lambda \mathrm{ind}_i^\lambda M \simeq (M \oplus \Pi M)^{\oplus c_i \tilde{\varphi}_i^\lambda(M)},$$
$$\mathrm{soc}\,\mathrm{ind}_i^\lambda \mathrm{res}_i^\lambda M \simeq (M \oplus \Pi M)^{\oplus c_i \varepsilon_i^\lambda(M)}.$$

Proof The statement about composition multiplicities follows from Theorem 19.4.1(i) and Lemma 19.7.1(i), taking into account how res_i^λ and ind_i^λ are related to e_i^λ and f_i^λ as explained in Lemmas 19.2.1 and 19.2.3. Now consider the statement about socles. We consider only $\mathrm{res}_i^\lambda \mathrm{ind}_i^\lambda M$, the other case being entirely similar but using results from Section 19.4 instead. By adjointness, it suffices to be able to compute

$$\mathrm{Hom}_{\mathcal{X}_{n+1}^\lambda}(\mathrm{ind}_i^\lambda N, \mathrm{ind}_i^\lambda M)$$

for any irreducible $\mathcal{X}_n^\lambda$-supermodule N. But in view of Lemma 19.2.3, this can be computed from knowledge of $\mathrm{Hom}_{\mathcal{X}_{n+1}^\lambda}(f_i^\lambda N, f_i^\lambda M)$, which is known by Corollary 19.7.2 (if $M \not\cong N$) and Lemma 19.7.1(ii) (if $M \cong N$). The details are similar to those in the proof of Lemma 19.7.3, so we omit them. □

Lemma 19.7.5 *Let M be an irreducible $\mathcal{X}_n^\lambda$-supermodule and $i \in I$. There are maps*

$$\mathrm{ind}_i^\lambda \mathrm{res}_i^\lambda M \xrightarrow{\psi} \mathrm{res}_i^\lambda \mathrm{ind}_i^\lambda M \xrightarrow{\mathrm{can}} \mathrm{res}_i^\lambda \mathrm{ind}_i^\lambda M / \mathrm{soc}\, \mathrm{res}_i^\lambda \mathrm{ind}_i^\lambda M,$$

whose composite is surjective.

Proof Let $k = \tilde{\varphi}_i^\lambda(M)$ and

$$\pi : \mathrm{ind}_{n,1}^{n+1} M \boxtimes \mathcal{R}_k(i) \twoheadrightarrow \mathrm{pr}^\lambda \mathrm{ind}_{n,1}^{n+1} M \boxtimes \mathcal{R}_k(i) = \mathrm{ind}_i^\lambda M$$

be the quotient map. Set $z = x_n^2 - q(i)$ (resp. $z = x_n$ if $i = 0$). Recall from Section 16.4 that viewed as an $\mathcal{X}_1'$-module, we have that $\mathcal{R}_k(i) \simeq \mathcal{X}_1'/(z^k)$. In particular, $\mathcal{R}_k(i)$ is a cyclic module generated by the image $\tilde{1}$ of $1 \in \mathcal{X}_1'$.

We first observe that for any $m \geq \varepsilon_i(M) + k$, z^m annihilates the vector

$$s_n \otimes (u \otimes v) \in \mathrm{ind}_{n,1}^{n+1} M \boxtimes \mathcal{R}_k(i)$$

for any $u \in M$, $v \in \mathcal{R}_k(i)$. This follows from the relations in $\mathcal{X}_{n+1}$, for example in the case where $i \neq 0$ we ultimately appeal to the facts that $(x_n^2 - q(i))^{\varepsilon_i(M)}$ annihilates u (see Theorem 19.4.1(ii)) and $(x_{n+1}^2 - q(i))^k$ annihilates v.

Therefore, for any $m \geq \varepsilon_i^\lambda(M) + k$, the following equality holds in $f_i^\lambda M$:

$$z^m \pi\big(s_n \otimes (u \otimes v)\big) = 0. \tag{19.21}$$

Next, it is not difficult to check that there exists a unique $\mathcal{X}_{n-1,1}$-homomorphism

$$(\mathrm{res}_i M) \boxtimes \mathcal{X}_1' \to \mathrm{res}_{n-1,1}^n \mathrm{res}_i \mathrm{ind}_i^\lambda M, \quad u \otimes 1 \mapsto \pi[s_n \otimes (u \otimes \tilde{1})]$$

for each $u \in \mathrm{res}_i M \subseteq M$. It follows from (19.21) that this homomorphism factors to induce a well-defined $\mathcal{X}_{n-1,1}$-module homomorphism

$$(\mathrm{res}_i M) \boxtimes \mathcal{R}_m(i) \to \mathrm{res}^n_{n-1,1} \mathrm{res}_i \mathrm{ind}_i^\lambda M.$$

We then get from Frobenius reciprocity an induced map

$$\psi_m : \mathrm{ind}^n_{n-1,1} (\mathrm{res}_i M) \boxtimes \mathcal{R}_m(i) \to \mathrm{res}_i \mathrm{ind}_i^\lambda M \tag{19.22}$$

for each $m \geq \varepsilon_i^\lambda(M) + k$. Each ψ_m factors through the quotient

$$\mathrm{pr}^\lambda \mathrm{ind}^n_{n-1,1} (\mathrm{res}_i M) \boxtimes \mathcal{R}_m(i),$$

so we get an induced map

$$\begin{aligned} \psi : \mathrm{ind}_i^\lambda \mathrm{res}_i M &= \varprojlim \mathrm{pr}^\lambda \mathrm{ind}^n_{n-1,1} (\mathrm{res}_i M) \boxtimes \mathcal{R}_m(i) \\ &\to \mathrm{res}_i \mathrm{ind}_i^\lambda M = \mathrm{res}_i^\lambda \mathrm{ind}_i^\lambda M. \end{aligned}$$

It remains to show that the composite of ψ with the canonical epimorphism from $\mathrm{res}_i^\lambda \mathrm{ind}_i^\lambda M$ to $\mathrm{res}_i^\lambda \mathrm{ind}_i^\lambda M / \mathrm{soc}\, \mathrm{res}_i^\lambda \mathrm{ind}_i^\lambda M$ is surjective.

By Mackey Theorem there exists an exact sequence

$$\begin{aligned} 0 \to M \boxtimes \mathcal{R}_k(i) &\to \mathrm{res}^{n+1}_{n,1} (\mathrm{ind}^{n+1}_{n,1} M \boxtimes \mathcal{R}_k(i)) \\ &\to \mathrm{ind}^{n,1}_{n-1,1,1} {}^{s_n}((\mathrm{res}^n_{n-1,1} M) \boxtimes \mathcal{R}_k(i)) \to 0. \end{aligned}$$

In other words, there is an even $\mathcal{X}_{n,1}$-isomorphism from

$$\mathrm{ind}^{n,1}_{n-1,1,1} {}^{s_n}((\mathrm{res}^n_{n-1,1} M) \boxtimes \mathcal{R}_k(i))$$

to

$$\mathrm{res}^{n+1}_{n,1} (\mathrm{ind}^{n+1}_{n,1} M \boxtimes \mathcal{R}_k(i)) / (M \boxtimes \mathcal{R}_k(i)),$$

which maps

$$h \otimes (u \otimes v) \mapsto h s_n \otimes u \otimes v + M \boxtimes \mathcal{R}_k(i)$$

for $h \in \mathcal{X}_n$, $u \in M$, $v \in \mathcal{R}_k(i)$, where $M \boxtimes \mathcal{R}_k(i)$ is embedded into $\mathrm{res}^{n+1}_{n,1} (\mathrm{ind}^{n+1}_{n,1} M \boxtimes \mathcal{R}_k(i))$ as $1 \otimes M \otimes \mathcal{R}_k(i)$. Recall from (16.11) that $\dim \mathcal{R}_k(i) = 2kc_i$, where c_i is as in Lemma 19.7.4. Hence, applying Lemma 19.7.4, we get

$$\mathrm{res}^{n,1}_n M \boxtimes \mathcal{R}_k(i) \simeq (M \oplus \Pi M)^{kc_i} \simeq \mathrm{soc}\, \mathrm{res}_i^\lambda \mathrm{ind}_i^\lambda M.$$

So applying the exact functor res_i to the isomorphism above we get an isomorphism

$$\begin{aligned} \mathrm{ind}^n_{n-1,1} (\mathrm{res}_i M) \boxtimes \mathcal{R}_k(i) &\xrightarrow{\sim} \mathrm{res}_i (\mathrm{ind}^{n+1}_{n,1} M \boxtimes \mathcal{R}_k(i)) / \mathrm{soc}\, \mathrm{res}_i^\lambda \mathrm{ind}_i^\lambda M, \\ h \otimes u \otimes v &\mapsto h T_n \otimes u \otimes v + \mathrm{soc}\, \mathrm{res}_i^\lambda \mathrm{ind}_i^\lambda M. \end{aligned}$$

It follows that there is a surjection

$$\theta : \mathrm{ind}^n_{n-1,1}(\mathrm{res}_i M) \boxtimes \mathcal{R}_k(i) \twoheadrightarrow \mathrm{res}_i^\lambda \mathrm{ind}_i^\lambda M / \mathrm{soc}\, \mathrm{res}_i^\lambda \mathrm{ind}_i^\lambda M.$$

such that the diagram

$$\begin{array}{ccc} \mathrm{ind}^n_{n-1,1}(\mathrm{res}_i M) \boxtimes \mathcal{R}_m(i) & \xrightarrow{\psi_m} & \mathrm{res}_i^\lambda \mathrm{ind}_i^\lambda M \\ \downarrow & & \downarrow{\scriptstyle \mathrm{can}} \\ \mathrm{ind}^n_{n-1,1}(\mathrm{res}_i M) \boxtimes \mathcal{R}_k(i) & \xrightarrow{\theta} & \mathrm{res}_i^\lambda \mathrm{ind}_i^\lambda M / \mathrm{soc}\, \mathrm{res}_i^\lambda \mathrm{ind}_i^\lambda M \end{array}$$

commutes for all $m \geq \varepsilon_i(M) + k$, where ψ_m is the map from (19.22) and the left-hand arrow is the natural surjection. Now surjectivity of θ immediately implies surjectivity of $\mathrm{can} \circ \psi_m$ and of $\mathrm{can} \circ \psi$. □

Lemma 19.7.6 *Let M be an irreducible $\mathcal{X}_n^\lambda$-supermodule with $\varepsilon_i^\lambda(M) > 0$. Then*

$$\tilde{\varphi}_i^\lambda(\tilde{e}_i^\lambda M) = \tilde{\varphi}_i^\lambda(M) + 1.$$

Proof Let us first show that

$$\tilde{\varphi}_i^\lambda(\tilde{e}_i^\lambda M) \geq \tilde{\varphi}_i^\lambda(M) + 1. \tag{19.23}$$

Recall $\varphi_i^\lambda(M) = 0$ if and only if $\tilde{\varphi}_i^\lambda(M) = 0$. Suppose first that $\varphi_i^\lambda(M) = 0$. Then $\varphi_i^\lambda(\tilde{e}_i^\lambda M) \neq 0$ in view of Lemma 17.2.2. Now, $\tilde{\varphi}_i^\lambda(M) = 0$ and $\tilde{\varphi}_i^\lambda(\tilde{e}_i^\lambda M) \neq 0$, so the conclusion certainly holds in this case. Next, assume that $\varphi_i^\lambda(M) > 0$, hence $\tilde{\varphi}_i^\lambda(M) > 0$. Note by Lemma 19.7.4,

$$\begin{aligned} [\mathrm{res}_i^\lambda \mathrm{ind}_i^\lambda M / \mathrm{soc}\, \mathrm{res}_i^\lambda \mathrm{ind}_i^\lambda M : M] &= 2c_i \varepsilon_i^\lambda(\tilde{f}_i^\lambda M) \tilde{\varphi}_i^\lambda(M) - 2c_i \tilde{\varphi}_i^\lambda(M) \\ &= 2c_i \varepsilon_i^\lambda(M) \tilde{\varphi}_i^\lambda(M) \neq 0. \end{aligned}$$

In particular, the map ψ in Lemma 19.7.5 is non-zero. So Lemma 19.7.5 implies that

$$[\mathrm{im}\ \psi : M] > 2c_i \varepsilon_i(M) \tilde{\varphi}_i^\lambda(M),$$

since at least one composition factor of $\mathrm{soc}\, \mathrm{im}\ \psi \subseteq \mathrm{soc}\, \mathrm{res}_i \mathrm{ind}_i^\lambda M$ must be sent to zero on composing with the second map can. Using another part of Lemma 19.7.4, this shows that

$$2c_i \varepsilon_i^\lambda(M) \tilde{\varphi}_i^\lambda(\tilde{e}_i^\lambda M) > 2c_i \varepsilon_i^\lambda(M) \tilde{\varphi}_i^\lambda(M)$$

and (19.23) follows.

Now using (19.23) and Lemma 19.7.4, we see that in the Grothendieck group,

$$[\mathrm{res}_i^\lambda \mathrm{ind}_i^\lambda M - \mathrm{ind}_i^\lambda \mathrm{res}_i^\lambda M : M] \le 2c_i(\tilde{\varphi}_i^\lambda(M) - \varepsilon_i^\lambda(M)),$$

with equality if and only if equality holds in (19.23). By central character considerations, for $i \neq j$,

$$[\mathrm{res}_i^\lambda \mathrm{ind}_j M : M] = [\mathrm{ind}_j^\lambda \mathrm{res}_i^\lambda M : M] = 0.$$

So using (19.5) we deduce that

$$[\mathrm{res}_{\mathcal{X}_n^\lambda}^{\mathcal{X}_{n+1}^\lambda} \mathrm{ind}_{\mathcal{X}_n^\lambda}^{\mathcal{X}_{n+1}^\lambda} M - \mathrm{ind}_{\mathcal{X}_{n-1}^\lambda}^{\mathcal{X}_n^\lambda} \mathrm{res}_{\mathcal{X}_{n-1}^\lambda}^{\mathcal{X}_n^\lambda} M : M]$$
$$\le 2(\tilde{\varphi}_0^\lambda(M) - \varepsilon_0^\lambda(M)) + 4\sum_{i=1}^{\ell}(\tilde{\varphi}_i^\lambda(M) - \varepsilon_i^\lambda(M))$$

with equality if and only if equality holds in (19.23) for all $i \in I$. Now Lemma 19.7.3 shows that the right-hand side equals $2\langle c, \lambda\rangle$, which does indeed equal the left-hand side, thanks to Theorem 15.5.2. □

Corollary 19.7.7 *For any irreducible $M \in \mathcal{X}_n^\lambda$-smod, $\varphi_i^\lambda(M) = \tilde{\varphi}_i^\lambda(M)$.*

Proof Similar to the proof of Corollary 8.5.7. □

Lemma 19.7.8 *Let M be an irreducible $\mathcal{X}_n^\lambda$-supermodule with central character χ_γ for $\gamma \in \Gamma_n$. Then $\varphi_i^\lambda(M) - \varepsilon_i^\lambda(M) = \langle h_i, \lambda - \gamma\rangle$.*

Proof In view of Corollary 19.6.4, it suffices to show that

$$(\varphi_0^\lambda(M) - \varepsilon_0^\lambda(M)) + 2\sum_{i=1}^{\ell}(\varphi_i^\lambda(M) - \varepsilon_i^\lambda(M)) = \langle c, \lambda\rangle.$$

But this is immediate from Lemma 19.7.3 and Corollary 19.7.7. □

Theorem 19.7.9 *Let $i \in I$ and M be an irreducible $\mathcal{X}_n^\lambda$-supermodule. Then:*

(i) *$[f_i^\lambda M] = \varphi_i^\lambda(M)[\tilde{f}_i^\lambda M] + \sum c_a[N_a]$ where the N_a are irreducibles with $\varphi_i^\lambda(N_a) < \varphi_i^\lambda(\tilde{f}_i^\lambda M) = \varphi_i^\lambda(M) - 1$;*
(ii) *$\varphi_i^\lambda(M)$ is the least $m \ge 0$ such that $f_i^\lambda M = \mathrm{pr}^\lambda \mathrm{ind}_{n,1}^{n+1} M \circledast L_m(i)$;*
(iii) *$\mathrm{End}_{\mathcal{X}_{n+1}}(f_i^\lambda M) \simeq \mathrm{End}_{\mathcal{X}_{n+1}}(\tilde{f}_i^\lambda M)^{\oplus \varphi_i^\lambda(M)}$ as vector superspaces.*

Proof Similar to the proof of Theorem 8.5.9. □

Lemma 19.7.10 *Let $\lambda \in P_+$, $i \in I$, $r \geq 1$, and M be an irreducible $\mathcal{X}_n^\lambda$-supermodule:*

(i) $(e_i^\lambda)^{(r)}M$ is non-zero if and only if $(\tilde{e}_i^\lambda)^r M \neq 0$.
(ii) $(f_i^\lambda)^{(r)}M$ is non-zero if and only if $(\tilde{f}_i^\lambda)^r M \neq 0$.

Proof Apply Lemma 19.3.2 and Theorems 19.4.1(i), 19.7.9(i). □

20
Construction of $U_{\mathbb{Z}}^{+}$ and irreducible modules

This chapter is parallel to Chapter 9.

20.1 Grothendieck groups revisited

Let us write

$$K(\infty) = \bigoplus_{n \geq 0} K(\mathrm{Rep}_I \mathcal{X}_n), \qquad K(\infty)_{\mathbb{Q}} = \mathbb{Q} \otimes_{\mathbb{Z}} K(\infty) \tag{20.1}$$

Thus $K(\infty)$ is a free $\mathbb{Z}$-module with canonical basis given by $B(\infty)$, the isomorphism classes of irreducible supermodules (see Section 19.2 for this notation), and $K(\infty)_{\mathbb{Q}}$ is the $\mathbb{Q}$-vector space on basis $B(\infty)$.

We let $K(\infty)^*$ denote the *restricted dual* of $K(\infty)$, namely, the set of functions $f: K(\infty) \to \mathbb{Z}$ such that f vanishes on all but finitely many elements of $B(\infty)$. Thus $K(\infty)^*$ is also a free $\mathbb{Z}$-module, with canonical basis

$$\{\delta_M \mid [M] \in B(\infty)\}$$

dual to the basis $B(\infty)$ of $K(\infty)$, that is $\delta_M([M]) = 1$, $\delta_M([N]) = 0$ for $[N] \in B(\infty)$ with $N \not\cong M$. Finally, we write $B(\infty)^*_{\mathbb{Q}} := \mathbb{Q} \otimes_{\mathbb{Z}} B(\infty)^*$.

Entirely similar definitions can be made for each $\lambda \in P_+$:

$$K(\lambda) = \bigoplus_{n \geq 0} K(\mathcal{X}_n^{\lambda}\text{-smod}) \tag{20.2}$$

denotes the Grothendieck groups of the categories $\mathcal{X}_n^{\lambda}$-smod for all n. Again, $K(\lambda)$ is a free $\mathbb{Z}$-module on the basis $B(\lambda)$ of isomorphism classes of irreducible supermodules. Moreover, infl^{λ} induces a canonical embedding $\mathrm{infl}^{\lambda} : K(\lambda) \hookrightarrow K(\infty)$ and $\mathrm{infl}^{\lambda} : B(\lambda) \hookrightarrow B(\infty)$. We will generally identify $K(\lambda)$ with its image under this embedding. We also define $K(\lambda)^*$ and $K(\lambda)_{\mathbb{Q}} = \mathbb{Q} \otimes_{\mathbb{Z}} K(\lambda)$ as above.

Recall the operators e_i and more generally the divided power operators $e_i^{(r)}$ for $r \geq 1$, defined on irreducible supermodules in $\mathrm{Rep}_I \mathcal{X}_n$ in (19.7) and (19.12) respectively. These induce linear maps

$$e_i^{(r)} : K(\infty) \to K(\infty) \tag{20.3}$$

for each $r \geq 1$. Similarly, the operators $(e_i^\lambda)^{(r)}$ and $(f_i^\lambda)^{(r)}$ from (19.13) and (19.14) respectively induce maps

$$(e_i^\lambda)^{(r)}, (f_i^\lambda)^{(r)} : K(\lambda) \to K(\lambda). \tag{20.4}$$

Recall by Lemma 19.3.2 that

$$e_i^r = (r!)e_i^{(r)}, \quad (e_i^\lambda)^r = (r!)(e_i^\lambda)^{(r)}, (f_i^\lambda)^r = (r!)(f_i^\lambda)^{(r)}. \tag{20.5}$$

Extending scalars, the maps $e_i^{(r)}$, $(e_i^\lambda)^{(r)}$, $(f_i^\lambda)^{(r)}$ induce linear maps on $K(\infty)_\mathbb{Q}$ and $K(\lambda)_\mathbb{Q}$.

20.2 Hopf algebra structure

Now we wish to give $K(\infty)$ the structure of a graded Hopf algebra over $\mathbb{Z}$. To do this, recall the canonical isomorphism

$$K(\mathrm{Rep}_I \mathcal{X}_m) \otimes_\mathbb{Z} K(\mathrm{Rep}_I \mathcal{X}_n) \to K(\mathrm{Rep}_I \mathcal{X}_{m,n}) \tag{20.6}$$

from (12.21), for each $m, n \geq 0$. The exact functor $\mathrm{ind}_{m,n}^{m+n}$ induces a well-defined map

$$\mathrm{ind}_{m,n}^{m+n} : K(\mathrm{Rep}_I \mathcal{X}_{m,n}) \to K(\mathrm{Rep}_I \mathcal{X}_{m+n}).$$

Composing with the isomorphism (20.6) and taking the direct sum over all $m, n \geq 0$, we obtain a homogeneous map

$$\diamond : K(\infty) \otimes_\mathbb{Z} K(\infty) \to K(\infty). \tag{20.7}$$

By transitivity of induction, this makes $K(\infty)$ into an associative graded $\mathbb{Z}$-algebra. By Corollary 17.3.2, τ induces the identity map on $K(\infty)$, so Theorem 14.7.1 implies that the multiplication $\diamond$ is commutative (in the usual unsigned sense). Moreover, there is a unit

$$\iota : \mathbb{Z} \to K(\infty), \ 1 \mapsto [\mathbf{1}] \in K(\mathrm{Rep}_I \mathcal{X}_0) \subset K(\infty) \tag{20.8}$$

The exact functor res_{n_1,n_2}^n induces a map

$$\mathrm{res}_{n_1,n_2}^n : K(\mathrm{Rep}_I \mathcal{X}_n) \to K(\mathrm{Rep}_I \mathcal{X}_{n_1,n_2}).$$

On composing with the isomorphism (20.6), we obtain maps

$$\Delta^n_{n_1,n_2} : K(\mathrm{Rep}_I \mathcal{X}_n) \to K(\mathrm{Rep}_I \mathcal{X}_{n_1}) \otimes_{\mathbb{Z}} K(\mathrm{Rep}_I \mathcal{X}_{n_2}),$$

$$\Delta^n = \sum_{n_1+n_2=n} \Delta^n_{n_1,n_2} : K(\mathrm{Rep}_I \mathcal{X}_n) \to \bigoplus_{n_1+n_2=n} K(\mathrm{Rep}_I \mathcal{X}_{n_1}) \otimes_{\mathbb{Z}} K(\mathrm{Rep}_I \mathcal{X}_{n_2}).$$

Now taking the direct sum over all $n \geq 0$ gives a homogeneous map

$$\Delta : K(\infty) \to K(\infty) \otimes_{\mathbb{Z}} K(\infty). \tag{20.9}$$

Transitivity of restriction implies that Δ is coassociative, while the homogeneous projection on to $K(\mathrm{Rep}_I \mathcal{X}_0) \cong \mathbb{Z}$ gives a counit

$$\varepsilon : K(\infty) \to \mathbb{Z}. \tag{20.10}$$

Thus $K(\infty)$ is also a graded coalgebra over $\mathbb{Z}$.

Theorem 20.2.1 *$(K(\infty), \diamond, \Delta, \iota, \varepsilon)$ is a commutative, graded Hopf algebra over $\mathbb{Z}$.*

Proof It just remains to check that Δ is an algebra homomorphism, which follows using the Mackey Theorem. Note in checking the details, we need to use Lemma 17.3.4 to take the definition of $\circledast$ into account correctly. □

We record the following lemma explaining how to compute the action of e_i on $K(\infty)$ explicitly in terms of Δ:

Lemma 20.2.2 *Let M be a supermodule in* $\mathrm{Rep}_I \mathcal{X}_n$*. Write*

$$\Delta^n_{n-1,1}[M] = \sum_a [M_a] \otimes [N_a]$$

for irreducible $\mathcal{X}_{n-1}$-supermodules M_a and irreducible $\mathcal{X}_1$-supermodules N_a. Then

$$e_i[M] = \sum_{a \text{ with } N_a \cong L(i)} [M_a].$$

Proof This is immediate from Lemma 19.2.1. □

Lemma 20.2.3 *The operators $e_i : K(\infty) \to K(\infty)$ satisfy the Serre relations, that is for $i, j \in I$:*

$$e_i e_j = e_j e_i, \quad \text{if } |i-j| > 1;$$

$$e_i^2 e_j + e_j e_i^2 = 2 e_i e_j e_i, \quad \text{if } |i-j| = 1, i \neq 0, j \neq \ell;$$

$$e_0^3e_1+3e_0e_1e_0^2=3e_0^2e_1e_0+e_1e_0^3, \text{ if } \ell\neq 1;$$
$$e_{\ell-1}^3e_\ell+3e_{\ell-1}e_\ell e_{\ell-1}^2=3e_{\ell-1}^2e_\ell e_{\ell-1}+e_\ell e_{\ell-1}^3, \text{ if } \ell\neq 1;$$
$$e_0^5e_1+5e_0e_1e_0^4+10e_0^3e_1e_0^2=10e_0^2e_1e_0^3+5e_0^4e_1e_0+e_1e_0^5, \text{ if } \ell=1,$$

Proof In view of Lemma 20.2.2 and coassociativity of Δ, this reduces to checking it on irreducible $\mathcal{X}_n$-supermodules for $n=2,3,4,4,6$ respectively. For this, the character information in Theorem 18.2.1 is sufficient. □

Next, $K(\lambda)$ has a natural structure as $K(\infty)$-comodule. The comodule structure map is the restriction of Δ to $K(\lambda)\subset K(\infty)$:

$$\Delta^\lambda: K(\lambda)\to K(\lambda)\otimes_{\mathbb{Z}} K(\infty).$$

The dual maps to $\diamond$, Δ, ι, ε induce on $K(\infty)^*$ the structure of a cocommutative graded Hopf algebra. Each $K(\lambda)$ is a left $K(\infty)^*$-module in the natural way: $f\in K(\infty)^*$ acts on the left on $K(\lambda)$ as the map $(\mathrm{id}\,\bar{\otimes} f)\circ\Delta^\lambda$. Similarly, $K(\infty)$ is itself a left $K(\infty)^*$-module, indeed in this case the action is even *faithful*.

Lemma 20.2.4 *The operator $e_i^{(r)}$ acts on $K(\infty)$ (resp. $(e_i^\lambda)^{(r)}$ on $K(\lambda)$ for any $\lambda\in P_+$) in the same way as the basis element $\delta_{L(i^r)}$ of $K(\infty)^*$.*

Proof The proof is similar to the proof of Lemma 9.2.5, but uses (20.5) and Lemma 20.2.2 instead of (9.6) and Lemma 9.2.3, respectively. □

Lemma 20.2.5 *There is a unique homomorphism $\pi: U_{\mathbb{Z}}^+\to K(\infty)^*$ of graded Hopf algebras such that $e_i^{(r)}\mapsto\delta_{L(i^r)}$ for each $i\in I$ and $r\geq 1$.*

Proof Similar to the proof of Lemma 9.2.6. □

20.3 Shapovalov form

Fix $\lambda\in P_+$. For a finite dimensional $\mathcal{X}_n^\lambda$-supermodule M, we let P_M denote its projective cover in the category $\mathcal{X}_n^\lambda$-smod. Since $\mathcal{X}_n^\lambda$ is a finite dimensional superalgebra, we can identify

$$K(\lambda)^*=\bigoplus_{n\geq 0}K(\mathcal{X}_n^\lambda\text{-proj}) \qquad (20.11)$$

so that the basis element δ_M corresponds to the isomorphism class $[P_M]$ for each irreducible $\mathcal{X}_n^\lambda$-supermodule M and each $n \geq 0$, see Section 12.2. Moreover, under this identification, the canonical pairing

$$(.,.) : K(\lambda)^* \times K(\lambda) \to \mathbb{Z} \tag{20.12}$$

satisfies

$$([P_M],[N]) = \begin{cases} \dim \operatorname{Hom}_{\mathcal{X}_n^\lambda}(P_M, N) & \text{if } M \text{ is of type } \mathtt{M}, \\ \frac{1}{2}\dim \operatorname{Hom}_{\mathcal{X}_n^\lambda}(P_M, N) & \text{if } M \text{ is of type } \mathtt{Q}, \end{cases} \tag{20.13}$$

for $\mathcal{X}_n^\lambda$-supermodules M, N with M irreducible (since the right-hand side clearly computes the composition multiplicity $[N: M]$).

There is a homogeneous map

$$\omega : K(\lambda)^* \to K(\lambda) \tag{20.14}$$

induced by the natural maps $K(\mathcal{X}_n^\lambda\text{-proj}) \to K(\mathcal{X}_n^\lambda\text{-smod})$ for each n.

As explained at the end of Section 19.3, we can define an action of $(e_i^\lambda)^{(r)}$ and $(f_i^\lambda)^{(r)}$ on the projective indecomposable supermodules, hence on $K(\lambda)^*$. We know by Lemma 19.3.3 that (20.5) holds for the operations on $K(\lambda)^*$ as well as on $K(\lambda)$. Also, the actions of $(e_i^\lambda)^{(r)}$ and $(f_i^\lambda)^{(r)}$ commute with ω by (19.15).

Lemma 20.3.1 *The operators e_i^λ, f_i^λ on $K(\lambda)^*$ and $K(\lambda)$ satisfy*

$$(e_i^\lambda x, y) = (x, f_i^\lambda y), \quad (f_i^\lambda x, y) = (x, e_i^\lambda y)$$

for each $x \in K(\lambda)^*$ and $y \in K(\lambda)$.

Proof Let M be an irreducible $\mathcal{X}_n^\lambda$-supermodule and N be an irreducible $\mathcal{X}_{n+1}^\lambda$-supermodule. We check that $(f_i^\lambda[P_M],[N]) = ([P_M], e_i^\lambda[N])$ in the special case that $i = 0$, M is of type $\mathtt{Q}$ and N is of type $\mathtt{M}$. In this case, by Lemmas 19.2.3, 19.1.1(i), and 19.2.1, we have

$$\begin{aligned}(f_i^\lambda[P_M],[N]) &= \frac{1}{2}(\operatorname{ind}_i^\lambda[P_M],[N]) = \frac{1}{2}\dim \operatorname{Hom}_{\mathcal{X}_{n+1}^\lambda}(\operatorname{ind}_i^\lambda P_M, N) \\ &= \frac{1}{2}\dim \operatorname{Hom}_{\mathcal{X}_n^\lambda}(P_M, \operatorname{res}_i^\lambda N) = ([P_M], \operatorname{res}_i^\lambda[N]) \\ &= ([P_M], e_i^\lambda[N]).\end{aligned}$$

All the other situations that need to be considered follow similarly. □

Corollary 20.3.2 *Suppose*

$$(e_i^\lambda)^{(r)}[M] = \sum_{[N]\in B(\lambda)} a_{M,N}[N], \quad (f_i^\lambda)^{(r)}[M] = \sum_{[N]\in B(\lambda)} b_{M,N}[N].$$

for $[M] \in B(\lambda)$. *Then*

$$(e_i^\lambda)^{(r)}[P_N] = \sum_{[M]\in B(\lambda)} b_{M,N}[P_M], \quad (f_i^\lambda)^{(r)}[P_N] = \sum_{[M]\in B(\lambda)} a_{M,N}[P_M]$$

for $[N] \in B(\lambda)$.

Lemma 20.3.3 *Let* M *be an irreducible supermodule in* $\mathcal{X}_n^\lambda$-smod. *Set* $\varepsilon = \varepsilon_i^\lambda(M)$, $\varphi = \varphi_i^\lambda(M)$. *Then, for any* $m \geq 0$,

$$(e_i^\lambda)^{(m)}[P_M] = \sum_{N \textit{ with } \varepsilon_i^\lambda(N)\geq m} a_N[P_{(\tilde{e}_i^\lambda)^m N}]$$

for coefficients $a_N \in \mathbb{Z}_{\geq 0}$. *Moreover, in case* $m = \varepsilon$,

$$(e_i^\lambda)^{(\varepsilon)}[P_M] = \binom{\varepsilon+\varphi}{\varepsilon}[P_{(\tilde{e}_i^\lambda)^\varepsilon M}] + \sum_{N \textit{ with } \varepsilon_i^\lambda(N)>\varepsilon} a_N[P_{(\tilde{e}_i^\lambda)^\varepsilon N}].$$

Proof Similar to the proof of Lemma 9.3.3, but using Corollary 20.3.2, Lemma 19.7.10(ii), (19.20), Lemma 19.7.1(i), and Theorem 19.7.9(i) instead of Corollary 9.3.2, Theorem 8.3.2(v), (8.17), Lemma 8.5.1(i), and Theorem 8.5.9(i), respectively. □

We also need:

Theorem 20.3.4 *Given an irreducible* $\mathcal{X}_n^\lambda$*-supermodule* M*, the element* $[P_M]$ *of* $K(\mathcal{X}_n^\lambda$-proj) *can be written as an integral linear combination of terms of the form* $(f_{i_1}^\lambda)^{(r_1)} \dots (f_{i_a}^\lambda)^{(r_a)}[\mathbf{1}_\lambda]$.

Proof Similar to the proof of Theorem 9.3.4, but using Corollary 20.3.2 and Theorem 19.4.1(i) instead of Corollary 9.3.2 and Theorem 5.5.1(i), respectively. □

Theorem 20.3.5 *The map* $\omega : K(\lambda)^* \to K(\lambda)$ *from (20.14) is injective.*

Proof Similar to the proof of Theorem 9.3.5, but using Lemma 20.3.3 instead of Lemma 9.3.3. □

In view of Theorem 20.3.5, we may identify $K(\lambda)^*$ with its image under ω, so $K(\lambda)^* \subseteq K(\lambda)$ are two different lattices in $K(\lambda)_{\mathbb{Q}}$. Extending scalars, the pairing (20.12) induces a bilinear form

$$(.,.) : K(\lambda)_{\mathbb{Q}} \times K(\lambda)_{\mathbb{Q}} \to \mathbb{Q} \tag{20.15}$$

with respect to which the operators e_i^λ and f_i^λ are adjoint.

Theorem 20.3.6 *The form $(.,.) : K(\lambda)_{\mathbb{Q}} \times K(\lambda)_{\mathbb{Q}} \to \mathbb{Q}$ is symmetric and non-degenerate.*

Proof Similar to the proof of Theorem 9.3.6 but using Theorem 20.3.4 instead of Theorem 9.3.4. □

20.4 Chevalley relations

Lemma 20.4.1 *The operators $e_i^\lambda, f_i^\lambda : K(\lambda) \to K(\lambda)$ satisfy the Serre relations (7.5) for $\mathfrak{g}$ as in Section 15.2.*

Proof Similar to the proof of Lemma 9.4.1. □

Now, for $i \in I$ and an irreducible $\mathcal{X}_n^\lambda$-supermodule M with central character χ_γ for $\gamma \in \Gamma_n$, define

$$h_i^\lambda[M] = \langle h_i, \lambda - \gamma \rangle [M]. \tag{20.16}$$

By Lemma 19.7.8 we have equivalently

$$h_i^\lambda[M] = (\varphi_i^\lambda(M) - \varepsilon_i^\lambda(M))[M].$$

More generally, define

$$\binom{h_i^\lambda}{r} : K(\lambda) \to K(\lambda), \qquad [M] \mapsto \binom{\varphi_i^\lambda(M) - \varepsilon_i^\lambda(M)}{r}[M]. \tag{20.17}$$

Extending linearly, each $\binom{h_i^\lambda}{r}$ can be viewed as a diagonal linear operator $K(\lambda) \to K(\lambda)$. The definition (20.16) implies immediately that:

Lemma 20.4.2 *As operators on $K(\lambda)$,*

$$[h_i^\lambda, e_j^\lambda] = \langle h_i, \alpha_j \rangle e_j^\lambda \quad \text{and} \quad [h_i^\lambda, f_j^\lambda] = -\langle h_i, \alpha_j \rangle f_j^\lambda$$

for all $i, j \in I$.

Lemma 20.4.3 *As operators on $K(\lambda)$, the relation*

$$[e_i^\lambda, f_j^\lambda] = \delta_{i,j} h_i^\lambda$$

holds for each $i, j \in I$.

Proof Let M be an irreducible $\mathcal{X}_n^\lambda$-supermodule. It follows immediately from Theorems 19.4.1(i) and 19.7.9(i) (together with central character considerations in case $i \neq j$) that $[M]$ appears in $e_i^\lambda f_j^\lambda [M] - f_j^\lambda e_i^\lambda [M]$ with multiplicity $\delta_{i,j}(\varphi_i^\lambda(M) - \varepsilon_i^\lambda(M))$. Therefore, it suffices simply to show that $e_i^\lambda f_j^\lambda [M] - f_j^\lambda e_i^\lambda [M]$ is a multiple of $[M]$. Let us show equivalently that $[\mathrm{res}_i^\lambda \mathrm{ind}_j^\lambda M] - [\mathrm{ind}_j^\lambda \mathrm{res}_i^\lambda M]$ is a multiple of $[M]$.

For $m \gg 0$, we have a surjection

$$\mathrm{ind}_{n,1}^{n+1} M \boxtimes \mathcal{R}_m(j) \twoheadrightarrow \mathrm{ind}_j^\lambda M.$$

Apply $\mathrm{pr}^\lambda \circ \mathrm{res}_i$ to get a surjection

$$\mathrm{pr}^\lambda \mathrm{res}_i \mathrm{ind}_{n,1}^{n+1} M \boxtimes \mathcal{R}_m(j) \twoheadrightarrow \mathrm{res}_i^\lambda \mathrm{ind}_j^\lambda M. \tag{20.18}$$

By the Mackey Theorem and (16.12), there is an exact sequence

$$\begin{aligned} 0 &\to (M \oplus \Pi M)^{\oplus \delta_{i,j} m c_j} \to \mathrm{res}_i \mathrm{ind}_{n,1}^{n+1} M \boxtimes \mathcal{R}_m(j) \\ &\to \mathrm{ind}_{n-1,1}^n (\mathrm{res}_i M) \boxtimes \mathcal{R}_m(j) \to 0, \end{aligned}$$

where c_j is as in Lemma 19.7.4. For sufficiently large m, we have

$$\mathrm{pr}^\lambda \mathrm{ind}_{n-1,1}^n (\mathrm{res}_i M) \boxtimes \mathcal{R}_m(j) = \mathrm{ind}_j^\lambda \mathrm{res}_i^\lambda M.$$

So on applying the right exact functor pr^λ and using the simplicity of M, this implies that there is an exact sequence

$$\begin{aligned} 0 &\to M^{\oplus m_1} \oplus \Pi M^{\oplus m_2} \to \mathrm{pr}^\lambda \mathrm{res}_i \mathrm{ind}_{n,1}^{n+1} M \boxtimes \mathcal{R}_m(j) \\ &\to \mathrm{ind}_j^\lambda \mathrm{res}_i^\lambda M \to 0, \end{aligned} \tag{20.19}$$

for some m_1, m_2. Now let N be any irreducible $\mathcal{X}_n^\lambda$-supermodule with $N \not\cong M$. Combining (20.18) and (20.19) shows that

$$[\mathrm{ind}_j^\lambda \mathrm{res}_i^\lambda M - \mathrm{res}_i^\lambda \mathrm{ind}_j^\lambda M : N] \geq 0 \tag{20.20}$$

Now summing over all i, j and using (19.5) gives that

$$[\mathrm{ind}\,\mathrm{res}\, M - \mathrm{res}\,\mathrm{ind}\, M : N] \geq 0.$$

But Theorem 15.5.2 shows that equality holds here, hence it must hold in (20.20) for all $i, j \in I$. This completes the proof. □

To summarize, we have shown in (20.5), Lemmas 20.4.1, 20.4.2, and 20.4.3 that:

Theorem 20.4.4 *The action of the operators $e_i^\lambda, f_i^\lambda, h_i^\lambda$ on $K(\lambda)$ satisfy the Chevalley relations (7.3), (7.4), and (7.5) for $\mathfrak{g}$ as in Section 15.2. Hence, the actions of $(e_i^\lambda)^{(r)}, (f_i^\lambda)^{(r)}$, and $\binom{h_i^\lambda}{r}$ for all $i \in I, r \geq 1$ make $K(\lambda)_{\mathbb{Q}}$ into a $U_{\mathbb{Q}}$-module so that $K(\lambda)^*, K(\lambda)$ are $U_{\mathbb{Z}}$-submodules.*

20.5 Identification of $K(\infty)^*$, $K(\lambda)^*$, and $K(\lambda)$

Theorem 20.5.1 *For any $\lambda \in P_+$:*

(i) *$K(\lambda)_{\mathbb{Q}}$ is precisely the integrable highest weight $U_{\mathbb{Q}}$-module of highest weight λ, with highest weight vector $[\mathbf{1}_\lambda]$;*
(ii) *the bilinear form $(.,.)$ from (20.15) on the highest weight module $K(\lambda)_{\mathbb{Q}}$ coincides with the usual Shapovalov form satisfying $([\mathbf{1}_\lambda], [\mathbf{1}_\lambda]) = 1$;*
(iii) *$K(\lambda)^* \subset K(\lambda)$ are integral forms of $K(\lambda)_{\mathbb{Q}}$ containing $[\mathbf{1}_\lambda]$, with $K(\lambda)^*$ being the minimal lattice $U_{\mathbb{Z}}^-[\mathbf{1}_\lambda]$ and $K(\lambda)$ being its dual under the Shapovalov form;*
(iv) *the classes $[M]$ of the irreducible supermodules $M \in \mathcal{X}_n^\lambda$-smod$[\gamma]$ form a basis of the $(\lambda-\gamma)$-weight space $V(\lambda)_{\lambda-\gamma}$. The same is true for the classes $[P_M]$ of projective indecomposable supermodules in $\mathcal{X}_n^\lambda$-smod$[\gamma]$.*

Proof It makes sense to think of $K(\lambda)_{\mathbb{Q}}$ as a $U_{\mathbb{Q}}$-module according to Theorem 20.4.4. The actions of e_i and f_i are locally nilpotent by Theorems 19.4.1(i) and 19.7.9(i). The action of h_i is diagonal by definition. Hence, $K(\lambda)_{\mathbb{Q}}$ is an integrable module. Clearly $[\mathbf{1}_\lambda]$ is a highest weight vector of highest weight λ. Moreover, $K(\lambda)_{\mathbb{Q}} = U_{\mathbb{Q}}^-[\mathbf{1}_\lambda]$ by Theorem 20.3.4. This completes the proof of (i), and (ii) follows immediately from Lemma 20.3.1. For (iii), we know already that $K(\lambda)^* \subset K(\lambda)$ are dual lattices of $K(\lambda)_{\mathbb{Q}}$, which are invariant under $U_{\mathbb{Z}}$. Moreover, Theorem 20.3.4 again shows that $K(\lambda)^* = U_{\mathbb{Z}}^-[\mathbf{1}_\lambda]$. Finally, (iv) follows from (20.16). □

Theorem 20.5.2 *The map $\pi : U_{\mathbb{Z}}^+ \to K(\infty)^*$, which was constructed in Lemma 20.2.5 is an isomorphism.*

Proof Similar to the proof of Theorem 9.5.3. □

20.6 Blocks of cyclotomic Sergeev superalgebras

Fix $\lambda \in P_+$.

Theorem 20.6.1 *Let M and N be irreducible $\mathcal{X}_n^\lambda$-supermodules with $M \not\cong N$, and*

$$0 \longrightarrow \mathrm{infl}^\lambda M \longrightarrow X \longrightarrow \mathrm{infl}^\lambda N \longrightarrow 0 \tag{20.21}$$

be an exact sequence of $\mathcal{X}_n$-supermodules. Then $\mathrm{pr}^\lambda X = X$.

Proof Similar to the proof of Theorem 9.6.1 but using Corollary 19.4.2 instead of Corollary 5.5.2. □

Recalling the definitions from Section 16.2 and using Theorem 20.5.1, we immediately deduce the following corollary which determines the blocks of $\mathcal{X}_n^\lambda$:

Corollary 20.6.2 *The blocks of cyclotomic Sergeev superalgebras $\mathcal{X}_n^\lambda$ are precisely the subcategories $\mathcal{X}_n^\lambda$-mod$[\gamma]$ for $\gamma \in \Gamma_n$. Moreover, the subcategory $\mathcal{X}_n^\lambda$-mod$[\gamma]$ is non-trivial if and only if the $(\lambda - \gamma)$-weight space of the highest weight module $K(\lambda)_{\mathbb{Q}}$ is non-zero.*

21
Identification of the crystal

This chapter is very short since essentially *no* changes need to be made to the statements and the proofs in Chapter 10. We only state the main results.

Recall the sets of isomorphism classes of irreducible supermodules $B(\infty)$ and $B(\lambda)$ from Section19.2. We can make them into crystals in the sense of Section10.2. For $B(\lambda)$, we use the operators $\tilde{e}_i^\lambda$, $\tilde{f}_i^\lambda$ from (19.10), (19.11) and functions ε_i^λ, φ_i^λ from (19.19), (19.20) to define the maps $\tilde{e}_i, \tilde{f}_i, \varepsilon_i, \varphi_i$, respectively. For the corresponding functions on $B(\infty)$ use $\tilde{e}_i$, $\tilde{f}_i$ from (17.10), (17.9), functions ε_i from (17.5), and φ_i defined below.

For the weight functions on $B(\infty)$ and $B(\lambda)$ set

$$\mathrm{wt}(M) = -\gamma, \tag{21.1}$$

for an irreducible $M \in \mathrm{Rep}_I\, \mathcal{X}_n[\gamma]$, and

$$\mathrm{wt}^\lambda(N) = \lambda - \gamma, \tag{21.2}$$

for an irreducible $N \in \mathcal{X}_n^\lambda\text{-smod}[\gamma]$, respectively. Finally, for $[M] \in B(\infty)$, define

$$\varphi_i(M) = \varepsilon_i(M) + \langle h_i, \mathrm{wt}(M)\rangle. \tag{21.3}$$

Lemma 21.0.3 *The tuples*

$$(B(\infty), \varepsilon_i, \varphi_i, \tilde{e}_i, \tilde{f}_i, \mathrm{wt})$$

and

$$(B(\lambda), \varepsilon_i^\lambda, \varphi_i^\lambda, \tilde{e}_i^\lambda, \tilde{f}_i^\lambda, \mathrm{wt}^\lambda)$$

for $\lambda \in P_+$ are crystals in the sense of Kashiwara.

Theorem 21.0.4 *The crystal $B(\infty)$ is isomorphic to Kashiwara's crystal $B(\infty)$ associated to the crystal base of $U_{\mathbb{Q}}^{-}$.*

Theorem 21.0.5 *For each $\lambda \in P_+$, the crystal $B(\lambda)$ is isomorphic to Kashiwara's crystal $B(\lambda)$ associated to the integrable highest weight $U_{\mathbb{Q}}$-module of highest weight λ.*

22
Double covers

In this chapter we specialize to the case $\lambda = \Lambda_0$. In this case $\mathcal{X}_n^\lambda \cong \mathcal{Y}_n$, see Remark 15.4.7. So, in view of Chapter 13, we will be getting results on spin representations of symmetric groups. The fact that we deal with the category of supermodules will usually provide important additional insights; for example, this will allow us to treat spin representation theory of *alternating groups* without any extra work. We will obtain a classification of the irreducible spin representations of S_n and A_n over an algebraically closed field F of characteristic $p \geq 0$, describe how the irreducibles split into blocks (spin version of "Nakayama's Conjecture"), and prove some of the branching rules. As $\lambda = \Lambda_0$ is fixed throughout the chapter, we will not use the superscripts and just write e_i for $e_i^{\Lambda_0}$, $\tilde{f}_i$ for $\tilde{f}_i^{\Lambda_0}$, etc. These should not be confused with the corresponding notions for the affine Sergeev algebra $\mathcal{X}_n$.

22.1 Description of the crystal graph

Kang [Kg] has given a convenient combinatorial description of the crystal $B(\Lambda_0)$ in terms of Young diagrams, which we now explain.

For any $n \geq 0$, let $\lambda = (\lambda_1, \lambda_2, \dots)$ be a partition of n. Recall $\ell \in \mathbb{Z}_{>0} \cup \{\infty\}$ from (15.1). We call λ a *p-strict partition* if p divides λ_r whenever $\lambda_r = \lambda_{r+1}$ for $r \geq 1$. If $p = 0$, we interpret this as a *strict partition*, that is a partition whose non-zero parts are all distinct. We say that a p-strict partition λ is *p-restricted* if in addition

$$\begin{cases} \lambda_r - \lambda_{r+1} < p & \text{if } p \mid \lambda_r, \\ \lambda_r - \lambda_{r+1} \leq p & \text{if } p \nmid \lambda_r \end{cases}$$

for each $r \geq 1$. Let $\mathcal{RP}_p(n)$ denote the set of all p-restricted p-strict partitions of n, and $\mathcal{RP}_p := \bigcup_{n \geq 0} \mathcal{RP}_p(n)$.

Let λ be a p-strict partition. As in Chapter 1, we identify λ with its *Young diagram*, but the way we are going to label the nodes of λ with *residues* is different from the one used in Chapter 1. Residues are now the elements of the set $I = \{0, 1, \ldots, \ell\}$, see (15.2). The labeling depends only on the column and follows the repeating pattern

$$0, 1, \ldots, \ell - 1, \ell, \ell - 1, \ldots, 1, 0,$$

starting from the first column and going to the right, see Example 22.1.1 below. The residue of the node A is denoted $\operatorname{res} A$. Define the *residue content* of λ to be the tuple

$$\operatorname{cont}(\lambda) = (\gamma_i)_{i \in I} \tag{22.1}$$

where for each $i \in I$, γ_i is the number of nodes of residue i contained in the diagram λ.

Let λ be a p-strict partition and $i \in I$ be some fixed residue. A node $A = (r, s) \in \lambda$ is called *i-removable* (for λ) if one of the following holds:

(R1) $\operatorname{res} A = i$ and $\lambda_A := \lambda - \{A\}$ is again a p-strict partition;
(R2) the node $B = (r, s+1)$ immediately to the right of A belongs to λ, $\operatorname{res} A = \operatorname{res} B = i$, and both $\lambda_B = \lambda - \{B\}$ and $\lambda_{A,B} := \lambda - \{A, B\}$ are p-strict partitions.

Similarly, a node $B = (r, s) \notin \lambda$ is called *i-addable* (for λ) if one of the following holds:

(A1) $\operatorname{res} B = i$ and $\lambda^B := \lambda \cup \{B\}$ is again a p-strict partition;
(A2) the node $A = (r, s-1)$ immediately to the left of B does not belong to λ, $\operatorname{res} A = \operatorname{res} B = i$, and both $\lambda^A = \lambda \cup \{A\}$ and $\lambda^{A,B} := \lambda \cup \{A, B\}$ are p-strict partitions.

We note that (R2) and (A2) above are only possible in case $i = 0$.

Now label all i-addable nodes of the diagram λ by $+$ and all i-removable nodes by $-$. Then, the *i-signature* of λ is the sequence of pluses and minuses obtained by going along the rim of the Young diagram from bottom left to top right and reading off all the signs. The *reduced i-signature* of λ is obtained from the i-signature by successively erasing all neighboring pairs of the form $+-$.

Note the reduced i-signature always looks like a sequence of $-$s followed by $+$s. Nodes corresponding to a $-$ in the reduced i-signature are called *i-normal*, nodes corresponding to a $+$ are called *i-conormal*. The rightmost i-normal node (corresponding to the rightmost $-$ in the reduced i-signature) is called *i-good*, and the leftmost i-conormal node (corresponding to the leftmost $+$ in the reduced i-signature) is called *i-cogood*.

A node is called *removable* (resp. *addable*, *normal* , *conormal*, *good*, *cogood*) if it is i-removable (resp. i-addable, i-normal, i-conormal, i-good, i-cogood) for some i.

Example 22.1.1 Let $p = 5$, so $\ell = 2$. The partition

$$\lambda = (16, 11, 10, 10, 9, 5, 1)$$

belongs to $\mathcal{RP}_5$, and its residues are as follows:

0	1	2	1	0	0	1	2	1	0	0	1	2	1	0	0
0	1	2	1	0	0	1	2	1	0	0					
0	1	2	1	0	0	1	2	1	0						
0	1	2	1	0	0	1	2	1	0						
0	1	2	1	0	0	1	2	1							
0	1	2	1	0											
0															

The 0-addable and 0-removable nodes are as labeled in the diagram:

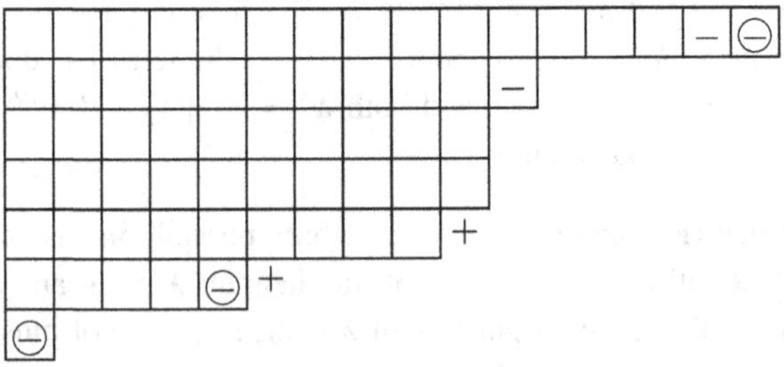

Hence, the 0-signature of λ is

$$-, -, +, +, -, -, -$$

and the reduced 0-signature is

$$-, -, -$$

Note the nodes corresponding to the $-$s in the reduced 0-signature have been circled in the above diagram. So, there are three 0-normal nodes, the rightmost of which is 0-good; there are no 0-conormal or 0-cogood nodes.

We define

$$\begin{aligned}\varepsilon_i(\lambda) &= \sharp\{i\text{-normal nodes in } \lambda\}\\ &= \sharp\{-\text{'s in the reduced } i\text{-signature of } \lambda\},\end{aligned} \tag{22.2}$$

$$\begin{aligned}\varphi_i(\lambda) &= \sharp\{i\text{-conormal nodes in } \lambda\}\\ &= \sharp\{+\text{'s in the reduced } i\text{-signature of } \lambda\}.\end{aligned} \tag{22.3}$$

Also set

$$\tilde{e}_i(\lambda) = \begin{cases} \lambda_A & \text{if } \varepsilon_i(\lambda) > 0 \text{ and } A \text{ is the } i\text{-good node,}\\ 0 & \text{if } \varepsilon_i(\lambda) = 0,\end{cases} \tag{22.4}$$

$$\tilde{f}_i(\lambda) = \begin{cases} \lambda^B & \text{if } \varphi_i(\lambda) > 0 \text{ and } B \text{ is the } i\text{-cogood node,}\\ 0 & \text{if } \varphi_i(\lambda) = 0.\end{cases} \tag{22.5}$$

Finally define

$$\mathrm{wt}(\lambda) = \Lambda_0 - \sum_{i\in I} \gamma_i \alpha_i \tag{22.6}$$

where $\mathrm{cont}(\lambda) = (\gamma_i)_{i\in I}$.

The definitions imply that $\tilde{e}_i(\lambda), \tilde{f}_i(\lambda)$ are p-restricted (or zero) in case λ is itself p-restricted. So we have now defined a datum

$$(\mathcal{RP}_p, \varepsilon_i, \varphi_i, \tilde{e}_i, \tilde{f}_i, \mathrm{wt})$$

which makes the set $\mathcal{RP}_p$ of all p-restricted p-strict partitions into a crystal in the sense of Section 10.2 (a combinatorial exercise, but follows from the theorem below anyway). We can now state the main result of Kang [Kg, 7.1] for type $A_{2\ell}^{(2)}$ (if $p = 0$ this is an easier result for type B_∞):

Theorem 22.1.2 *The set* $\mathcal{RP}_p$ *equipped with* $\varepsilon_i, \varphi_i, \tilde{e}_i, \tilde{f}_i, \mathrm{wt}$ *as above is isomorphic (in the unique way) to the crystal* $B(\Lambda_0)$ *associated to the integrable highest weight* $U_{\mathbb{Q}}$*-module of fundamental highest weight* Λ_0.

Example 22.1.3 The crystal graph of $\mathcal{RP}_3 = B(\Lambda_0)$, up to degree 10, is as follows:

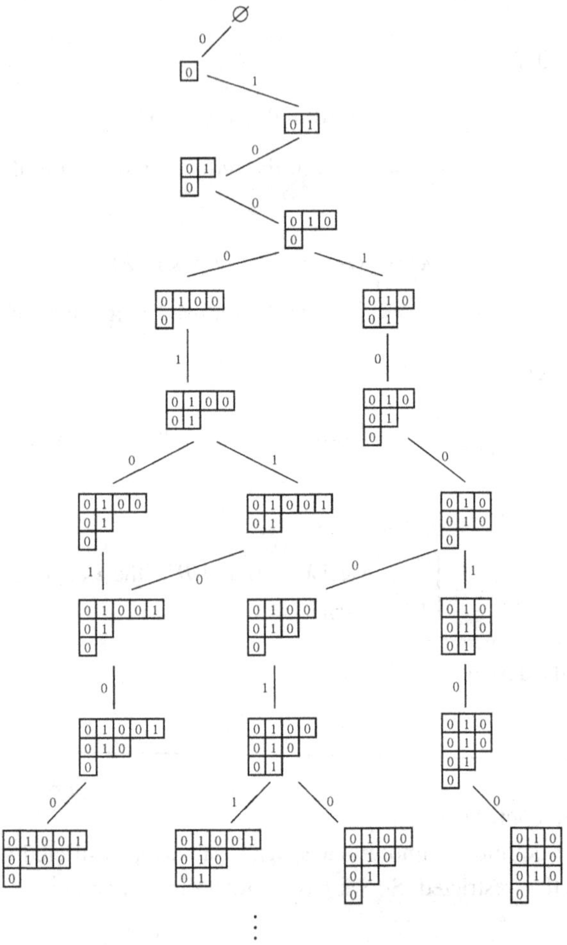

Finally, we discuss here the extension of Morris' notion of p-bar core [Mo] to an arbitrary p-strict partition λ. By a *p-bar* of λ, we mean one of the following:

(B1) the rightmost p nodes of row i of λ if $\lambda_i \geq p$ and either $p|\lambda_i$ or λ has no row of length $(\lambda_i - p)$;

(B2) the set of nodes in rows i and j of λ if $\lambda_i + \lambda_j = p$.

If λ has no p-bars, it is called a *p-bar core*. In general, the p-bar core $\tilde{\lambda}$ of λ is obtained by successively removing p-bars, reordering the rows each time so that the result still lies in $\mathcal{P}_p$, until it is reduced to a core. The *p-bar weight* of λ, denoted $w(\lambda)$, is then the total number of p-bars that get removed. There is a notion of a p-bar abacus due to Morris and Yassin [MoY$_1$], which implies easily that for p-strict partitions μ, λ of n we have

$$\mathrm{cont}(\mu) = \mathrm{cont}(\lambda) \text{ if and only if } \tilde{\mu} = \tilde{\lambda}. \tag{22.7}$$

This was first observed in [LT$_2$, Section 4].

The Lie theoretic interpretation of these combinatorial notions is exactly the same as in the classical situation, see the end of Section 11.1. In particular,

$$w(\lambda) = \begin{cases} \gamma_0\gamma_1 + \cdots + \gamma_{\ell-2}\gamma_{\ell-1} + 2\gamma_{\ell-1}\gamma_\ell & \text{if } \ell > 1, \\ -\frac{1}{2}\gamma_0(\gamma_0 - 1) - \gamma_1^2 - \cdots - \gamma_{\ell-1}^2 - 2\gamma_\ell^2 & \\ 2\gamma_0\gamma_1 - \frac{1}{2}\gamma_0(\gamma_0 - 1) - 2\gamma_1^2 & \text{if } \ell = 1, \end{cases}$$

if $\mathrm{cont}(\lambda) = (\gamma_0, \gamma_1, \ldots, \gamma_\ell)$. Also, bearing in mind Theorem 22.1.2, we can state Kac' formula [Kc, (12.13.5)] for the character of the highest weight $U_{\mathbb{Q}}$-module of highest weight Λ_0 as follows: for $\lambda \in \mathcal{RP}_p(n)$,

$$\sharp\{\mu \in \mathcal{RP}_p(n) \mid \mathrm{cont}(\mu) = \mathrm{cont}(\lambda)\} = \mathrm{Par}_\ell(w(\lambda)), \tag{22.8}$$

where $\mathrm{Par}_\ell(N)$ denotes the number of partitions of N as a sum of positive integers of ℓ different colors.

22.2 Representations of Sergeev superalgebras

Now that we have an explicit description of the crystal $B(\Lambda_0)$, we formulate a more combinatorial description of our main results for the representation theory of the Sergeev superalgebras $\mathcal{Y}_n$. Recall from Remark 15.4.7 that this is precisely the cyclotomic Sergeev superalgebra $\mathcal{X}_n^{\Lambda_0}$.

By Theorems 22.1.2 and 21.0.5, we can identify $B(\Lambda_0)$ with $\mathcal{RP}_p$. In other words, we can use the set $\mathcal{RP}_p(n)$ of p-restricted p-strict partitions of n to parametrize the irreducible $\mathcal{Y}_n$-supermodules for each $n \geq 0$. Let us write $M(\lambda)$ for the irreducible $\mathcal{Y}_n$-supermodule corresponding to $\lambda \in \mathcal{RP}_p(n)$. To be precise,

$$M(\lambda) := L(i_1, \ldots, i_n)$$

if $\lambda = \tilde{f}_{i_n} \ldots \tilde{f}_{i_1}\varnothing$. Here the operator $\tilde{f}_i$ is as defined in (22.5), corresponding under the identification $\mathcal{RP}_p(n) = B(\Lambda_0)$ to the crystal operator denoted $\tilde{f}_i^{\Lambda_0}$ in (19.11), and $\varnothing$ denotes the empty partition, corresponding to $[\mathbf{1}_\lambda] \in B(\Lambda_0)$.

For $\lambda \in \mathcal{RP}_p(n)$, we also define

$$b(\lambda) := \sharp\{r \geq 1 \mid p \nmid \lambda_r > 0\}, \tag{22.9}$$

the number of (non-zero) parts of λ that are not divisible by p. The definition of residues immediately gives that

$$b(\lambda) \equiv \gamma_0 \pmod{2}, \tag{22.10}$$

where γ_0 denotes the number of 0s in the residue content of λ.

Theorem 22.2.1 *The supermodules $\{M(\lambda) \mid \lambda \in \mathcal{RP}_p(n)\}$ form a complete set of pairwise non-isomorphic irreducible $\mathcal{Y}_n$-supermodules. Moreover, for $\lambda, \mu \in \mathcal{RP}_p(n)$:*

(i) *$M(\lambda) \cong M(\lambda)^\tau$;*
(ii) *$M(\lambda)$ is of type* M *if $b(\lambda)$ is even, type* Q *if $b(\lambda)$ is odd;*
(iii) *$M(\mu)$ and $M(\lambda)$ belong to the same block if and only if* $\operatorname{cont}(\mu) = \operatorname{cont}(\lambda)$*;*
(iv) *$M(\lambda)$ is projective if and only if λ is a p-bar core.*

Proof We have already discussed the first statement of the theorem, being a consequence of our main results combined with Theorem 22.1.2. For the rest, (i) follows from Corollary 17.3.2, (ii) is a special case of Lemma 17.3.4 combined with (22.10), and (iii) is a special case of Corollary 20.6.2. For (iv), note that if $M(\lambda)$ is projective then it is the only irreducible in its block, hence by (22.8), $\operatorname{Par}_\ell(w(\lambda)) = 1$. So either $w(\lambda) = 0$, or $\ell = 1$ and $w(\lambda) = 1$. Now if $w(\lambda) = 0$ then λ is a p-bar core so the Shapovalov form on the (1-dimensional) $\operatorname{wt}(\lambda)$-weight space of $K(\Lambda_0)_{\mathbb{Z}}$ is 1 (since $\operatorname{wt}(\lambda)$ is conjugate to Λ_0 under the action of the affine Weyl group). Hence, $M(\lambda)$ is projective by Theorem 20.5.1(ii). To rule out the remaining possibilty $\ell = 1$ and $w(\lambda) = 1$, we check in that case that the Shapovalov form on the $\operatorname{wt}(\lambda)$-weight space of $K(\Lambda_0)_{\mathbb{Z}}$ is 3. □

The next two theorems summarize earlier results concerning restriction and induction.

Theorem 22.2.2 *Let $\lambda \in \mathcal{RP}_p(n)$. There exist $\mathcal{Y}_{n-1}$-supermodules $e_i M(\lambda)$ for each $i \in I$, unique up to isomorphism, such that:*

(i) *$\operatorname{res}^{\mathcal{Y}_n}_{\mathcal{Y}_{n-1}} M(\lambda)$ is isomorphic to*

$$\begin{cases} 2e_0 M(\lambda) \oplus 2e_1 M(\lambda) \oplus \cdots \oplus 2e_\ell M(\lambda) & \text{if } b(\lambda) \text{ is odd,} \\ e_0 M(\lambda) \oplus 2e_1 M(\lambda) \oplus \cdots \oplus 2e_\ell M(\lambda) & \text{if } b(\lambda) \text{ is even;} \end{cases}$$

(ii) *for each $i \in I$, $e_i M(\lambda) \neq 0$ if and only if λ has an i-good node A, in which case $e_i M(\lambda)$ is a self-dual indecomposable supermodule with irreducible socle and head isomorphic to $M(\lambda_A)$.*

Moreover, if $i \in I$ and λ has an i-good node A, then:

(iii) *the multiplicity of $M(\lambda_A)$ in $e_i M(\lambda)$ is $\varepsilon_i(\lambda)$, $\varepsilon_i(\lambda_A) = \varepsilon_i(\lambda) - 1$, and $\varepsilon_i(\mu) < \varepsilon_i(\lambda) - 1$ for all other composition factors $M(\mu)$ of $e_i M(\lambda)$;*
(iv) $\mathrm{End}_{\mathcal{Y}_{n-1}}(e_i M(\lambda)) \simeq \mathrm{End}_{\mathcal{Y}_{n-1}}(M(\lambda_A))^{\oplus \varepsilon_i(\lambda)}$ *as a vector superspace;*
(v) $\mathrm{Hom}_{\mathcal{Y}_{n-1}}(e_i M(\lambda), e_i M(\mu)) = 0$ *for all $\mu \in \mathcal{RP}_p(n)$ with $\mu \neq \lambda$;*
(vi) *$e_i M(\lambda)$ is irreducible if and only if $\varepsilon_i(\lambda) = 1$. Hence, the restriction $\mathrm{res}^{\mathcal{Y}_n}_{\mathcal{Y}_{n-1}} M(\lambda)$ is completely reducible if and only if $\varepsilon_i(\lambda) \leq 1$ for every $i \in I$.*

Proof The existence of such supermodules $e_i M(\lambda)$ follows from (19.5), Lemma 19.2.1 and Theorem 19.2.5(i), combined as usual with Theorem 22.1.2. Uniqueness follows from Krull–Schmidt and the block classification from Theorem 22.2.1(iii). For the remaining properties, (iii), (iv), and (v) follow from Theorem 19.4.1 and Corollary 19.4.2. Finally, (vi) follows from (iii) as $e_i M(\lambda)$ is a module with irreducible socle and head, both isomorphic to $M(\lambda_A)$. □

Theorem 22.2.3 *Let $\lambda \in \mathcal{RP}_p(n)$. There exist $\mathcal{Y}_{n+1}$-supermodules $f_i M(\lambda)$ for each $i \in I$, unique up to isomorphism, such that:*

(i) $\mathrm{ind}^{\mathcal{Y}_{n+1}}_{\mathcal{Y}_n} M(\lambda)$ *is isomorphic to*

$$\begin{cases} 2f_0 M(\lambda) \oplus 2f_1 M(\lambda) \oplus \cdots \oplus 2f_\ell M(\lambda) & \text{if } b(\lambda) \text{ is odd,} \\ f_0 M(\lambda) \oplus 2f_1 M(\lambda) \oplus \cdots \oplus 2f_\ell M(\lambda) & \text{if } b(\lambda) \text{ is even;} \end{cases}$$

(ii) *for each $i \in I$, $f_i M(\lambda) \neq 0$ if and only if λ has an i-cogood node B, in which case $f_i M(\lambda)$ is a self-dual indecomposable supermodule with irreducible socle and head isomorphic to $M(\lambda^B)$.*

Moreover, if $i \in I$ and λ has an i-cogood node B, then:

(iii) *the multiplicity of $M(\lambda^B)$ in $f_i M(\lambda)$ is $\varphi_i(\lambda)$, $\varphi_i(\lambda^B) = \varphi_i(\lambda) - 1$, and $\varphi_i(\mu) < \varphi_i(\lambda) - 1$ for all other composition factors $M(\mu)$ of $f_i M(\lambda)$;*
(iv) $\mathrm{End}_{\mathcal{Y}_{n+1}}(f_i M(\lambda)) \simeq \mathrm{End}_{\mathcal{Y}_{n+1}}(M(\lambda^B))^{\oplus \varphi_i(\lambda)}$ *as a vector superspace;*
(v) $\mathrm{Hom}_{\mathcal{Y}_{n+1}}(f_i M(\lambda), f_i M(\mu)) = 0$ *for all $\mu \in \mathcal{RP}_p(n)$ with $\mu \neq \lambda$;*
(vi) *$f_i M(\lambda)$ is irreducible if and only if $\varphi_i(\lambda) = 1$. Hence, the induction $\mathrm{ind}^{\mathcal{Y}_{n+1}}_{\mathcal{Y}_n} M(\lambda)$ is completely reducible if and only if $\varphi_i(\lambda) \leq 1$ for every $i \in I$.*

Proof The argument is the same as Theorem 22.2.2, but using (19.5), Lemma 19.2.3, Theorem 19.2.5(ii), Corollary 19.7.2, and Theorem 19.7.9. □

There is one $\mathcal{Y}_n$-supermodule that deserves special mention, the so-called *basic spin supermodule*. Recall from Section 13.2 that the subalgebra of $\mathcal{Y}_n$ generated by $s_1, \dots, s_{n-1}$ is isomorphic to the group algebra $\mathcal{G}_n$ of the symmetric group S_n. It has the trivial 1-dimensional module denoted $\mathbf{1}$, on which each s_i acts as multiplication by 1. For $n \geq 1$, we define

$$I(n) := \operatorname{ind}_{\mathcal{G}_n}^{\mathcal{Y}_n} \mathbf{1}, \tag{22.11}$$

giving a $\mathcal{Y}_n$-supermodule of dimension 2^n. Also define the p-restricted p-strict partition

$$\omega_n := \begin{cases} (p^a, b) & \text{if } b \neq 0, \\ (p^{a-1}, p-1, 1) & \text{if } b = 0, \end{cases} \tag{22.12}$$

where $n = ap + b$ with $0 \leq b < p$.

Lemma 22.2.4 *If $p \nmid n$ then $I(n) \cong M(\omega_n)$; if $p | n$ then $I(n)$ is an indecomposable module with two composition factors both isomorphic to $M(\omega_n)$. In particular,*

$$\dim M(\omega_n) = \begin{cases} 2^n & \text{if } p \nmid n, \\ 2^{n-1} & \text{if } p \mid n. \end{cases}$$

Proof This is obvious if $n = 1, 2$ and easy to check directly if $n = 3$. Now for $n > 3$ we proceed by induction using Theorem 22.2.2 together with the observation that

$$\operatorname{res}^{\mathcal{Y}_n}_{\mathcal{Y}_{n-1}} I(n) \simeq I(n-1) \oplus \Pi I(n-1).$$

We consider the four cases $n \equiv 0, 1$ or $2 \pmod p$ and $n \not\equiv 0, 1, 2 \pmod p$ separately.

Suppose first that $n \not\equiv 0, 1, 2 \pmod p$. Considering the crystal graph shows that $\tilde{f}_i \omega_{n-1} \neq 0$ only for $i = 0$ and for one other $i \in I$, for which $\tilde{f}_i \omega_{n-1} = \omega_n$. By the induction hypothesis, $\operatorname{res}^{\mathcal{Y}_n}_{\mathcal{Y}_{n-1}} I(n) \cong 2M(\omega_{n-1})$. Hence by Theorem 22.2.2, $I(n)$ can only contain $M(\omega_n)$ and $M(\tilde{f}_0 \omega_{n-1})$ as composition factors. But the latter case cannot hold since by Theorem 22.2.2 again, $\operatorname{res}^{\mathcal{Y}_n}_{\mathcal{Y}_{n-1}} M(\tilde{f}_0 \omega_{n-1})$ is not isotypic. Hence all composition factors of $I(n)$ are $\cong M(\omega_n)$, and we easily get that in fact $I(n) \cong M(\omega_n)$ by a dimension argument.

Next suppose that $n \equiv 0 \pmod p$. This time, $\tilde{f}_0 \omega_{n-1} = \omega_n$ and all other $\tilde{f}_i \omega_{n-1}$ are zero. Hence, by the induction hypothesis and the branching rules, $I(n)$ only involves $M(\omega_n)$ as a constituent. But we have that $\operatorname{res}^{\mathcal{Y}_n}_{\mathcal{Y}_{n-1}} M(\omega_n) = e_0 M(\omega_n) \cong M(\omega_{n-1})$ so in fact that $I(n)$ must have $M(\omega_n)$ as a constituent with multiplicity two. Further consideration of the endomorphism ring of $I(n)$ shows moreover that it is an indecomposable module.

The argument in the remaining two cases $n \equiv 1 \pmod p$ and $n \equiv 2 \pmod p$ is entirely similar. □

22.3 Spin representations of S_n

Here we consider representation theory of the twisted group algebra $\mathcal{T}_n$, see Section 13.1. By Theorem 22.2.1, we have a parametrization $\{M(\lambda) \mid \lambda \in \mathcal{RP}_p(n)\}$ of the irreducible $\mathcal{Y}_n$-supermodules. Proposition 13.2.2 shows that the functors $\mathfrak{F}_n$ and $\mathfrak{G}_n$ set up a natural correspondence between classes of irreducible $\mathcal{T}_n$ and $\mathcal{Y}_n$-supermodules, type-preserving if n is even and type-reversing if n is odd. Hence we have a parametrization

$$\{D(\lambda) \mid \lambda \in \mathcal{RP}_p(n)\}$$

of the irreducible $\mathcal{T}_n$-supermodules, letting $D(\lambda)$ be an irreducible $\mathcal{T}_n$-supermodule corresponding to $M(\lambda)$ under the correspondence. Also, recalling the definition (22.9), define

$$a(\lambda) := n - b(\lambda) \tag{22.13}$$

for $\lambda \in \mathcal{RP}_p(n)$. We observe by (22.10) that

$$a(\lambda) \equiv \gamma_1 + \cdots + \gamma_\ell \pmod 2, \tag{22.14}$$

where $\gamma_1 + \cdots + \gamma_\ell$ counts the number of nodes in the Young diagram λ of residue *different from* 0. Finally, it is an easy combinatorial exercise to see that

$$a(\lambda) \equiv n - h_{p'}(\lambda) \pmod 2, \tag{22.15}$$

where the *p'-height* $h_{p'}(\lambda)$ of λ is the number of parts in λ not divisible by p.

Theorem 22.3.1 *The supermodules $\{D(\lambda) \mid \lambda \in \mathcal{RP}_p(n)\}$ form a complete set of pairwise non-isomorphic irreducible $\mathcal{T}_n$-supermodules. Moreover, for $\lambda, \mu \in \mathcal{RP}_p(n)$,*

(i) $D(\lambda) \cong D(\lambda)^\tau$;

(ii) *$D(\lambda)$ is of type* M *if $a(\lambda)$ is even, type* Q *if $a(\lambda)$ is odd;*

(iii) *$D(\mu)$ and $D(\lambda)$ belong to the same block if and only if* $\mathrm{cont}(\mu) = \mathrm{cont}(\lambda)$;

(iv) *$D(\lambda)$ is projective if and only if λ is a p-bar core.*

Proof Observe that (i)–(iv) follow directly from Theorem 22.2.1 using Proposition 13.2.2. □

Remark 22.3.2 The p-blocks of the ordinary irreducible spin representations of S_n were described by Humphreys [H], in terms of the notion of p-bar core. However, unlike the case of S_n, Humphreys' result does not imply Theorem 22.3.1(iii) because of the lack of information on decomposition numbers. See, however, Remark 22.3.20.

Recall the definition of $\omega_n \in \mathcal{RP}_p(n)$ in (22.12). We call the irreducible $\mathcal{T}_n$-supermodule $D(\omega_n)$ the *basic spin supermodule*. The following result is closely related to [W].

Lemma 22.3.3 *$D(\omega_n)$ is of dimension $2^{\lfloor n/2 \rfloor}$, unless $p|n$ when its dimension is $2^{\lfloor (n-1)/2 \rfloor}$. Moreover, $D(\omega_n)$ is equal to the reduction modulo p of the basic spin module $D((n))_{\mathbb{C}}$ of $(\mathcal{T}_n)_{\mathbb{C}}$ over $\mathbb{C}$, except if $p|n$ and n is even when the reduction modulo p of $D((n))_{\mathbb{C}}$ has two composition factors, both isomorphic to $D(\omega_n)$.*

Proof The statement about dimension is immediate from Lemma 22.2.4 and Proposition 13.2.2. The final statement is easily proved by working in terms of $\mathcal{Y}_n$ and using the explicit construction given in (22.11). □

To motivate the next two theorems, note that the map $[D(\lambda)] \mapsto [M(\lambda)]$ for each $\lambda \in \mathcal{RP}_p(n)$ extends linearly to an isomorphism

$$K(\mathcal{T}_n\text{-smod}) \xrightarrow{\sim} K(\mathcal{Y}_n\text{-smod})$$

of Grothendieck groups. Using this identification, we can lift the operators e_i and f_i on $K(\Lambda_0) = \bigoplus_{n\geq 0} K(\mathcal{Y}_n\text{-smod})$ defined earlier to define similar operators on $\bigoplus_{n\geq 0} K(\mathcal{T}_n\text{-smod})$. Then all our earlier results about $K(\Lambda_0)$, for instance Theorems 20.4.4 and 20.5.1, could be restated purely in terms of the representations of $\mathcal{T}_n$ instead of $\mathcal{Y}_n$. In fact, we can do slightly better and define the operators e_i and f_i on irreducible $\mathcal{T}_n$-supermodules, not just on the Grothendieck group.

Theorem 22.3.4 *Let $\lambda \in \mathcal{RP}_p(n)$. There exist $\mathcal{T}_{n-1}$-supermodules $e_i D(\lambda)$ for each $i \in I$, unique up to isomorphism, such that:*

(i) $\mathrm{res}^{\mathcal{T}_n}_{\mathcal{T}_{n-1}} D(\lambda)$ *is isomorphic to*

$$\begin{cases} e_0 D(\lambda) \oplus 2e_1 D(\lambda) \oplus \cdots \oplus 2e_\ell D(\lambda) & \text{if } a(\lambda) \text{ is odd,} \\ e_0 D(\lambda) \oplus e_1 D(\lambda) \oplus \cdots \oplus e_\ell D(\lambda) & \text{if } a(\lambda) \text{ is even;} \end{cases}$$

(ii) *for each $i \in I$, $e_i D(\lambda) \neq 0$ if and only if λ has an i-good node A, in which case $e_i D(\lambda)$ is a self-dual indecomposable supermodule with irreducible socle and head isomorphic to $D(\lambda_A)$.*

Moreover, if $i \in I$ and λ has an i-good node A, then:

(iii) *the multiplicity of $D(\lambda_A)$ in $e_i D(\lambda)$ is $\varepsilon_i(\lambda)$, $\varepsilon_i(\lambda_A) = \varepsilon_i(\lambda) - 1$, and $\varepsilon_i(\mu) < \varepsilon_i(\lambda) - 1$ for all other composition factors $D(\mu)$ of $e_i D(\lambda)$;*

(iv) $\mathrm{End}_{\mathcal{T}_{n-1}}(e_i D(\lambda)) \simeq \mathrm{End}_{\mathcal{T}_{n-1}}(D(\lambda_A))^{\oplus \varepsilon_i(\lambda)}$ *as a vector superspace;*

(v) $\mathrm{Hom}_{\mathcal{T}_{n-1}}(e_i D(\lambda), e_i D(\mu)) = 0$ *for all $\mu \in \mathcal{RP}_p(n)$ with $\mu \neq \lambda$;*

(vi) *$e_i D(\lambda)$ is irreducible if and only if $\varepsilon_i(\lambda) = 1$. Hence, $\mathrm{res}^{\mathcal{T}_n}_{\mathcal{T}_{n-1}} D(\lambda)$ is completely reducible if and only if $\varepsilon_i(\lambda) \leq 1$ for every $i \in I$.*

Proof If n is odd, we simply define $e_i D(\lambda) := \mathfrak{G}_{n-1}(e_i M(\lambda))$ for each $i \in I$, $\lambda \in \mathcal{RP}_p(n)$. If n is even, take

$$e_i D(\lambda) := \begin{cases} \mathfrak{G}_{n-1}(e_i M(\lambda)) & \begin{array}{l}\text{if } a(\lambda) \text{ is even and } i \neq 0,\\ \text{or } a(\lambda) \text{ is odd and } i = 0,\end{array} \\ \overline{\mathfrak{G}}_{n-1}(e_i M(\lambda)) & \begin{array}{l}\text{if } a(\lambda) \text{ is even and } i = 0,\\ \text{or } a(\lambda) \text{ is odd and } i \neq 0.\end{array} \end{cases}$$

We need to explain the notation $\overline{\mathfrak{G}}_{n-1}$ used in the last two cases: here, $e_i M(\lambda)$ admits an odd involution in view of Remark 19.2.6 and Theorem 22.2.1(ii), and also the Clifford supermodule U_{n-1} has an odd involution since n is even. So in exactly the same way as in the definition of (19.6), we can introduce the space

$$\overline{\mathfrak{G}}_{n-1}(e_i M(\lambda)) := \overline{\mathrm{Hom}}_{\mathcal{C}_{n-1}}(U_{n-1}, e_i M(\lambda)).$$

It is then the case that

$$\mathfrak{G}_{n-1}(e_i M(\lambda)) \simeq \overline{\mathfrak{G}}_{n-1}(e_i M(\lambda)) \oplus \Pi \overline{\mathfrak{G}}_{n-1}(e_i M(\lambda)).$$

Equivalently, by Proposition 13.2.2(ii) $e_i D(\lambda)$ can be characterized by

$$e_i M(\lambda) \cong \mathfrak{F}_{n-1}(e_i D(\lambda))$$

if $a(\lambda)$ is even and $i = 0$, or $a(\lambda)$ is odd and $i \neq 0$.

With these definitions, it is now a straightforward matter to prove (i)–(vi) using Theorem 22.2.2 and Proposition 13.2.2. Finally, the uniqueness statement is immediate from Krull–Schmidt and the description of blocks from Theorem 22.3.1(iii). □

Theorem 22.3.5 *Let $\lambda \in \mathcal{RP}_p(n)$. There exist $\mathcal{T}_{n+1}$-supermodules $f_i D(\lambda)$ for each $i \in I$, unique up to isomorphism, such that:*

(i) $\mathrm{ind}_{\mathcal{T}_n}^{\mathcal{T}_{n+1}} D(\lambda)$ *is isomorphic to*

$$\begin{cases} f_0 D(\lambda) \oplus 2f_1 D(\lambda) \oplus \cdots \oplus 2f_\ell D(\lambda) & \text{if } a(\lambda) \text{ is odd,} \\ f_0 D(\lambda) \oplus f_1 D(\lambda) \oplus \cdots \oplus f_\ell D(\lambda) & \text{if } a(\lambda) \text{ is even;} \end{cases}$$

(ii) *for each $i \in I$, $f_i D(\lambda) \neq 0$ if and only if λ has an i-cogood node B, in which case $f_i D(\lambda)$ is a self-dual indecomposable supermodule with irreducible socle and head isomorphic to $D(\lambda^B)$.*

Moreover, if $i \in I$ and λ has an i-cogood node B, then:

(iii) *the multiplicity of $D(\lambda^B)$ in $f_i D(\lambda)$ is $\varphi_i(\lambda)$, $\varphi_i(\lambda^B) = \varphi_i(\lambda) - 1$, and $\varphi_i(\mu) < \varphi_i(\lambda) - 1$ for all other composition factors $D(\mu)$ of $f_i D(\lambda)$;*

(iv) $\mathrm{End}_{\mathcal{T}_{n+1}}(f_i D(\lambda)) \simeq \mathrm{End}_{\mathcal{T}_{n+1}}(D(\lambda^B))^{\oplus \varphi_i(\lambda)}$ *as a vector superspace;*

(v) $\mathrm{Hom}_{\mathcal{T}_{n+1}}(f_i D(\lambda), f_i D(\mu)) = 0$ *for all $\mu \in \mathcal{RP}_p(n)$ with $\mu \neq \lambda$;*

(vi) *$f_i D(\lambda)$ is irreducible if and only if $\varphi_i(\lambda) = 1$. Hence the induction $\mathrm{ind}_{\mathcal{T}_n}^{\mathcal{T}_{n+1}} D(\lambda)$ is completely reducible if and only if $\varphi_i(\lambda) \leq 1$ for every $i \in I$.*

Proof This is deduced from Theorem 22.2.3 by similar argument to the proof of Theorem 22.3.4. □

Remark 22.3.6 Over $\mathbb{C}$, the branching rules in the preceeding two theorems are the same as Morris' branching rules, see [Mo]. Using this observation, we can show that our labeling of irreducibles over $\mathbb{C}$ agrees with the standard labeling.

It is possible to define the formal characters of $\mathcal{T}_n$-supermodules, as well as operators e_i and f_i in terms of the Jucys–Murphy elements $M_k \in \mathcal{T}_n$ (13.6) without referring to the Sergeev superalgebra $\mathcal{Y}_n$. (Of course, the *proofs* do depend on the representation theory of $\mathcal{Y}_n$ developed so far). Recall the notation $\ell = (p-1)/2$ (resp. $\ell = \infty$ if $p = 0$) and $I = \{0, 1, \dots, \ell\}$. Given a tuple $\underline{i} = (i_1, \dots, i_n) \in I^n$ and an $\mathcal{T}_n$-supermodule V, we define the *$\underline{i}$-weight space* of V:

$$V_{\underline{i}} = \left\{ v \in V \mid \left(M_k^2 - \frac{i_k(i_k+1)}{2}\right)^N v = 0 \text{ for } N \gg 0 \text{ and } k = 1, \dots, n \right\}.$$

Lemma 22.3.7 *Any $V \in \mathcal{T}_n$-smod decomposes as $V = \bigoplus_{\underline{i} \in I^n} V_{\underline{i}}$.*

Proof A similar statement for $\mathcal{Y}_n$ with L_k^2 in place of M_k^2 and $q(i) = i(i+1)$ in place of $\frac{i(i+1)}{2}$ follows from Lemmas 16.1.2, 15.3.1, and Remark 15.4.7. Now the desired fact follows from Lemma 13.2.5(ii). □

Now fix $\underline{i} \in I^n$, and let $\gamma = \sum_{i \in I} \gamma_i \alpha_i$ be the content of $\underline{i}$, see (16.2). Consider the Clifford–Grassman superalgebra $\mathcal{A}(\underline{i})$ from Example 12.1.4. By Example 12.2.14, it has a unique irreducible supermodule $U(\underline{i})$ of dimension $2^{\lfloor \frac{n-\gamma_0+1}{2} \rfloor}$. Now suppose that M is a $\mathcal{T}_n$-supermodule. The weight space $M_{\underline{i}}$ is obviously invariant under the action of the subalgebra $\mathfrak{H}$ of $\mathcal{T}_n$ generated by the JM-elements M_k. Put

$$j_k := \frac{i_k(i_k+1)}{2} \qquad (1 \leq k \leq n)$$

Then the M_k satisfy the relations

$$M_k^2 = j_k \qquad (1 \leq k \leq n)$$
$$M_k M_l = -M_l M_k \qquad (1 \leq k \neq l \leq n)$$

on every $\mathfrak{H}$-irreducible constituent of $M_{\underline{i}}$. Note that, since $i_k \in I$, we have $j_k = 0$ if and only if $i_k = 0$, so $\mathcal{A}(\underline{j}) \cong \mathcal{A}(\underline{i})$ and

$$\dim U(\underline{i}) = \dim U(\underline{j}) = 2^{\lfloor \frac{n-\gamma_0+1}{2} \rfloor}.$$

This shows that $\dim M_{\underline{i}}$ is divisible by $\dim U(\underline{i})$. Now define the *formal character* of M by

$$\operatorname{ch} M := \sum_{\underline{i} \in I^n} \frac{\dim M_{\underline{i}}}{\dim U(\underline{i})} e^{\underline{i}}, \tag{22.16}$$

an element of the free $\mathbb{Z}$-module on basis $\{e^{\underline{i}} \mid \underline{i} \in I^n\}$. The next lemma explains why we want to divide by $\dim U(\underline{i})$: we set this definition up so that the formal characters of the $\mathcal{Y}_n$-module $M(\lambda)$ and the corresponding $\mathcal{T}_n$-module $D(\lambda)$ were essentially the same.

Lemma 22.3.8 *Let $\lambda \in \mathcal{RP}_p(n)$, and*

$$\operatorname{ch} M(\lambda) = \sum_{\underline{i} \in I^n} a_{\underline{i}} [L(i_1) \circledast \cdots \circledast L(i_n)],$$
$$\operatorname{ch} D(\lambda) = \sum_{\underline{i} \in I^n} b_{\underline{i}} e^{\underline{i}}.$$

Then $a_{\underline{i}} = b_{\underline{i}}$ for all $\underline{i} \in I^n$.

Proof Let $\underline{i} \in I^n$. We calculate the dimension of $\mathfrak{G}_n(M(\lambda))_{\underline{i}}$. We know that $\dim U_n = 2^{\lfloor \frac{n+1}{2} \rfloor}$ and U_n is of type M if and only if n is even. Now, by Lemma 13.2.5(ii), $\varphi(M_k^2) = L_k^2/2$, and so the $\varphi(M_k^2)$ commute with the Clifford generators c_l in $\mathcal{Y}_n$, whence

$$\dim \mathfrak{G}_n(M(\lambda))_{\underline{i}} = \frac{a_{\underline{i}} \dim L(i_1) \circledast \cdots \circledast L(i_n)}{\dim U_n} 2^{\delta_n}$$
$$= a_{\underline{i}} 2^{n - \lfloor \frac{\gamma_0}{2} \rfloor - \lfloor \frac{n+1}{2} \rfloor + \delta_n},$$

where $\delta_n := n \pmod 2$. Now, $D(\lambda) = \mathfrak{G}_n(M(\lambda))$, unless both n and γ_0 are odd, in which case $\mathfrak{G}_n(M(\lambda)) \cong D(\lambda) \oplus D(\lambda)$. So the formula above implies that

$$\dim M(\lambda)_{\underline{i}} = 2^{\lfloor \frac{n-\gamma_0+1}{2} \rfloor} a_{\underline{i}} = \dim U(\underline{i}) a_{\underline{i}},$$

which implies the desired result. □

Corollary 22.3.9 *The characters of the pairwise inequivalent irreducible $\mathcal{T}_n$-supermodules are linearly independent.*

Proof Follows from Lemma 22.3.8 and Theorem 17.3.1. □

Recall the notation Γ_n from Section 16.2. Given $\gamma = \sum \gamma_i \alpha_i \in \Gamma_n$ and a $\mathcal{T}_n$-supermodule M, we set

$$M[\gamma] := \sum_{\underline{i} \in I^n \text{ with } \mathrm{cont}(\underline{i}) = \gamma} M_{\underline{i}}.$$

Corollary 22.3.10 *The decomposition $M = \bigoplus_{\gamma \in \Gamma_n} M[\gamma]$ is precisely the decomposition of M into blocks as an $\mathcal{T}_n$-supermodule.*

Proof By (16.3), (21.2), and (22.6), we have $\mathrm{cont}(\lambda) = \mathrm{cont}(\underline{i})$ for any $\underline{i} \in I^n$ with $[L(i_1) \circledast \cdots \circledast L(i_n)]$ appearing in $\mathrm{ch}\, M(\lambda)$. So the result follows from Lemma 22.3.8 and Theorems 22.2.1, 22.3.1(iii). □

For $\gamma \in \Gamma_n$ and a $\mathcal{T}_n$-supermodule M, we say that M *belongs to the block* γ if $M = M[\gamma]$. It is also now clear that the type of an irreducible $\mathcal{T}_n$-supermodule D can be read off from its formal character:

Corollary 22.3.11 *Let D be an irreducible $\mathcal{T}_n$-supermodule belonging to the block γ. Then D is of type* M *if $(n - \gamma_0)$ is even, type* Q *if $(n - \gamma_0)$ is odd.*

Next, we proceed to define i-induction and i-restriction purely in terms of $\mathcal{T}_n$. Let M be a $\mathcal{T}_n$-supermodule belonging to the block $\gamma \in \Gamma_n$. Given $i \in I$, define

$$\mathrm{res}_i M := (\mathrm{res}^{\mathcal{T}_n}_{\mathcal{T}_{n-1}} M)[\gamma - \alpha_i], \tag{22.17}$$

$$\mathrm{ind}_i M := (\mathrm{ind}^{\mathcal{T}_{n+1}}_{\mathcal{T}_n} M)[\gamma + \alpha_i], \tag{22.18}$$

where $\mathrm{res}_i M$ is interpreted as zero in case $\gamma_i = 0$. These definitions extend in an obvious way to give exact functors res_i and ind_i, which are adjoint to each other. We note in particular that $\mathrm{res}_i M$ is the generalized eigenspace of eigenvalue $i(i+1)/2$ for the action of M_n^2. Hence Lemma 22.3.7 and adjointness imply:

Lemma 22.3.12 *For an $S(n)$-supermodule M,*

$$\mathrm{res}^{\mathcal{T}_n}_{\mathcal{T}_{n-1}} M \cong \bigoplus_{i\in I} \mathrm{res}_i M, \qquad \mathrm{ind}^{\mathcal{T}_{n+1}}_{\mathcal{T}_n} M \cong \bigoplus_{i\in I} \mathrm{ind}_i M.$$

The next lemma shows how to define of the operators e_i and f_i without referring to the Sergeev superalgebra.

Lemma 22.3.13 *Let D be an irreducible $\mathcal{T}_n$-supermodule, and $i \in I$.*

(i) *There is a $\mathcal{T}_{n-1}$-supermodule $e_i D$, unique up to isomorphism, such that*

$$\mathrm{res}_i D \cong \begin{cases} e_i D \oplus e_i D & \text{if } i \neq 0 \text{ and } D \text{ is of type } \mathtt{Q}, \\ e_i D & \text{if } i = 0 \text{ or } D \text{ is of type } \mathtt{M}. \end{cases}$$

(ii) *There is an $\mathcal{T}_{n+1}$-supermodule $f_i D$, unique up to isomorphism, such that*

$$\mathrm{ind}_i D \cong \begin{cases} f_i D \oplus f_i D & \text{if } i \neq 0 \text{ and } D \text{ is of type } \mathtt{Q}, \\ f_i D & \text{if } i = 0 \text{ or } D \text{ is of type } \mathtt{M}. \end{cases}$$

Moreover operators e_i and f_i defined in (i) and (ii) agree with the operators e_i and f_i defined in Theorems 22.3.4 and 22.3.5, respectively.

Proof Consider for example (i) for $D = D(\lambda)$ with $a(\lambda)$ odd. It follows from the description of blocks of $\mathcal{T}_n$ and (22.7) that $e_0 D(\lambda)$ and $2e_i D(\lambda)$ for $i \neq 0$, as defined in Theorem 22.3.4 are precisely $\mathrm{res}_0 D(\lambda)$ and $\mathrm{res}_i D(\lambda)$, respectively, as defined in (22.17). Now the result follows from Theorem 22.3.4 and Krull–Schmidt. □

Note again that the e_i, f_i $(i \in I)$ are operators on irreducible $\mathcal{T}_n$-supermodules, but note they are *not* functors defined on arbitrary supermodules. However, extending linearly, they induce operators also denoted e_i, f_i at the level of characters or equivalently on the level of Grothendieck groups, see Corollary 22.3.9. The effect of e_i on characters is exactly the same as in the classical case (cf. (5.9)):

Lemma 22.3.14 *If*

$$\operatorname{ch} M = \sum_{\underline{i}\in I^n} c_{\underline{i}} e^{\underline{i}}$$

then

$$\operatorname{ch}(e_i M) = \sum_{\underline{i}\in I^{n-1}} c_{(i_1,\dots,i_{n-1},i)} e^{\underline{i}}.$$

Proof This follows from Lemma 22.3.8 and Corollary 19.2.2. □

There are also divided power operators $e_i^{(r)}$ and $f_i^{(r)}$. Again we just state a lemma characterizing them uniquely, rather than giving their explicit definition:

Lemma 22.3.15 *Let D be an irreducible $\mathcal{T}_n$-supermodule, and $i \in I$.*

(i) *There is an $\mathcal{T}_{n-r}$-supermodule $e_i^{(r)} D$, unique up to isomorphism, such that*

$$(\mathrm{res}_i)^r D \cong \begin{cases} (e_i^{(r)} D)^{\oplus r!} & \text{if } i = 0,\\ (e_i^{(r)} D)^{\oplus 2^{\lfloor r/2\rfloor} r!} & \text{if } i \neq 0 \text{ and } D \text{ is of type } \mathtt{M},\\ (e_i^{(r)} D)^{\oplus 2^{\lfloor (r+1)/2\rfloor} r!} & \text{if } i \neq 0 \text{ and } D \text{ is of type } \mathtt{Q}.\end{cases}$$

(ii) *There is an $\mathcal{T}_{n+r}$-supermodule $f_i^{(r)} D$, unique up to isomorphism, such that*

$$(\mathrm{ind}_i)^r D \cong \begin{cases} (f_i^{(r)} D)^{\oplus r!} & \text{if } i = 0,\\ (f_i^{(r)} D)^{\oplus 2^{\lfloor r/2\rfloor} r!} & \text{if } i \neq 0 \text{ and } D \text{ is of type } \mathtt{M},\\ (f_i^{(r)} D)^{\oplus 2^{\lfloor (r+1)/2\rfloor} r!} & \text{if } i \neq 0 \text{ and } D \text{ is of type } \mathtt{Q}.\end{cases}$$

Proof Use Lemma 19.3.1. Details omitted. □

Note comparing Lemmas 22.3.12 and 22.3.15, we see that

$$e_i^r = r!e_i^{(r)}, \qquad f_i^r = r!f_i^{(r)} \tag{22.19}$$

at the level of characters.

Finally, we "desuperize", that is deduce results about the usual (ungraded) $\mathcal{T}_n$-modules from our theory of supermodules. At the same time we get results on $\mathcal{U}_n$-modules, where $\mathcal{U}_n = (\mathcal{T}_n)_{\bar{0}}$ is the twisted group algebra of the alternating group A_n, see Section 13.1. Using Theorem 22.3.1, Corollary 12.2.10

and Proposition 12.2.11, we get the following. If $a(\lambda)$ is odd then $D(\lambda)$ decomposes as an ungraded module as

$$D(\lambda) = D(\lambda, +) \oplus D(\lambda, -)$$

for two non-isomorphic irreducible $\mathcal{T}_n$-modules $D(\lambda, +)$ and $D(\lambda, -)$. Moreover, the restrictions $\operatorname{res}^{\mathcal{T}_n}_{\mathcal{U}_n} D(\lambda, +)$ and $\operatorname{res}^{\mathcal{T}_n}_{\mathcal{U}_n} D(\lambda, -)$ are irreducible and isomorphic to each other. We denote

$$E(\lambda, 0) := \operatorname{res}^{\mathcal{T}_n}_{\mathcal{U}_n} D(\lambda, \pm).$$

If $a(\lambda)$ is even, then $D(\lambda)$ is irreducible viewed as an ungraded $\mathcal{T}_n$-module, but we denote it instead by $D(\lambda, 0)$ to make it clear that we are no longer considering a $\mathbb{Z}_2$-grading. Moreover,

$$\operatorname{res}^{\mathcal{T}_n}_{\mathcal{U}_n} D(\lambda, 0) \cong E(\lambda, +) \oplus E(\lambda, -)$$

for two non-isomorphic irreducible $\mathcal{U}_n$-modules $E(\lambda, +)$ and $E(\lambda, -)$. Finally:

Theorem 22.3.16

$$\{D(\lambda, 0) \mid \lambda \in \mathcal{RP}_p(n), a(\lambda) \textit{ even}\} \cup$$
$$\{D(\lambda, +), D(\lambda, -) \mid \lambda \in \mathcal{RP}_p(n), a(\lambda) \textit{ odd}\}$$

is a complete set of pairwise non-isomorphic irreducible $\mathcal{T}_n$-modules, and

$$\{E(\lambda, 0) \mid \lambda \in \mathcal{RP}_p(n), a(\lambda) \textit{ odd}\} \cup$$
$$\{E(\lambda, +), E(\lambda, -) \mid \lambda \in \mathcal{RP}_p(n), a(\lambda) \textit{ even}\}$$

is a complete set of pairwise non-isomorphic irreducible $\mathcal{U}_n$-modules.

Remark 22.3.17 We would like to emphasize that, as it is usual for spin representations of symmetric groups even in characteristic 0, the parametrizations obtained in Theorem 22.3.16 are "defective" in the sense that we cannot effectively distinguish between $D(\lambda, +)$ and $D(\lambda, -)$. So we believe that when working with spin representations of S_n and A_n, it is more natural to work with the category of supermodules for as long as possible, and then to "desuperize" in the last moment. Examples of that are given by Theorem 22.3.16, as well as Theorems 22.3.18, 22.3.19, and 22.3.20 below.

As another illustration of "desuperization" procedure, we give the solution to a problem important for group theory, namely description of irreducible restrictions from $\mathcal{T}_n$ to $\mathcal{T}_{n-1}$ and from $\mathcal{U}_n$ to $\mathcal{U}_{n-1}$.

Theorem 22.3.18 *Let $\lambda \in \mathcal{RP}_p(n)$. Then:*

(i) *If $a(\lambda)$ is even, $\mathrm{res}^{\mathcal{T}_n}_{\mathcal{T}_{n-1}} D(\lambda, 0)$ is irreducible if and only if*

$$\varepsilon_0(\lambda) = \sum_{i \in I} \varepsilon_i(\lambda) = 1.$$

(ii) *If $a(\lambda)$ is odd, $\mathrm{res}^{\mathcal{T}_n}_{\mathcal{T}_{n-1}} D(\lambda, \pm)$ is irreducible if and only if*

$$\sum_{i \in I} \varepsilon_i(\lambda) = 1.$$

Proof (i) $D(\lambda, 0)$ is just $D(\lambda)$ considered as an ungraded module. So, by Theorem 22.3.4(i), the restriction to $\mathrm{res}^{\mathcal{T}_n}_{\mathcal{T}_{n-1}} D(\lambda, 0)$ is irreducible only if $\varepsilon_i(\lambda) = 1$ for some $i \in I$ and $\varepsilon_j(\lambda) = 0$ for all $j \neq i$. Moreover, in this case the restriction is $e_i D(\lambda)$, considered as an ungraded module. In view of Theorem 22.3.4(vi), $e_i D(\lambda)$ is an irreducible supermodule of type $\mathtt{M}$ if $i = 0$ and type $\mathtt{Q}$ otherwise, see (22.14) and Theorem 22.3.1. So the restriction $\mathrm{res}^{\mathcal{T}_n}_{\mathcal{T}_{n-1}} D(\lambda, 0)$ is irreducible if and only if $i = 0$.

(ii) In view of Corollary 12.2.10, we have

$$D(\lambda, +) \cong D(\lambda, -) \otimes \mathbf{sgn},$$

where **sgn** is the 1-dimensional sign representation of $\mathcal{T}_n$. It follows that $\mathrm{res}^{\mathcal{T}_n}_{\mathcal{T}_{n-1}} D(\lambda, +)$ is irreducible if and only if $\mathrm{res}^{\mathcal{T}_n}_{\mathcal{T}_{n-1}} D(\lambda, -)$ is irreducible. So $\mathrm{res}^{\mathcal{T}_n}_{\mathcal{T}_{n-1}} D(\lambda, +)$ is irreducible if and only if $\mathrm{res}^{\mathcal{T}_n}_{\mathcal{T}_{n-1}} D(\lambda)$ has two composition factors when considered as an ungraded module. It is now easy to see using Theorems 22.3.4 and 22.3.1 that this happens if and only if $\sum_{i \in I} \varepsilon_i(\lambda) = 1$. □

Theorem 22.3.19 *Let $\lambda \in \mathcal{RP}_p(n)$. Then:*

(i) *If $a(\lambda)$ is even, $\mathrm{res}^{\mathcal{U}_n}_{\mathcal{U}_{n-1}} E(\lambda, \pm)$ is irreducible if and only if*

$$\sum_{i \in I} \varepsilon_i(\lambda) = 1.$$

(ii) *If $a(\lambda)$ is odd, $\mathrm{res}^{\mathcal{U}_n}_{\mathcal{U}_{n-1}} E(\lambda, 0)$ is irreducible if and only if*

$$\varepsilon_0(\lambda) = \sum_{i \in I} \varepsilon_i(\lambda) = 1.$$

Proof Similar to the proof of Theorem 22.3.18. □

Finally, we give a description of the ordinary (ungraded) blocks of $\mathcal{T}_n$. This does not quite follow from the description of the "superblocks" in

Theorem 22.3.1(iii), unless we invoke the work of Humphreys [H]; in fact, all we need from [H] is to know the *number* of ordinary blocks.

Theorem 22.3.20 *Let $D(\lambda, \varepsilon)$ and $D(\mu, \delta)$ be ungraded irreducible $\mathcal{T}_n$-modules, see Theorem 22.3.16. Then, with one exception, $D(\lambda, \varepsilon)$ and $D(\mu, \delta)$ lie in the same block if and only if λ and μ have the same p-bar core. The exception is if $\lambda = \mu$ is a p-bar core, $a(\lambda)$ is odd and $\varepsilon = -\delta$, when $D(\lambda, \varepsilon)$ and $D(\mu, \delta)$ are in different blocks.*

Proof If the p-bar core of λ is different from the p-bar core of μ, then, by Theorem 22.3.1(iii), the supermodules $D(\lambda)$ and $D(\mu)$ belong to different blocks of the superalgebra $\mathcal{T}_n$. This means that they correspond to two mutually orthogonal even central idempotents in $\mathcal{T}_n$, whence $D(\lambda, \varepsilon)$ and $D(\mu, \delta)$ belong to different ungraded blocks.

Next assume that $\lambda = \mu$ is a p-bar core and $a(\lambda)$ is odd. By Theorem 22.3.1(iv), $D(\lambda)$ is a projective supermodule, whence $D(\lambda, +)$ and $D(\lambda, -)$ are projective ungraded modules. So they are in different blocks.

Thus, the combinatorial conditions in the statement of the theorem separate the irreducible $\mathcal{T}_n$-modules into classes which are unions of blocks. In view of [H, 1.1], the amount of these classes is equal to the number of blocks of $\mathcal{T}_n$. So each class must comprise exactly one block. □

Remark 22.3.21 Using the same method as the one explained in Remark 11.2.29, we can prove the following (see [BK_6] for details): *if B is a superblock of p-bar weight w of $\mathcal{T}_n$, then the determinant of the Cartan matrix of B is p^N where N equals*

$$\sum \frac{2r_1+2r_3+2r_5+\dots}{p-1}\binom{\frac{p-3}{2}+r_1}{r_1}\binom{\frac{p-3}{2}+r_2}{r_2}\binom{\frac{p-3}{2}+r_3}{r_3}\dots,$$

the sum being over all partitions $\lambda = (1^{r_1}2^{r_2}\dots)$ of w.

In order to "desuperize" this result, note that for blocks of type M this same formula gives the Cartan determinant of the corresponding ungraded block of $\mathcal{T}_n$. We have conjectured in [BK_5] that the same is true for blocks of type Q (in this case the ungraded block has twice as many irreducibles as in the corresponding superblock). Recently Bessenrodt and Olsson confirmed this conjecture.

References

[AF] F. Anderson and K. Fuller, *Rings and Categories of Modules*, Springer-Verlag, 1974.

[ABO$_1$] G. E. Andrews, C. Bessenrodt, and J. B. Olsson, Partition identities and labels for some modular characters, *Trans. Amer. Math. Soc.* **344** (1994), 597–615.

[ABO$_2$] G. E. Andrews, C. Bessenrodt, and J. B. Olsson, A refinement of a partition identity and blocks of some modular characters, *Arch. Math.* (Basel) **66** (1996), no. 2, 101–113.

[A$_1$] S. Ariki, On the decomposition numbers of the Hecke algebra of type $G(m, 1, n)$, *J. Math. Kyoto Univ.* **36** (1996), 789–808.

[A$_2$] S. Ariki, On the classification of simple modules for cyclotomic Hecke algebras of type $G(m, 1, n)$ and Kleshchev multipartitions, to appear in *Osaka J. Math.*.

[A$_3$] S. Ariki, *Representations of Quantum Algebras and Combinatorics of Young Tableaux*, University Lecture Series, 26. American Mathematical Society, Providence, RI, 2002.

[AK] S. Ariki and K. Koike, A Hecke algebra of $(Z/rZ) \wr S_n$ and construction of its irreducible representations, *Advances Math.* **106** (1994), 216–243.

[AM] S. Ariki and A. Mathas, The number of simple modules of the Hecke algebras of type $G(r, 1, n)$, *Math. Z.* **233** (2000), 601–623.

[Be] G. Bergman, The diamond lemma for ring theory, *Advances Math.* **29** (1978), 178–218.

[BZ] I. Bernstein and A. Zelevinsky, Induced representations of reductive p-adic groups, I, *Ann. Sci. Ecole Norm. Sup.* **10** (1977), 441–472.

[BeKa] A. Berenstein and D. Kazhdan, Perfect bases and crystal bases, preprint, University of Oregon, 2004.

[BMO] C. Bessenrodt, A. O. Morris, and J. B. Olsson, Decomposition matrices for spin characters of symmetric groups at characteristic 3, *J. Algebra* **164** (1994), 146–172.

[BO$_1$] C. Bessenrodt and J. B. Olsson, On residue symbols and the Mullineux conjecture, *J. Algebraic Combin.* **7** (1998), 227–251.

[BO$_2$] C. Bessenrodt and J. B. Olsson, A note on Cartan matrices for symmetric groups, *Arch. Math.* **81** (2003), 497–504.

[B] J. Brundan, Modular branching rules and the Mullineux map for Hecke algebras of type **A**, *Proc. London Math. Soc.* **77** (1998), 551–581.

[BK_1] J. Brundan and A. Kleshchev, Translation functors for general linear and symmetric groups, *Proc. London Math. Soc.* **80** (2000), 75–106.

[BK_2] J. Brundan and A. Kleshchev, Projective representations of the symmetric group via Sergeev duality, *Math. Z.* **239** (2002), 27–68.

[BK_3] J. Brundan and A. Kleshchev, Hecke–Clifford superalgebras, crystals of type $A_{2\ell}^{(2)}$, and modular branching rules for $\hat{S}_n$, *Repr. Theory* **5** (2001), 317–403.

[BK_4] J. Brundan and A. Kleshchev, Representations of the symmetric group which are irreducible over subgroups, *J. Reine Angew. Math.* **530** (2001), 145–190.

[BK_5] J. Brundan and A. Kleshchev, Representation theory of symmetric groups and their double covers, *Groups, Combinatorics and Geometry (Durham, 2001)*, pp. 31–53, World Scientific, Publishing, River Edge, NJ, 2003.

[BK_6] J. Brundan and A. Kleshchev, Cartan determinants and Shapovalov forms, *Math. Ann.* **324** (2002), 431–449.

[BM] M. Broué and G. Malle, Zyklotomische Heckealgebren, *Astérisque* **212** (1993), 119–189.

[Ch] I. Cherednik, A new interpretation of Gel'fand-Tzetlin bases, *Duke Math. J.* **54** (1987), 563–577.

[CK] J. Chuang and R. Kessar, Symmetric groups, wreath products, Morita equivalences, and Broué's abelian defect group conjecture, *Bull. London Math. Soc.* **34** (2002), 174–184.

[CR] J. Chuang and R. Rouquier, Derived equivalences for symmetric groups and $\mathfrak{sl}_2$-categorifications, preprint, Inst. Math. Jussieu, Prepub. 376, 2004.

[CKK] C. De Concini, V. Kac, and D. Kazhdan, Boson–Fermion correspondence over $\mathbb{Z}$, in Infinite-dimensional Lie algebras and groups (Luminy-Marseille, 1988), *Adv. Ser. Math. Phys.* **7** (1989), 124–137.

[At] J. H. Conway, R. T. Curtis, S. P. Norton, R. A. Parker, and Wilson, *Atlas of Finite Groups*, Clarendon Press, Oxford, 1985.

[DJ] R. Dipper and G. James, Representations of Hecke algebras of general linear groups, *Proc. London Math. Soc.* **52** (1986), 20–52.

[DG] P. Diaconis and C. Greene, Applications of Murphy's elements, *Stanford University Technical Report* (1989), no. 335.

[D] V. G. Drinfeld, Degenerate affine Hecke algebras and Yangians, *Functional Anal. Appl.* **20** (1986), 62–64.

[E] M. Enguehard, Isométries parfaites entre blocs de groupes symétriques, *Astérisque* **181–182** (1990), 157–171.

[FK] B. Ford and A. Kleshchev, A proof of the Mullineux conjecture, *Math. Z.* **226** (1997), 267–308.

[G_1] I. Grojnowski, Representations of affine Hecke algebras (and affine quantum GL_n) at roots of unity, *Internat. Math. Res. Notices*, 1994, no. 5, 215 ff., approx. 3 pp.

[G_2] I. Grojnowski, Affine $\widehat{\mathfrak{sl}}_p$ controls the modular representation theory of the symmetric group and related Hecke algebras, preprint, math.RT/9907129, 1999.

[G_3] I. Grojnowski, Blocks of the cyclotomic Hecke algebra, preprint, 1999 (private communication).

[GV] I. Grojnowski and M. Vazirani, Strong multiplicity one theorems for affine Hecke algebras of type A, *Transform. Groups* **6** (2001), 143–155.

[H] J. F. Humphreys, Blocks of projective representations of the symmetric groups, *J. London Math. Soc.* **33** (1986), 441–452.

[J] G. D. James, *The Representation Theory of the Symmetric Groups*, Springer-Verlag, Berlin–Heidelberg–New York, 1978.

[JK] G. James and A. Kerber, *The Representation Theory of the Symmetric Groups*, Addison-Wesley, London, 1980.

[Ju] A. Jucys, Symmetric polynomials and the center of the symmetric group ring, *Reports Math. Phys.* **5** (1974), 107–112.

[MAt] C. Jansen, K. Lux, R. A. Parker, and R. A. Wilson, *An Atlas of Brauer Characters*, Oxford University Press, Oxford, 1995.

[JS] J. C. Jantzen and G. M. Seitz, On the representation theory of the symmetric groups, *Proc. London Math. Soc.* **65** (1992), 475–504.

[JN] A. Jones and M. Nazarov, Affine Sergeev algebra and q-analogues of the Young symmetrizers for projective representations of the symmetric group, *Proc. London Math. Soc.* **78** (1999), 481–512.

[Jos] T. Józefiak, Semisimple superalgebras, in *Algebra – Some Current Trends (Varna, 1986)*, pp. 96–113, Lecture Notes in Math. **1352**, Springer, Berlin–New York, 1988.

[Kc] V. Kac, *Infinite Dimensional Lie Algebras*, Cambridge University Press, third edition, 1995.

[KR] V. G. Kac and A. K. Raina, *Bombay Lectures on Highest Weight Representations of Infinite-Dimensional Lie Algebras*, Advanced Series in Mathematical Physics, 2, World Scientific Publishing, Teaneck, NJ, 1987.

[Kg] S.-J. Kang, Crystal bases for quantum affine algebras and combinatorics of Young walls, *Proc. London Math. Soc. (3)* **86** (2003), 29–69.

[Ka] M. Kashiwara, On crystal bases, in Representations of groups (Banff 1994), *CMS Conf. Proc.* **16** (1995), 155–197.

[KS] M. Kashiwara and Y. Saito, Geometric construction of crystal bases, *Duke Math. J.* **89** (1997), 9–36.

[Kt] S. Kato, Irreducibility of principal series representations for Hecke algebras of affine type, *J. Fac. Sci. Univ. Tokyo Sect. IA Math.* **28** (1981), 929–943.

[K_1] A. Kleshchev, Branching rules for modular representations of symmetric groups II, *J. Reine Angew. Math.* **459** (1995), 163–212.

[K_2] A. Kleshchev, Branching rules for modular representations of symmetric groups III: some corollaries and a problem of Mullineux, *J. London Math. Soc.* **54** (1996), 25–38.

[K_3] A. Kleshchev, Completely splittable representations of symmetric groups, *J. Algebra* **181** (1996), 584–592.

[K_4] A. Kleshchev, On decomposition numbers and branching coefficients for symmetric and special linear groups, *Proc. London Math. Soc.* **75** (1997), 497–558.

[K_5] A. Kleshchev, Branching rules for modular representations of symmetric groups IV, *J. Algebra* **201** (1998), 547–572.

[K_6] A. Kleshchev, Branching rules for symmetric groups and applications, in *Algebraic Groups and Their Representations*, R. W. Carter and J. Saxl, editors, NATO ASI Series C, Vol. 517, pp. 103–130, Kluwer Academic, Dordrecht–Boston–London, 1998.

[KR] C. Kriloff and A. Ram, Representations of graded Hecke algebras, *Represent. Theory* **6** (2002), 31–69.

[La] P. Landrock, *Finite group algebras and their modules*, Lond. Math. Soc. Lecture Note Series, 84, Cambridge University Press, Cambridge, 1983.

[LLT] A. Lascoux, B. Leclerc, and J.-Y. Thibon, Hecke algebras at roots of unity and crystal bases of quantum affine algebras, *Comm. Math. Phys.* **181** (1996), 205–263.

[LM] B. Leclerc and H. Miyachi, Some closed formulas for canonical bases of Fock spaces, *Represent. Theory* **6** (2002), 290–312 (electronic).

[LNT] B. Leclerc, M. Nazarov, and J.-Y. Thibon, Induced representations of affine Hecke algebras and canonical bases of quantum groups, *Studies in Memory of Issai Schur (Chevaleret/Rehovot, 2000)*, pp. 115–153, Progr. Math., **210**, Birkhäuser, Boston, MA, 2003.

[LT_1] B. Leclerc and J.-Y. Thibon, Canonical bases of q-deformed Fock spaces, *Internat. Math. Res. Notices* **9** (1996), 447–456.

[LT_2] B. Leclerc and J.-Y. Thibon, q-Deformed Fock spaces and modular representations of spin symmetric groups, *J. Phys. A* **30** (1997), 6163–6176.

[Le] D. A. Leites, Introduction to the theory of supermanifolds, *Russian Math. Surveys* **35** (1980), 1–64.

[L] G. Lusztig, Affine Hecke algebras and their graded version, *J. Amer. Math Soc.* **2** (1989), 599–635.

[ML] S. MacLane, *Categories for the Working Mathematician*, Graduate Texts in Mathematics 5, Springer-Verlag, Berlin,1971.

[Man] Yu I. Manin, *Gauge Field Theory and Complex Geometry*, Grundlehren der mathematischen Wissenschaften 289, second edition, Springer-Verlag, Berlin, 1997.

[Ma] A. Marcus, On equivalences between blocks of group algebras: reduction to the simple components, *J. Algebra*, **184** (1996), 372–396.

[M] O. Mathieu, On the dimension of some modular irreducible representations of the symmetric group, *Lett. Math. Phys.* **38** (1996), 23–32.

[MiMi] K. Misra and T. Miwa, Crystal base for the basic representation of $U_q(\mathfrak{sl}(n))$, *Comm. Math. Phys.* **134** (1990), 79–88.

[Mo] A. O. Morris, The spin representations of the symmetric group, *Canad. J. Math.* **17** (1965), 543–549.

[MoY_1] A.O. Morris and A.K. Yaseen, Some combinatorial results involving involving shifted Young diagrams, *Math. Proc. Camb. Phil. Soc.* **99** (1986), 23–31.

[MoY_2] A. O. Morris and A. K. Yaseen, Decomposition matrices for spin characters of symmetric groups, *Proc. Roy. Soc. Edinburgh Sect. A* **108** (1988), no. 1–2, 145–164.

[Mu_1] G. Murphy, A new construction of Young's seminormal representation of the symmetric group, *J. Algebra* **69** (1981), 287–291.

[Mu_2] G. Murphy, The idempotents of the symmetric group and Nakayama's conjecture, *J. Algebra* **81** (1983), 258–265.

[N] M. Nazarov, Young's symmetrizers for projective representations of the symmetric group, *Advances Math.* **127** (1997), 190–257.

[OV] A. Okounkov and A. Vershik, A new approach to representation theory of symmetric groups. *Selecta Math. (N.S.)* **2** (1996), 581–605.

[O_1] G. Olshanski, Extension of the algebra $U(g)$ for finite dimensional classical Lie algebras g and the Yangians $Y(gl(m))$, *Soviet Math. Doklady* **36** (1988), 569–573.

[O_2] G. Olshanski, Quantized universal enveloping superalgebra of type Q and a super-extension of the Hecke algebra, *Lett. Math. Phys.* **24** (1992), 93–102.

[R] A. Ram, Affine Hecke algebras and generalized standard Young tableaux, *J. Algebra* **260** (2003), 367–415.

[Sch] O. Schiffmann, The Hall algebra of a cyclic quiver and canonical bases of Fock spaces, *Internat. Math. Res. Notices* **8** (2000), 413–440.

[Sc] J. Scopes, Cartan matrices and Morita equivalence for blocks of the symmetric groups, *J. Algebra* **142** (1991), 441–455.

[S_1] A. N. Sergeev, Tensor algebra of the identity representation as a module over the Lie superalgebras $GL(n, m)$ and $Q(n)$, *Math. USSR Sbornik* **51** (1985), 419–427.

[S_2] A. N. Sergeev, The Howe duality and the projective representations of symmetric groups, *Represent. Theory* **3** (1999), 416–434.

[St] R. Steinberg, On a theorem of Pittie, *Topology* **14** (1975), 173–177.

[Su] M. Suzuki, *Group Theory I*, Springer-Verlag, Berlin–Heidelberg–New York, 1982.

[VV] M. Varagnolo and E. Vasserot, On the decomposition matrices of the quantized Schur algebra, *Duke Math. J.* **100** (1999), 267–297.

[V_1] M. Vazirani, Irreducible modules over the affine Hecke algebra: a strong multiplicity one result, Ph.D. thesis, UC Berkeley, 1999.

[V_2] M. Vazirani, Filtrations on the Mackey decomposition for cyclotomic Hecke algebras, *J. Algebra* **252** (2002), 205–227.

[W] D. Wales, Some projective representations of S_n, *J. Algebra* **61** (1979), 37–57.

[Z_1] A. Zelevinsky, Induced representations of reductive p-adic groups, II, *Ann. Sci. Ecole Norm. Sup.* **13** (1980), 165–210.

[Z_2] A. Zelevinsky, *Representations of Finite Classical Groups*, Lecture Notes in Math. 869, Springer-Verlag, Berlin, 1981.

Index